The New Medicine

ENDOCRINOLOGY

The New Medicine

An Integrated System of Study

The New Medicine

An Integrated System of Study
Series Editors R. Harden and A. Marcus

Volume 2

ENDOCRINOLOGY

Edited by I.R. Hart and R.W. Newton

The New Medicine Series has been
produced in collaboration with
Update Publications Limited

1983 **MTP PRESS LIMITED**
a member of the KLUWER ACADEMIC PUBLISHERS GROUP
BOSTON / THE HAGUE / DORDRECHT / LANCASTER

Published by
MTP Press Limited
Falcon House
Lancaster, England

British Library Cataloguing in Publication Data

Endocrinology.—(New medicine)
  1. Endocrinology
I. Newton, R.W. II. Hart, I.R. III. Series
  616.4   RC648

ISBN-13: 978-0-85200-401-2   e-ISBN-13: 978-94-010-9298-2
DOI: 10.1007/978-94-010-9298-2

Typeset by Servis Filmsetting Ltd., Manchester
Colour origination by Speedlith Photo Litho Ltd., Manchester

CONTENTS

LIST OF CONTRIBUTORS

EDITORS

Dr I.R. Hart
Division of Metabolism and Endocrinology
Ottawa Civic Hospital
Ottawa, Ontario
Canada

Dr R.W. Newton
Wards 1 and 2
Ninewells Hospital and Medical School
Dundee, Scotland

CONTRIBUTORS

Ms M. Browning
Department of Biochemical Medicine
Ninewells Hospital and Medical School
Dundee, Scotland

Prof. J. Crooks
Department of Pharmacology and Therapeutics
Ninewells Hospital and Medical School
Dundee, Scotland

Dr C. Forsyth
Department of Child Health
Ninewells Hospital and Medical School
Dundee, Scotland

Mr A. Gunn
Department of Surgery
University of Dundee
Ninewells Hospital and Medical School
Dundee, Scotland

Dr T. Isles
Department of Biochemical Medicine
Ninewells Hospital and Medical School
Dundee, Scotland

Professor A. Klopper
Department of Obstetrics and Gynaecology
Aberdeen Royal Infirmary
Foresterhill
Aberdeen, Scotland

Dr J. McKellican
Rye Hill Health Centre
Dundee, Scotland

Dr C. Paterson
Department of Clinical Chemistry
Ninewells Hospital and Medical School
Dundee, Scotland

Dr S. Scott
Department of Radiotherapy
University of Dundee
Ninewells Hospital and Medical School
Dundee, Scotland

Prof. J. Stowers
Diabetic Department
City Hospital
Urquhart Road
Aberdeen, Scotland

Dr W. Wilson Downie
Boehringer Ingelheim AB
127 21 Skarholmen
Stockholm
Sweden

SERIES EDITORS

Professor R. McG. Harden
Centre for Medical Education
Ninewells Hospital and Medical School
University of Dundee
Scotland

Dr A. Marcus
Chairman
Update Publications Ltd
London
England

CO-ORDINATING EDITOR
Dr R. Cairncross
Centre for Medical Education
Ninewells Hospital and Medical School
University of Dundee
Scotland

INTRODUCTION

The need for a new approach to textbooks

Many books have been written for students of medicine. The conventional textbook, however, imposes many constraints upon the reader and the author. While a considerable effort has been put into developing newer, more sophisticated methods of learning such as television, audio-tape and slides, and computers, few attempts have been made to improve the more traditional approach – the book. The aim of this series of textbooks is to minimize the limitations of the standard text and to maximize the usefulness of the book as an aid to learning.

We believe that in a number of ways this series is unique. It is the first textbook to be produced as a collaborative project between a publisher and a University Department of Medical Education. The intenton has been to produce a series of textbooks which take into account three significant trends in medical education: a move towards a more integrated approach to teaching, an increased emphasis on student-centred learning, and greater use of problem-based learning.

A more integrated text

Firstly, there is a general move to a more integrated approach to learning, a trend reflected in the curricula of many schools. This involves a shift from subject- or discipline-based teaching where the emphasis is on the individual subjects or disciplines such as medicine, surgery and therapeutics, to a multi-disciplinary or integrated approach where the student is encouraged to take a more holistic view of medicine and to learn the appropriate medicine, surgery or therapeutics in relation to each system such as the cardiovascular system, respiratory system, etc. Unfortunately, textbooks, in general, have not kept pace with these developments and many textbooks still look at medicine from the point of view of each separate discipline. Patients, however, present to the doctor with symp-

toms such as abdominal pain, or swellings in the neck, and don't come neatly labelled as a surgical case or a medical case. The examination of the patient, and his further investigation and management, must take into account both 'medical' and 'surgical' pathologies. The advantages and indications for medical treatment must be reviewed alongside those of surgical intervention. This series of medical textbooks presents such an integrated view of medicine and has been written by a multi-disciplinary team.

One approach to the production of an integrated textbook is to ask a series of specialists from different backgrounds each to prepare a chapter or section looking at the subject from his own view point. Unfortunately, such a strategy frequently results in a disjointed look at the subject and the juxtaposition of sections written by a surgeon, a physician and a general practitioner is a poor substitute for a truly integrated book.

In this book the contributors have worked together as a team, planning and writing the book under the direction of an editor. As a group they have taken overall responsibility for its contents. It is hoped that the result will be a more meaningful integration of the subjects.

A useful aid to the student

A second trend in medical education is the move towards more student-centred learning where the emphasis is on the student and what he learns rather than what he is taught. This is a move away from a more teacher-centred approach when the emphasis is on the teacher and what he teaches.

A student-centred approach results in more effective learning and prepares the student better for his continuing education or life-long learning. This series of books has been designed to provide the teacher and the student with an effective resource for learning. It can be used as a basis for a course where the emphasis is on independent learning, as a resource to provide background information for small group work and as a text for use in relation to a lecture course.

Each volume contains questions relating to the content of the volume. The reader can use these to assess his knowledge of the subject. They can be used either before or after he reads the relevant sections of the book. The reader by trying to answer the questions can obtain an indication as to the extent to which he has mastered the subject and to which further reading is necessary.

A more problem-based approach to medicine

A third trend in medical education is the move towards a more problem-based approach to learning. In the past the emphasis in medical education has been placed on the teaching of facts about patients and their diseases rather than on the application of the facts and the use of the information to solve problems relating to patients. To take account of this trend, each volume in the series contains a section which looks at how patients present with problems relating to the system under consideration. It is hoped that this will encourage a more problem-based approach to medicine and provide a resource which can be used in more problem-based curricula.

Format of books

The volumes in the series have a standard format. Each volume has five sections and each section tackles the subject from a different direction. Section one presents appropriate background information and briefly reviews the relevant general anatomy, bio-chemistry, pathology and epidemiology. Section two considers how to take a history from a patient and conduct a physical examination in relation to the system under consideration. Section three discusses the investigation and management of the common clinical presentations and leads to a series of differential diagnoses. Section four considers the diseases relating to the system and discusses their management. Section five covers in more detail some aspects of the pharmacology and therapeutics.

For whom is the text intended?

Undergraduate students can use the books in this series as they work their way through the curriculum. The series will be of value not only in schools with integrated curricula but in more traditional schools. The texts will provide the necessary information on each subject while at the same time encouraging the student to relate the various subjects he is studying one to another. While the books will be of particular value in the later years of the cirriculum, they can also serve to introduce students to medicine in the earlier years. Many teachers have attempted in recent years to introduce a more clinical approach in the early phases of the medical school curriculum and to relate the basic and paramedical sciences to clinical medicine. This series has been designed to encourage the student to relate the medical sciences to the practice of medicine.

The series also has a place in postgraduate and continuing education. For postgraduate students the series can serve as introductory texts in each area. For doctors who have completed their vocational training, it can provide a useful and up-to-date review of medicine. While participation in courses, attendance at meetings, reading of journals and interaction with colleagues are all useful in continuing medical education, a readily available reference source is also necessary. This series of books can be used for this purpose.

SECTION I

Background to Endocrine Disease

EPIDEMIOLOGY

Apart from obesity, the most commonly occurring endocrine/metabolic disorders are those of the thyroid gland and diabetes mellitus. A survey carried out by the Royal College of General Practitioners found that the average general practitioner is likely to see seven new cases of thyroid disease and ten patients with a new diagnosis of diabetes each year. This compares to 25 patients with asthma and 35 with chronic bronchitis.

Of the 40 patients seen with obesity in the same period, few, if any, are found to have an identifiable endocrine cause. He would see only one new patient with thyroid cancer every 20 years, compared to two new patients with lung cancer every year and one new case of gastric carcinoma every 2 years.

Some diseases of the endocrine glands are rare. The average general practitioner for example, with a practice of 2500 patients, may not see patients with pituitary or adrenal dysfunction in a professional lifetime.

World Health Statistics reveal that direct endocrine causes of death are uncommon, being directly associated with only 1%–4% of deaths (Table 1).

However, disorders affecting the endocrine system and their diseases are more important than these figures might suggest, since:

- endocrine disorders may present in many different ways and affect different systems in the body

  Unless there is an awareness of them, much needless investigation and unnecessary treatment may result.

- patients with endocrine disorders are frequently in the young or middle age groups

- treatment is usually very effective and usually the quality of life for the patient is restored to normal

- hormones are used extensively for the management of a wide range of conditions

  Such therapy is frequently associated with side-effects.

The figures given in Table 2 show the episodes of disease and consultation rates per 1000 population and were published by The Royal College of General Practitioners, Office of Population Censuses and Surveys, Department of Health and Social Security, 1974 (morbidity status from *General Practitioners' Second National Study 1970–1971*. London: HMSO).

Table 2. Episodes of endocrine disease

Disease	Episodes of disease	Consultation rate
Total endocrine nutrition and metabolism	17.0	38.7
Goitre and thyrotoxicosis	0.6	1.7
Cretinism and myxoedema	0.7	0.9
Other thyroid	0.1	0.3
Diabetes	5.1	15.7
Vitamin deficiencies and other nutritional disorders	0.4	0.8

PATHOLOGY

A wide variety of diseases may be associated with

Table 1. Deaths per 100 000 population – endocrine causes

	Canada	USA	Japan	Philippines	Belgium	England & Wales	N. Ireland	Scotland
Total endocrine, nutritional and metabolic disorders	17.3 (2.33%)	21.6 (2.36%)	9.4 (1.49%)	33.2 (4.85%)	38.9 (3.27%)	14.6 (1.23%)	12.2 (1.09%)	15.1 (1.22%)
Non-toxic goitre	–	–	–	0.11 (0.01%)	0.1 (0.01%)	0.1 (0.01%)	–	0.1
Hyperthyroidism	–	–	0.1 (0.02%)	0.1 (0.01%)	0.4 (0.03%)	0.5 (0.04%)	0.5 (0.05%)	0.2 (0.02%)
Diabetes mellitus	14.9 (1.9%)	17.7 (1.93%)	8.1 (1.29%)	2.7 (0.39%)	32.8 (2.75%)	10.5 (0.88%)	7.7 (0.69%)	12.1 (0.98%)
Avitaminosis and other nutritional deficiencies	0.8 (0.1%)	1.2 (0.13%)	0.4 (0.07%)	30.2 (4.41%)	3.9 (0.32%)	0.5 (0.04%)	1.0 (0.09%)	0.5 (0.04%)
Other endocrine and metabolic diseases	2.4 (0.32%)	2.6 (0.28%)	0.7 (0.11%)	0.2 (0.03%)	1.8 (0.15%)	3.1 (0.26%)	3.0 (0.27%)	2.2 (0.17%)

Showing deaths per 100 000 population with percentage of all causes of deaths in brackets, as reported in the World Health Statistics (Geneva: WHO, 1977); and showing geographical differences between the UK causes disturbances of the endocrine system and alterations in one gland may affect the function of another. Disorders of the function of endocrine glands may be reflected by changes in every organ in the body. The endocrine disorders can be classified as disorders associated with:

- excessive hormone action
- inadequate hormone action

Excessive hormone action

Excessive hormone action may be the result of:

- excessive gland function due to hypertrophy, hyperplasia or tumour
- excessive trophic stimulation of the gland
- increased target organ sensitivity
  Here there is normal hormone production but an increase in peripheral tissue response.

Inadequate hormone action

Inadequate hormone action may be:

- The result of primary hypofunction of the gland due to:
  congenital absence or hypoplasia

  abnormalities of hormone biosynthesis
  therapeutic destruction by surgical removal or radiotherapy
  autoimmune destruction
  infarction
  haemorrhage
  infiltration by non-endocrine tissue, e.g. neoplasia, sarcoid
- secondary to lack of trophic hormone
- the result of decreased target organ sensitivity
  This means normal hormone production but a decrease in peripheral tissue response.

Examples of the various types of disorders of the endocrine glands are summarized in Table 3. There are others, discussed in Section IV of the present volume.

THE HYPOTHALAMUS AND THE PITUITARY GLAND

Anatomy and embryology

Hypothalamus

The hypothalamus lies at the base of the brain around the third ventricle. The small mass of cerebral tissue which comprises the hypothalamus (approximately 2.5 g in the adult human) contains both well-defined

Table 3. Various types of endocrine disorders

	Overproduction of hormone	Underproduction of hormone
Hypothalamus	Cushing's disease (some cases)	Isolated GH, LH and FSH deficiencies*
Pituitary	Acromegaly Cushing's disease	Hypopituitarism Short stature
Thyroid	Hyperthyroidism	Hypothyroidism
Parathyroid	Hyperparathyroidism	Hypoparathyroidism
Adrenal	Cushing's syndrome Primary hyperaldosteronism Phaeochromocytoma	Hypoadrenalism (Addison's disease)
Pancreas	Insulinoma Glucagonoma	Diabetes mellitus
Ovary	Rare tumours	Menopause Gonadal dysgenesis
Testes	Rare tumours	Trauma Orchitis

*GH = growth hormone; LH = luteinizing hormone; FSH = follicle stimulating hormone

nuclei and others whose outline is less easily determined. The most easily defined are the

- supraoptic nuclei – composed almost wholly of large nerve cells
- paraventricular nuclei – composed of large nerve cells plus a number of smaller cells

See LD-1a,b and LD-2a,b.

The large cells of these nuclei synthesize:

- vasopressin
- oxytocin
- neurophysins which are specific binding proteins for vasopressin and oxytocin

These pass down the axons of the cells into the hypothalamic–neurohypophyseal tract and are secreted into the capillaries of the posterior pituitary gland.

The nuclei in the basal part of the hypothalamus are concerned mainly with the production of releasing and inhibiting hormones and with the factors which control the activities of the anterior pituitary gland. The efferent nerve fibres from these nuclei are non-myelinated. They join the fibres of the hypothalamic–neurohypophyseal tract to end on the coiled capillaries which form the primary capillary bed of the long and short portal vessels (LD-1a,b and 2a,b).

Large numbers of both afferent and efferent non-myelinated nerve fibres connect the hypothalamic nerve cells with various parts of the cerebral hemisphere and brain stem. They form a capsule of nerve fibres around the hypothalamus.

The fornix, a massive tract of myelinated fibres, brings impulses from each temporal lobe to the ipsilateral mamillary body. The mamillothalamic tract is more medially situated and is composed of myelinated fibres. It connects each mamillary body with the ipsilateral anterior nucleus of the thalamus from which impulses are relayed to the frontal lobes. Efferent hypothalamic autosomal tracts pass down the spinal cord directly to the preganglionic nuclei of the sympathetic and parasympathetic systems.

Pituitary

The anterior pituitary gland or adenohypophysis is formed early in embryonic life. An evagination from the roof of the primitive oral region, *Rathke's pouch*, extends upward toward the base of the brain, and is met by an outpouching of the floor of the third ventricle destined to become the neurohypophysis (posterior pituitary) and the neural portion of the pituitary stalk.

A pair of lateral buds arises from Rathke's pouch and extends superiorly to invest the neural stalk with a thin cloak of cells which later become the pars tuberalis. In the neural tissue lie the various coiled capillary vessels on which end the nerve fibres derived from the cells in the hypothalamus.

The neurohormones coming down these nerve fibres are secreted from the endings of the fibres into the blood passing through the coiled capillaries and thus into the portal vessels.

The epithelial cells forming the anterior lobe of the pituitary secrete their hormones directly into the blood flowing through the sinusoids that run between the cells (LD-2a,b).

The superior hypophysial arteries supply the arterial ring and thereafter the afferent arterioles to the coiled capillaries. The long portal vessels are derived from there and run down the pituitary stalk to supply the larger part of the anterior lobe.

The inferior hypophysial arteries eventually supply the coiled capillaries that form the short portal vessels. These supply a restricted part of the anterior lobe adjacent to the lower infundibular stem.

Physiology and biochemistry

The hormones produced by the anterior pituitary gland are:

- growth hormone (GH)
- prolactin
- thyroid stimulating hormone (thyrotrophin, TSH)
- luteinizing hormone (LH)
- follicle stimulating hormone (FSH) } gonadotrophins
- adrenocorticotrophic hormone (corticotrophin, ACTH)

The secretion of the anterior pituitary hormones is controlled in the case of the trophic hormones (TSH, LH, FSH, and ACTH) by a combination of negative feedback and hypothalamic releasing hormones. The secretion of the non-trophic hormones (growth hormone and prolactin) is controlled by releasing and inhibitory hypothalamic hormones. Hypothalamic releasing hormones form the first part of a cascading amplifier which controls adenocortical, thyroid and gonadal function.

The human hypothalamus contains only about 1 ng of any one releasing hormone and the amount released in a short period into the portal capillaries is extremely small. However, the amount of trophic hormone released from the anterior pituitary gland in response to the signal from the hypothalamus is about a million times greater.

Further amplification, of a similar magnitude, occurs when the hormone acts on the target gland. The amplifier thus has three stages with a very precise sensing system for feedback control at hypothalamic and pituitary level.

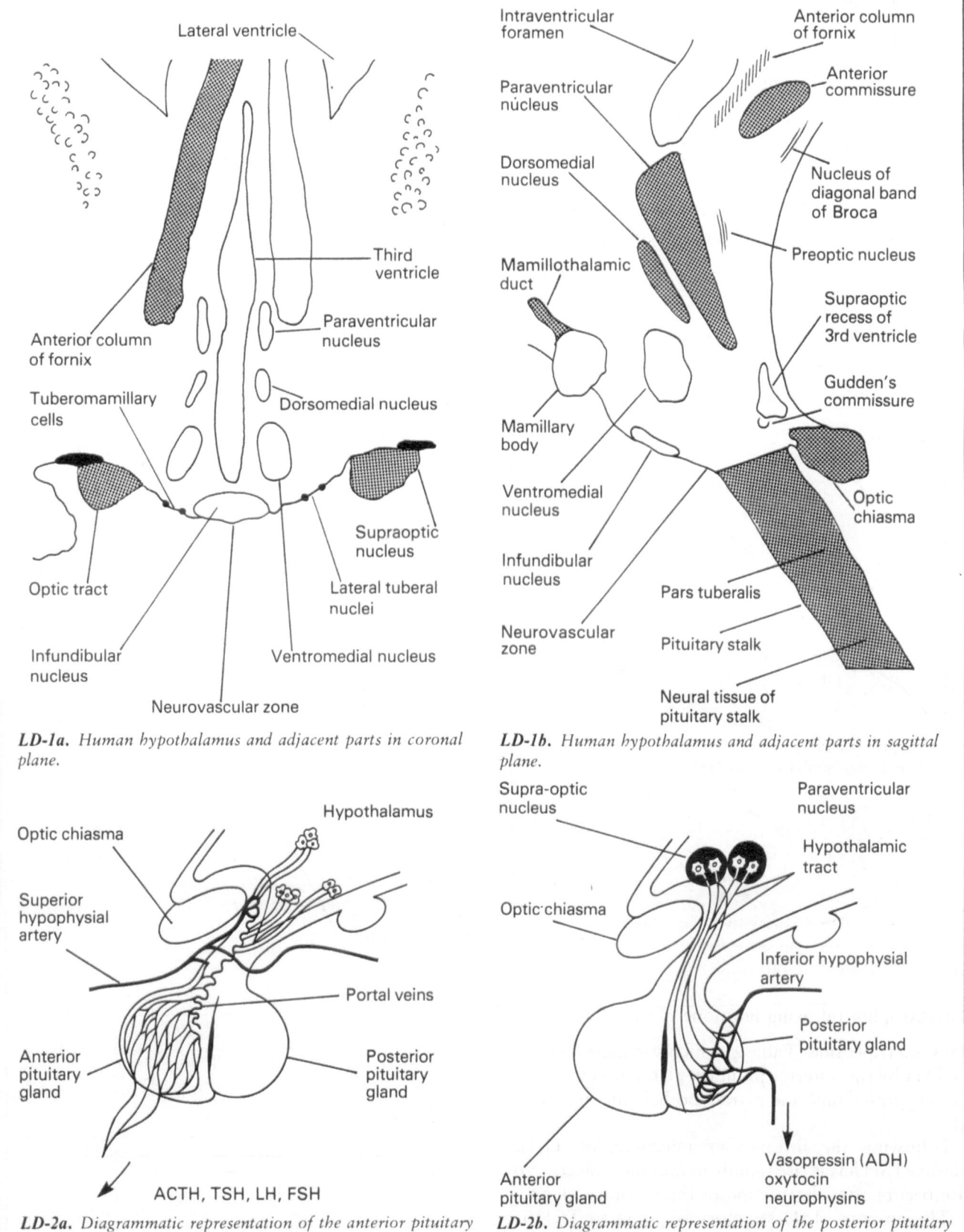

LD-1a. *Human hypothalamus and adjacent parts in coronal plane.*

Lateral ventricle

Anterior column of fornix

Tuberomamillary cells

Optic tract

Infundibular nucleus

Third ventricle

Paraventricular nucleus

Dorsomedial nucleus

Supraoptic nucleus

Lateral tuberal nuclei

Ventromedial nucleus

Neurovascular zone

LD-1b. *Human hypothalamus and adjacent parts in sagittal plane.*

Intraventricular foramen

Paraventricular nucleus

Dorsomedial nucleus

Mamillothalamic duct

Mamillary body

Ventromedial nucleus

Infundibular nucleus

Neurovascular zone

Anterior column of fornix

Anterior commissure

Nucleus of diagonal band of Broca

Preoptic nucleus

Supraoptic recess of 3rd ventricle

Gudden's commissure

Optic chiasma

Pars tuberalis

Pituitary stalk

Neural tissue of pituitary stalk

LD-2a. *Diagrammatic representation of the anterior pituitary and its main hypothalamic connections.*

Optic chiasma

Hypothalamus

Superior hypophysial artery

Portal veins

Anterior pituitary gland

Posterior pituitary gland

ACTH, TSH, LH, FSH

LD-2b. *Diagrammatic representation of the posterior pituitary and its main hypothalamic connections.*

Supra-optic nucleus

Paraventricular nucleus

Hypothalamic tract

Optic chiasma

Inferior hypophysial artery

Posterior pituitary gland

Anterior pituitary gland

Vasopressin (ADH) oxytocin neurophysins

Hypothalamic releasing and inhibiting hormones

Corticotrophin releasing factor (CRF) (LD-3)

This was the first of the hypothalamic regulating hormones to be described; its structure is still not known. A substance which has the ability to release ACTH has been isolated from the hypothalamus. Vasopressin has some of the properties of CRF but evidence suggests that it is not the physiological CRF.

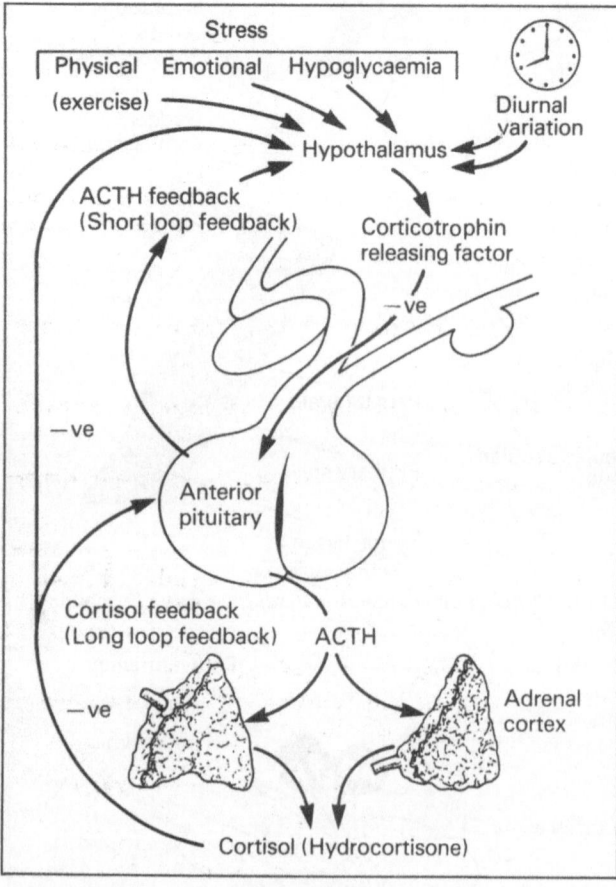

LD-3. *Regulation of ACTH release.*

Thyrotrophin releasing hormone (TRH) (LD-4)

This is a tripeptide (Table 4) which stimulates release of TSH by the anterior pituitary gland. It is found in both hypothalamic and extrahypothalamic sites (LD-4).

Following the intravenous injections of TRH, plasma TSH levels rise rapidly in normal subjects. No rise occurs after destruction of the pituitary gland.

The pituitary TSH secretory response to TRH is modified by the thyroid status of the individual. Excess thyroid hormone inhibits pituitary response

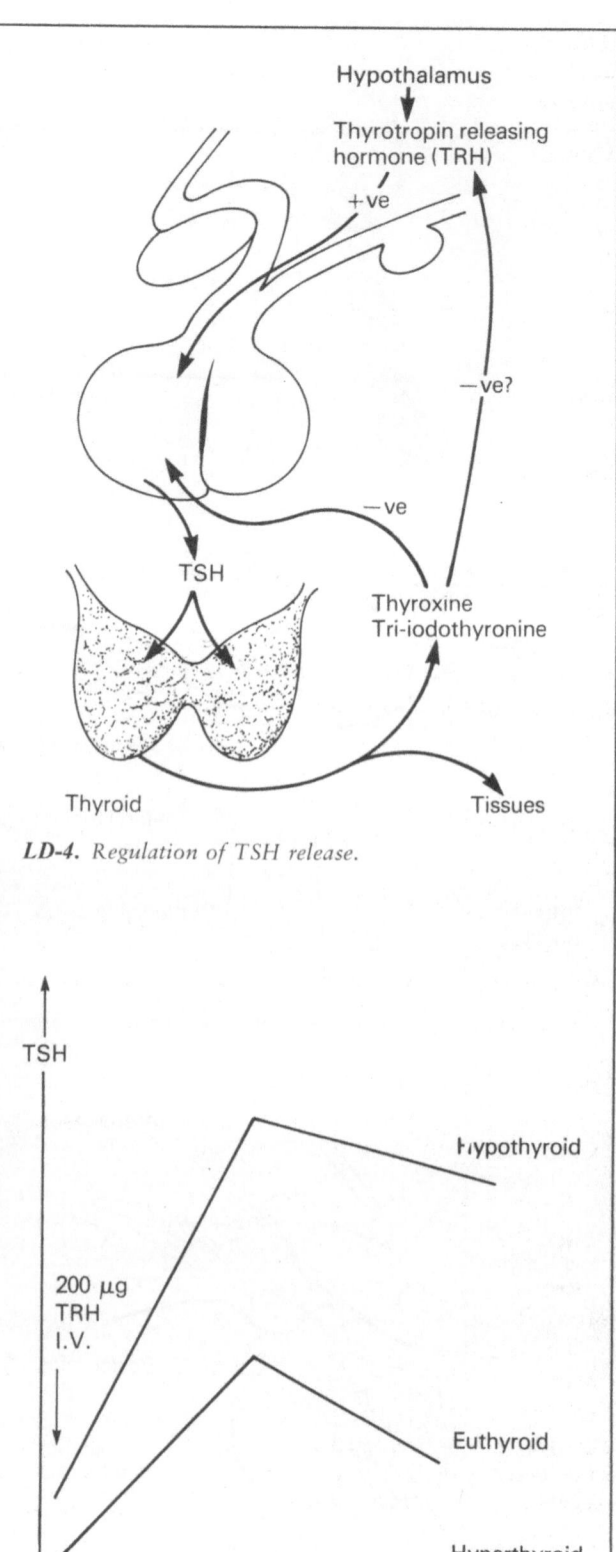

LD-4. *Regulation of TSH release.*

LD-5. *Pituitary TSH secretory response.*

Table 4. Pituitary and hypothalamic hormone sequences

Hormone	Structure
Vasopressin (antidiuretic hormone)	CYS–TYR–PHE–GLN–ASN–CYS–PRO–ARG–GLY–NH$_2$ 1 2 3 4 5 6 7 8 9
	CYS–TYR–ILE–GLN–ASN–CYS–PRO–LEU–GLY–NH$_2$ 1 2 3 4 5 6 7 8 9
Thyrotrophin releasing hormone (TRH)	GLU–HIS–PRO–NH$_2$ 1 2 3
Gonadotrophin releasing hormone (LHRH)	GLU–HIS–TRP–SER–TYR–GLY–LEU–ARG–PRO–GLY–NH$_2$ 1 2 3 4 5 6 7 8 9 10
Somatostatin	H·ALA–GLY–CYS–LYS–ASN–PHE–PHE–TRP–LYS–THR–PHE–THR–SER–CYS 1 2 3 4 5 6 7 8 9 10 11 12 13 14

to TRH and following destruction of the thyroid the pituitary is sensitized to TRH (LD-5).

The administration of TRH also increases the production of prolactin by the anterior pituitary gland. In those disease states where there is increased production of TRH (hypothyroidism) circulating prolactin levels are elevated (see LD-7).

This effect of TRH on prolactin is pharmacological and TRH does not appear to be involved in the physiological control of prolactin.

Within the hypothalamus TRH containing cell bodies are found in the paraventricular region with neurons projecting to the external layer of the median eminence. Extrahypothalamically other nuclei in the ventral horn and the intermediolateral cell column of the spinal cord appear to contain TRH.

Gonadotrophin releasing hormone (GnRH, LHRH)

It is now believed that one hormone is responsible for the release of both the gonadotrophins:

- luteinizing hormone (LH)
- follicle stimulating hormone (FSH)

Gonadotrophin releasing hormone (GnRH, commonly called LHRH) is a decapeptide (Table 4).

The response of the pituitary to LHRH varies throughout life. **N.B.** Before puberty both males and females show only a small rise in the LH response to LHRH. In the early stages of puberty the response to FSH increases and is similar to that seen in adults. As puberty progresses the response to LH increases until it reaches adult levels.

In women there is also a variation in response throughout the menstrual cycle, much greater response being found in the luteal phase than in the follicular phase of the cycle.

The cause of the increase in production of LHRH around midcycle and the increased sensitivity of the anterior pituitary to it is not clear. As the follicular phase progresses more oestradiol-17β is secreted by the ovary producing a positive feedback effect on the anterior pituitary and presumably also on the hypothalamus. Under the influence of a slowly rising oestradiol-17β the pituitary becomes more sensitive to LHRH.

Somatostatin and growth hormone releasing factor (LD-6)

Growth hormone secretion is controlled by means of two hypothalamic factors:

- growth hormone releasing factor – the structure of this is unknown

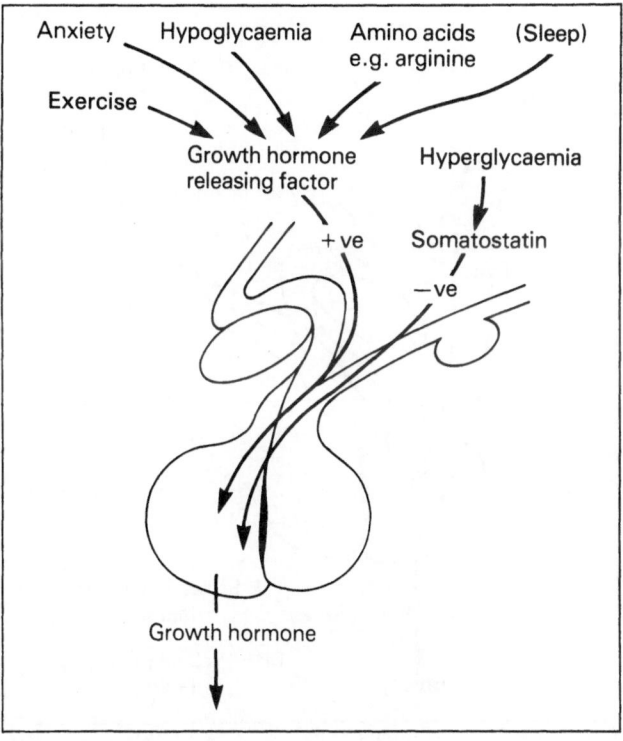

LD-6. Regulation of growth hormone release.

- growth hormone releasing inhibiting factor (somatostatin)

Somatostatin is a tetradecapeptide (Table 4). It has been detected in high concentrations in the median eminence and the hypothalamus, and in lower but significant concentrations in other areas of the brain. It has also been detected in the pancreas and the gastrointestinal tract.

Other actions of somatostatin include the reduction of both basal and stimulated insulin, glucagon and gastrin levels.

Prolactin inhibitory factors (LD-7)

Prolactin is unique among the anterior pituitary hormones in that its secretion is under tonic inhibitory control by the hypothalamus. The principal inhibitory factor is believed to be dopamine, though prolactin production is also reduced by noradrenaline.

Dopamine terminals abut directly on to the portal capillary bed, suggesting that dopamine is released directly into the hypophyseal portal capillaries. Dopamine receptors are found in prolactin secreting cells. There is evidence to suggest that there may be other prolactin inhibitory factors.

Prolactin releasing factors (LD-7)

Thyrotrophin releasing hormone does stimulate the release of prolactin but it is not believed to be of physiological importance.

An additional prolactin releasing factor has been postulated but not identified. Breast stimulation increases prolactin secretion probably by decreasing dopamine production.

The hormones of the anterior pituitary gland

These are:

- luteinizing hormone (LH)
- follicle stimulating hormone (FSH)
- prolactin
- thyroid stimulating hormone (TSH)
- adrenocorticotrophic hormone (ACTH) and related compounds
- growth hormone (GH)

Gonadotrophins (LD-8a,b)

These are:

- luteinizing hormone (LH)
- follicle stimulating hormone (FSH)

Structure

The gonadotrophins are both glycoproteins, with molecular weights of 29 000 and 32 000 respectively. Each is composed of two subunits.

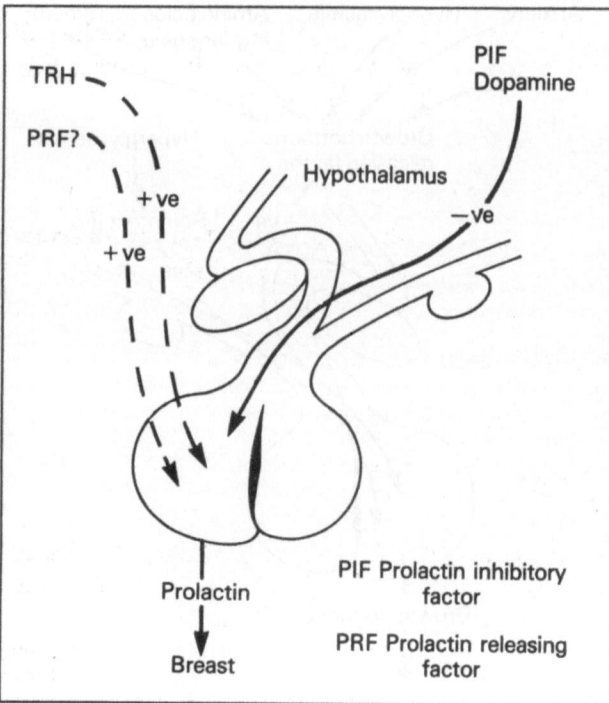

LD-7. Control of prolactin secretion.

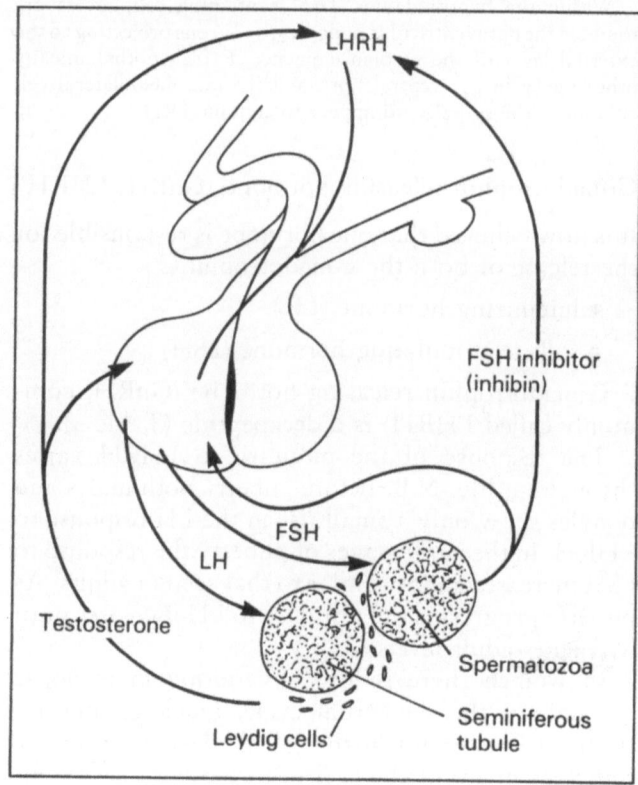

LD-8a. Gonadotrophin releasing hormone Gn RH/LHRH and FSH-LH release in the male.

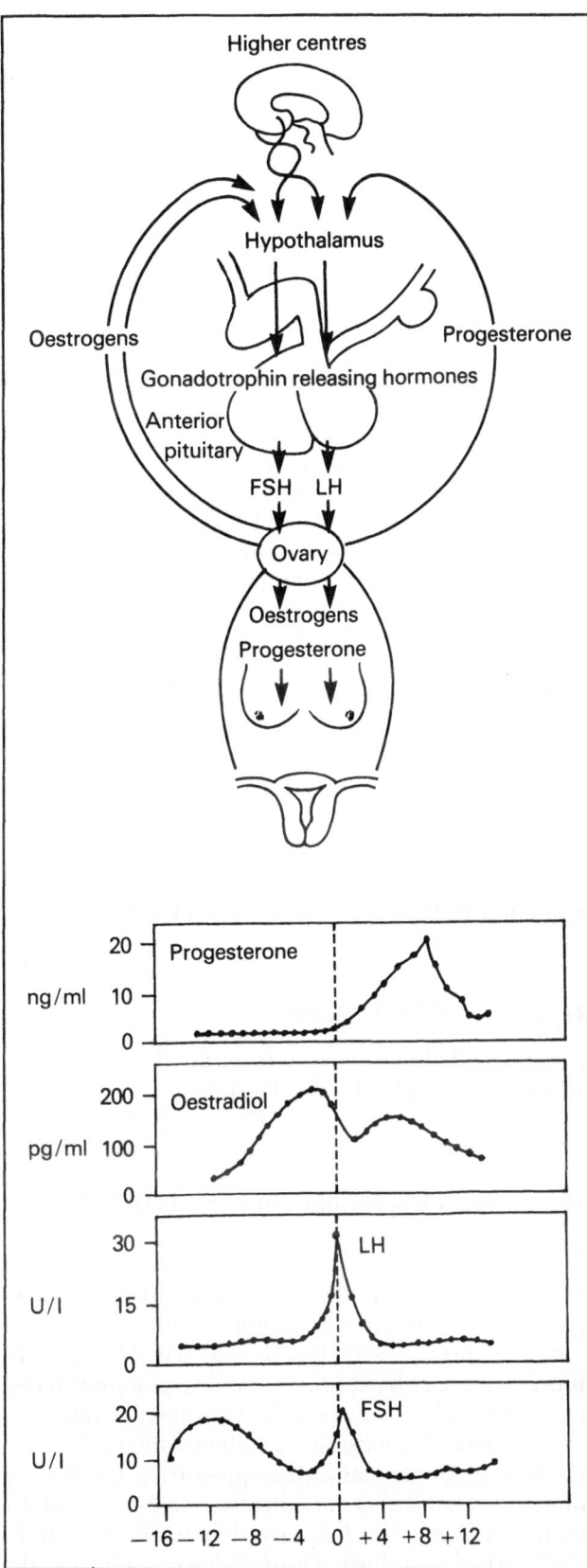

LD-8b. Gonadotrophin releasing hormone Gn RH/LHRH and FSH-LH release in the female.

One, the beta-subunit, is unique for each hormone, but the alpha-subunit is common to both LH and FSH as well as to thyroid stimulating hormone (TSH) and human chorionic gonadotrophin (HCG).

Biological activity

The release of luteinizing and follicle stimulating hormones is governed by the release of LHRH from the hypothalamus.

LH and FSH act synergistically in *the female* to induce (LD-8b):

- ovarian growth
- oestrogen biosynthesis
- oestrogen secretion

FSH is primarily responsible in the female for follicular development and this effect is augmented by oestradiol-17β and LH.

In *the male* (LD-8a) FSH is mainly responsible for spermatogenesis, an effect augmented by testosterone and LH.

LH affects the ovary by inducing:

- a rapid preovulation enlargement of one or more selected follicles
- ovulation
- transformation of granulosa cells to luteal cells and thus formation of the corpus luteum

It is also responsible for maintenance of the corpus luteum and secretion of progesterone.

In *the male* LH is responsible for the morphological development and stimulation of androgen secretion by the Leydig cell.

Pattern of secretion

The secretion of the gonadotrophins at different stages in the development of gonadal function

- infancy
- puberty
- adult life
- menopause
- old age

is qualitatively and quantitatively different.

In the *adult* state the secretion of LH and, to a lesser degree, FSH is pulsatile without there being a true circadian rhythm.

The levels of both LH and FSH are low in infancy and childhood with FSH always greater than LH in children of both sexes. Prepubertally the administration of LHRH gives a greater rise in FSH than in LH.

During puberty the levels of both hormones increase.

In boys FSH levels increase rapidly during the early phases but during the final stages of sexual maturation FSH levels stabilize to a plateau whereas LH and testosterone show a progressive and steady increase throughout puberty.

In girls FSH levels rise during the early stages of puberty and then plateau, whereas LH levels rise steadily during sexual maturation with a rapid rise in the final stages prior to the onset of menstruation.

The menstrual cycle

In all ovulatory cycles in the adult female there is a peak of LH and FSH at midcycle. In addition to these regular midcycle peaks the gonadotrophin levels are generally higher in the preovulatory period than during the luteal phase. In particular FSH levels are higher during the first few days of the follicular phase and subsequently decrease during the days that precede ovulation (LD-8b).

The elevated levels of FSH at the beginning of the cycle are less than those at midcycle. During the cycle the secretion of LH and perhaps also of FSH occurs in an episodic manner with short periods of release followed by periods when there is little if any secretion. Oestrogen levels reach their maximum in the 48 hours before the ovulatory peak of LH and trigger the release of the gonadotrophins (LD-7).

The menopause and later life

Basal gonadotrophin levels in postmenopausal women are elevated with FSH levels always higher than those seen in the course of the normal menstrual cycle. LH levels are approximately equal to the maximum values seen during the menstrual cycle.

In men over the age of 50 years there is a reduction in spermatogenesis and in testicular production of testosterone with a consequent increase in circulating FSH and LH levels.

The administration of testosterone to males is followed by a decrease in plasma LH levels; the effects on FSH are less well defined.

A substance, probably produced by the graafian follicle, called 'inhibin' selectively inhibits the secretion of FSH.

In the female, oestrogens usually inhibit FSH secretion whereas they exert a biphasic action on LH – first negative then positive. The magnitude and time course of the responses depend upon the stage of the menstrual cycle at which they were administered.

Prolactin (LD-7)

Structure

Prolactin is a single chain polypeptide with a molecular weight of approximately 21 500 and is very similar in structure to growth hormone (GH) and to human placental lactogen (HPL).

Biological activity

In females its principal actions are to stimulate the growth and differentiation of breast tissue and milk production.

In males no functional role for prolactin has yet been found.

In excess, prolactin blocks the actions of LH and FSH on gonadal tissue in both males and females.

Thyroid stimulating hormone (TSH) (LD-4)

Structure

TSH is a glycoprotein with a molecular weight of approximately 26 000. It is composed of two chemically distinct subunits – α and β which are linked by covalent bonds. The structure of the α subunit is similar to that of LH, FSH and HCG.

Biological activity

TSH stimulates the uptake of iodide by the thyroid gland as well as the synthesis and release of thyroid hormones. Continued stimulation of the thyroid by TSH results in hypertrophy of the gland and an increased blood flow through it.

TSH deficiency results in glandular atrophy and a reduction in thyroid hormone production.

Regulation of TSH release

Control of TSH production is outlined in the section on the thyroid gland (see LD-4).

Adrenocorticotrophic hormone (ACTH) (LD-3)

Structure

ACTH is a polypeptide with a molecular weight of 4500. It is composed of 39 amino acids.

It has now been shown that ACTH and β-lipotrophin (a polypeptide containing 91 amino acids) are produced by the same cells in the anterior pituitary cell and probably exist in a combined molecule. This has been given the name pro-opiocortin. The first 13 amino acids of ACTH are identical to α-MSH whilst amino acids 41–58 of β-lipotrophin are identical to β-MSH. Also contained within β-lipotrophin are the amino acid sequences which form metenkephalin (61–65) and β-endorphin (61–91).

| 1 | ACTH | 39 | 1 | β-lipotrophin | 91 |

| 1 α-MSH 13 18 | CLIP | 39 | 41 | β-MSH | 58 |

61 65
metenkephalin

61 β-endorphin 91

Biological activity

The principal physiological action of ACTH is to stimulate the adrenal cortex to produce cortisol. Other adrenal steroids including corticosterone and the 'adrenal androgens' are also produced in response to stimulation by ACTH. ACTH also promotes adrenal blood flow and hypertrophy of the gland.

Regulation of ACTH release

A reduction in circulating cortisol will, in the normal individual, result in a rise in ACTH output. Conversely, a rise in circulating cortisol will result in a fall in ACTH output. This may be due, in part, to a direct action of cortisol on the anterior pituitary gland but is probably mainly mediated through the hypothalamus via corticotrophin releasing factor (LD-3).

The output of ACTH by the pituitary is not constant throughout the day but is greatest in the morning and at its lowest level in the late evening. This nyctohemeral rhythm appears to be related to activity rather than light and dark and may be reversed in those who work prolonged periods of night shift. The output of ACTH is also increased by both emotional and physical stress.

β-Lipotrophin, enkephalins, endorphins and related compounds

α-MSH and corticotrophin-like intermediate lobe peptide (CLIP) are only found in the human pituitary gland during fetal life and during pregnancy, when a distinct intermediate lobe can be found. γ-Lipotrophin, the first 58 amino acids of the β-lipotrophin, is also secreted by the human pituitary but β-MSH has never been found in the human, and material reported as human β-MSH is an artefact of the extraction procedure.

Enkephalins were discovered when pharmacological studies into the effects of opiates revealed that the brain itself contained a morphine-like factor localized in the synaptosomal fraction. The morphine-like factor was isolated and its structure elucidated. It was found to be a mixture of two pentapeptides, metenkephalin and leu-enkephalin. The structure of metenkephalin is contained in β-lipotrophin, forming the first part of the molecule remaining after cleavage to give γ-lipotrophin. The opiate action of the entire 31 amino acid fragment was then studied and found to be more potent than metenkephalin.

Endorphins is the general name given to the peptides with opiate properties related to β-lipotrophin. The 31 amino acid fragment is known as endorphin. The release of β-endorphin from the pituitary has been shown to be regulated in parallel with ACTH but the function of β-endorphin is still not known; it is possible that these peptides may be involved in the control of mood.

Growth hormone (LD-6)

Structure

Growth hormone is a polypeptide and has a molecular weight of approximately 22 000, and is structurally similar to prolactin and to human placental lactogen.

Biological activity

It was originally thought to be solely concerned with growth in the early years of life. It has now been shown to exert physiological actions throughout life, including:

- facilitating amino acid transport and incorporation into protein

- mobilizing free fatty acids from peripheral fat stores
- reducing lipid synthesis

The stimulation by growth hormone of collagen and protein synthesis in cartilage is mediated by somatomedin, a low molecular weight plasma component which is produced by the liver in response to growth hormone.

As part of its general anabolic action growth hormone promotes retention of:

- calcium
- phosphorus
- nitrogen

It is responsible for the increased serum inorganic phosphate and alkaline phosphatase levels seen in children. Growth hormone opposes the actions of insulin, and elevated growth hormone levels (as in acromegaly) can produce glycosuria, impaired glucose tolerance and resistance to insulin.

Regulation of growth hormone release

The hypothalamus exerts a predominantly stimulatory effect on secretion of growth hormone in the mammal. This stimulatory factor has not been isolated although the structure of the inhibitory factor, somatostatin, is known (LD-6).

Plasma levels of growth hormone measured at frequent intervals throughout day and night show a pulsatile release with the largest peaks occurring during the first 2 hours of night sleep. The number and magnitude of spontaneous growth hormone bursts are age dependent, increasing during the growth spurt of adolescence and declining thereafter. The sleep-associated release of growth hormone is commonly absent in adults over 50 years of age.

Exercise and stress, both physical and emotional, also cause release of growth hormone.

Hyperglycaemia temporarily suppresses daytime surges of growth hormone secretion while insulin induced hypoglycaemia increases daytime growth hormone secretion.

Although the hypothalamus appears capable of maintaining basal growth hormone secretion, reflex growth hormone secretion associated with stress or sleep may require neural input from higher centres such as the hippocampus. Feedback regulation of growth hormone secretion may occur by the direct effect of the hormone on its own secretion or by the effects of somatomedins which are generated in the liver.

The hormones of the posterior pituitary gland

The hormones secreted by the posterior pituitary gland are:

- oxytocin
- vasopressin or antidiuretic hormone, ADH (see Table 4)

They are synthesized in the supraoptic and paraventricular nuclei. After synthesis ADH and oxytocin are transported along the axons of the neurohypophyseal tract and then stored in the posterior pituitary. Within the axons the hormones are contained in neurosecretory granules bound to their specific neurophysins.

Vasopressin (ADH) (LD-9)

Vasopressin is not only secreted by the neurons of the hypophyseal tract but also found in the hypophyseal portal blood supply. This suggests that in addition to its antidiuretic role it may also affect anterior pituitary function.

Under normal conditions ADH release is primarily regulated by osmoreceptors which are probably located in the supraoptic nuclei. There are also volume receptors in the left atrium and baroreceptors in the aorta and carotid arteries which relay impulses to the hypothalamus via the vagus nerves. These produce

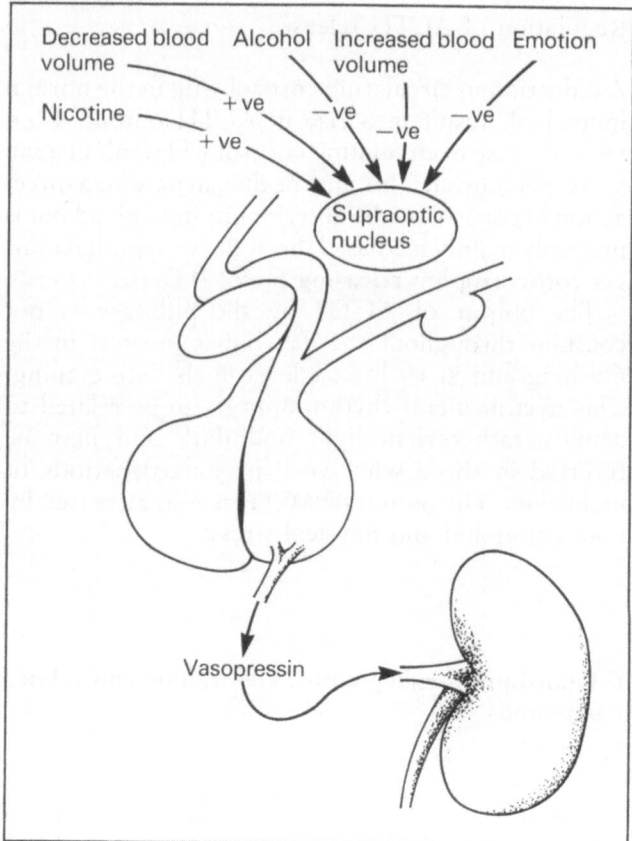

LD-9. Vasopressin (antidiuretic hormone).

tonic and episodic inhibition of vasopressin release. Vasopressin acts to conserve water and to concentrate urine. This action assists in maintaining the constancy of the osmolality and volume of body fluids.

Diabetes insipidus is due to impaired renal conservation of water. It results from low blood levels of ADH. Excessive production of ADH causes an inability to secrete or dilute urine. Ingested fluids are retained and a dilutional hyponatraemia develops.

Oxytocin (LD-10)

Oxytocin is also a nonapeptide, differing by only two amino acids from vasopressin (Table 4). Along with its specific neurophysin, oxytocin is released in response to manipulation or distension of the female genital tract or, postpartum, to suckling. It acts on the excitable membranes surrounding the myometrial and myo-epithelial cells of the uterus and results in an increased force of contraction. Sensitivity of the myometrium to oxytocin increases with the duration of pregnancy but the mechanism of the initiation and maintenance of labour is not known (LD-10).

Postpartum oxytocin continues to exert a contractile action on the myometrium as well as contracting the myo-epithelial cells of the mammary alveoli causing them to expel milk.

Despite the similarities in structure there is little

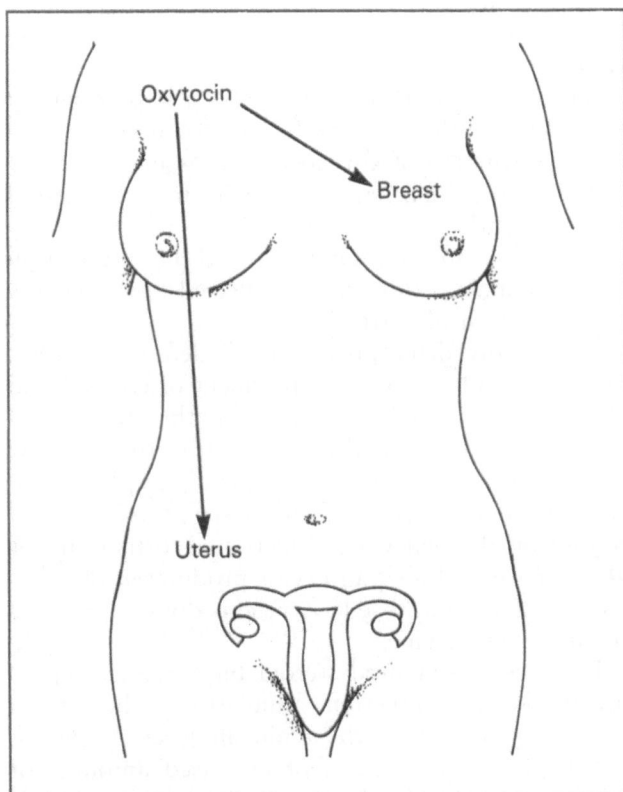

LD-10. *Oxytocin.*

crossover in the actions of ADH and oxytocin. The relative potencies of vasopressin and oxytocin are:

	Oxytocin	*Vasopressin*
Antidiuretic potency	1	200
Milk ejecting activity	100	1

THE ADRENAL GLANDS

The adrenal glands each consist of two separate endocrine structures:

- the adrenal cortex
- the adrenal medulla

These are embryologically, histologically and functionally distinct.

Embryology

The adrenal cortex is of mesodermal origin, while the adrenal medulla arises from ectodermal tissue. The fetal adrenal cortex is larger than that found in adult life and consists mainly of large fetal cells with a narrow rim of definitive cortical cells at the superior margin. ACTH is essential for the development of the fetal zone of the adrenal gland.

Anatomy

The adrenal glands are paired pyramidal structures adhering to the upper poles of the kidneys. The inner reddish part is the medulla while the outer adrenal cortex is yellow in colour. The normal adult gland weighs approximately 4–5 g.

The adrenal gland has a very rich vascular supply.

These vessels run along each of the three surfaces of the gland then break up into small vessels which penetrate the capsule to form a subcapsular plexus. Capillary loops then pass downwards between the cortical cells to enter the thin-walled vascular sinusoids of the zona reticularis. The blood then passes to the adrenal medulla and thence to the adrenal veins. On the left side the adrenal vein drains into the renal vein and on the right side the adrenal vein drains directly into the inferior vena cava. The blood supply to the adrenal medulla contains a high concentration of adrenocortical steroids.

The adrenal medulla is richly innervated by the autonomic nervous system but there is no innervation of the cells of the cortex.

The adrenal cortex consists of three layers or zones:

- zona glomerulosa producing aldosterone
- zona fasciculata
- zona reticularis producing glucocorticoids, androgens and to a much lesser extent oestrogens

In the centre of the gland, the muscle of the central vein is concentric, but away from the centre it is composed of eccentric longitudinal bundles between which pass the small venules draining the cortex. These fibres are innervated by postganglionic sympathetic nerves.

It appears therefore that control of secretion from the adrenal cortex may be partly neurogenic through the peculiar arrangement of the longitudinal muscle in the adrenal vein and partly humoral from the liberation of endogenous ACTH.

THE ADRENAL CORTEX

Physiology and biochemistry

All steroid hormones contain the basic perhydro-cyclopentone phenanthrene nucleus

and are derived from cholesterol.

The adrenal cortex can synthesize cholesterol from acetate or take it up from the circulation. The principal adrenal steroids in the human are:

- cortisol
- aldosterone

An adequate production of both of these steroids is essential for life.

Androgens are also secreted, especially:

- androstenedione
- 11β-hydroxyandrostenedione
- dehydroepiandrosterone
- dehydroepiandrosterone sulphate

The routes of biosynthesis and the enzyme systems involved in the biosynthesis of the adrenal steroids are shown in LD-11. Deficiencies in the production of any of these enzymes result in reduced synthesis of cortisol and/or aldosterone.

Regulation of steroid biosynthesis

The regulation of cortisol production by the adrenal cortex is under the control of ACTH (see LD-3). A normal adult secretes between 10 and 30 mg/day of cortisol (27.6–82.8 μmol/day) the higher levels being found in those with the greatest body mass.

Aldosterone production is controlled mainly by the renin–angiotensin system and by the plasma concentrations of sodium and potassium (LD-12). The secretion of aldosterone varies according to the sodium content of the diet but on an unrestricted sodium diet ranges from 62 to 27.5 μg/day (170–760 nmol/day). A reduction in sodium intake will, in normal individuals, lead to a marked rise in aldosterone secretion. An increase in serum potassium concentration stimulates and hypokalaemia inhibits aldosterone production (LD-13).

Impaired steroid biosynthesis

A low circulating level of cortisol and of aldosterone will stimulate production of ACTH and angiotensin respectively.

There are three possible reasons for impairment of steroid biosynthesis by the adrenal cortex:

1. Destruction of adrenocortical tissue.
2. Enzyme deficiencies which impair steroid biosynthesis.
3. Inadequate stimulus – i.e., if ACTH production is impaired this will be reflected in a very low secretion of cortisol.

Genetically determined enzyme defects are rare, but may result in severe impairment of cortisol and aldosterone production which is life threatening and must be diagnosed and treated within the first few days of life (see Section IV). Depending upon which enzyme is deficient, the presenting symptoms vary and result from deficiency of production of cortisol and/or aldosterone and also an excess production of other adrenal steroids whose biosynthesis does not require the deficient enzyme.

In cases of impaired steroid biosynthesis due to impaired enzyme function, stimulation of the adrenal cortex by ACTH or the renin–angiotensin system results in the production of increased amounts of steroid precursors in an attempt to achieve a satisfactory plasma level of cortisol and aldosterone. In very

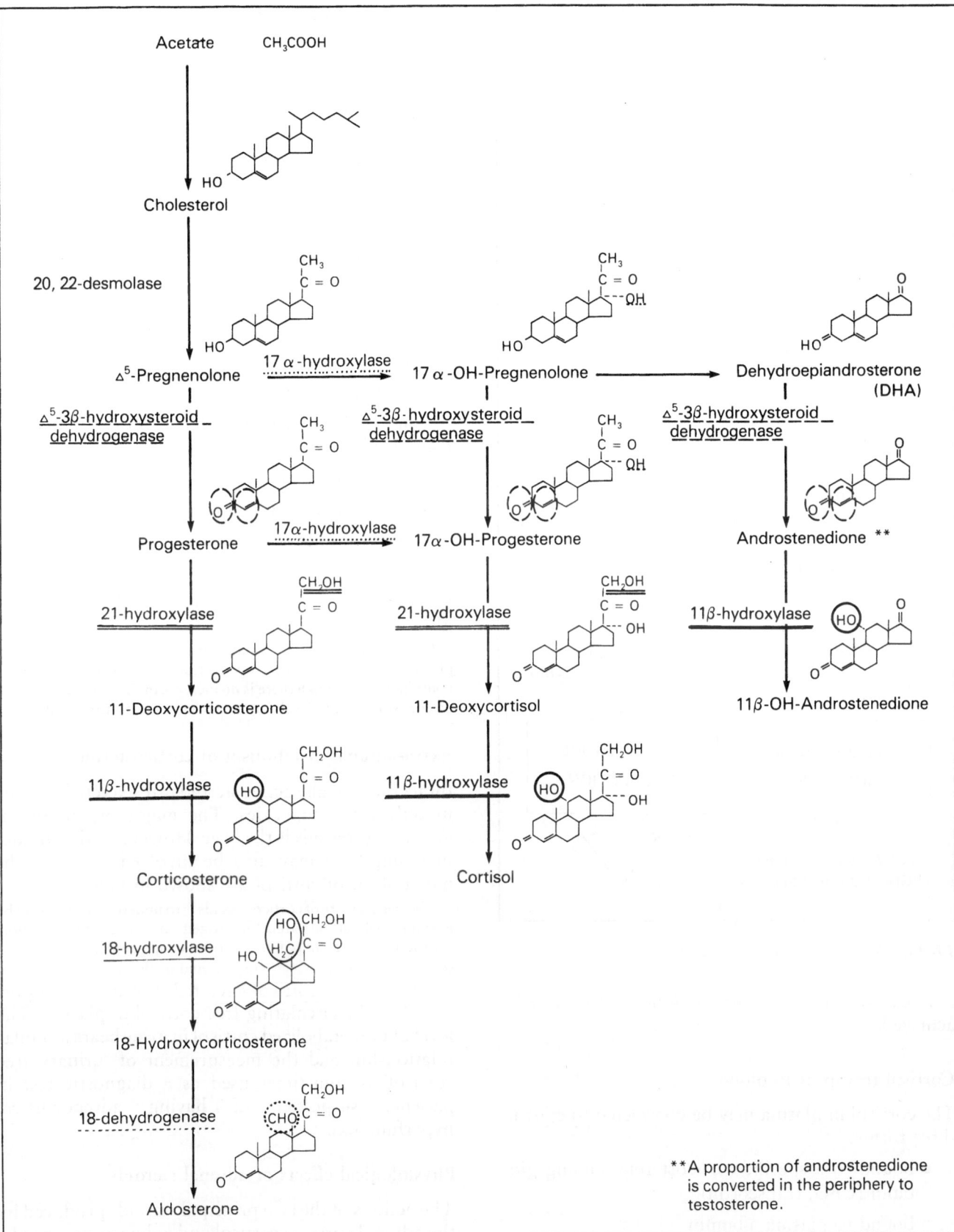

LD-11. *Biosynthesis of adrenal steroids.*

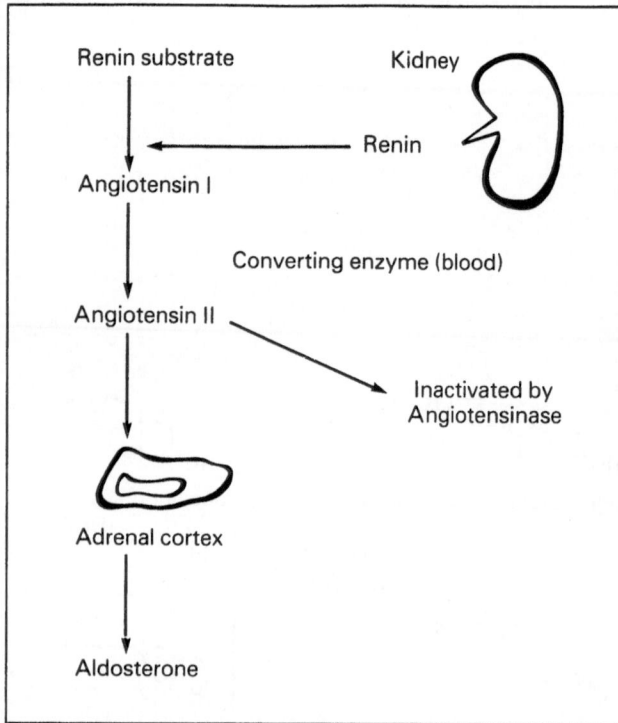

LD-12. *Aldosterone production.*

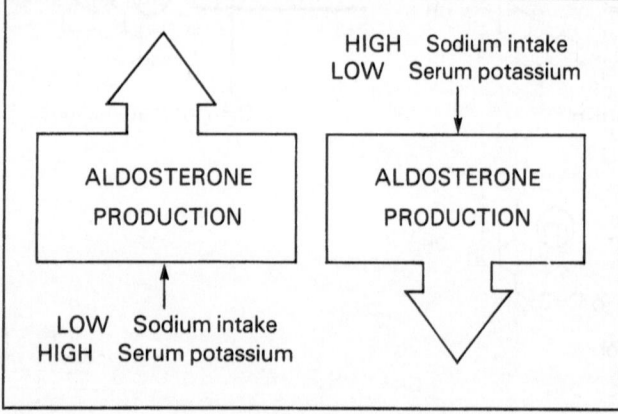

LD-13. *Effects of aldosterone production.*

mild cases of impaired enzyme function this may be achieved.

Cortisol transport in blood

The cortisol in plasma may be considered to exist in three forms:

- bound to a specific corticosteroid-binding globulin (CBG, transcortin)
- bound to plasma albumin
- unbound or free

Transcortin has a high affinity for cortisol but a relatively low capacity; conversely, albumin has a low affinity but an almost infinite capacity. The capacity of transcortin becomes saturated when the total plasma cortisol concentration reaches approximately 500 mmol/l. At this concentration approximately 5% (25 mmol/l) of cortisol is in the unbound or free form. At plasma cortisol concentrations in excess of 500 mmol/l the level of free (unbound) cortisol increases rapidly.

The protein binding of cortisol is completely reversible.

Transcortin also has a high affinity for:

- progesterone
- deoxycorticosterone
- corticosterone

Some other hormones have little affinity for transcortin:

- dehydroepiandrosterone (DHA)
- androstenedione
- testosterone
- oestradiol-17β

but the latter two are bound by a different specific globulin.

In normal pregnancy, and during treatment with supraphysiological doses of oestrogens, transcortin levels increase. In consequence although there is no increase in the adrenal secretion of cortisol, plasma levels are increased two- to threefold.

Extra-adrenal metabolism of corticosteroids

Intact biologically active steroids are excreted in urine in only trace quantities. The major organ for inactivating steroids is the liver. However, other tissues including skin may also be involved. Steps in the metabolism of cortisol are shown in LD-14.

The general fate of corticosteroids is to undergo inactivation by enzymes which introduce hydrogen atoms at one or more positions. The compounds are then conjugated to form water soluble derivatives which are excreted in the urine (LD-15).

The level of cortisol excreted in saliva is proportional to the circulating free cortisol in plasma. The level of unmetabolized cortisol in urine bears a similar relationship and the measurement of 'urinary free cortisol' is sometimes used as a diagnostic test in patients suspected of having adrenocortical hyperfunction.

Physiological effects of adrenal steroids

The actions of the two principal steroids produced by the adrenal cortex – cortisol and aldosterone – can be divided into two general categories:

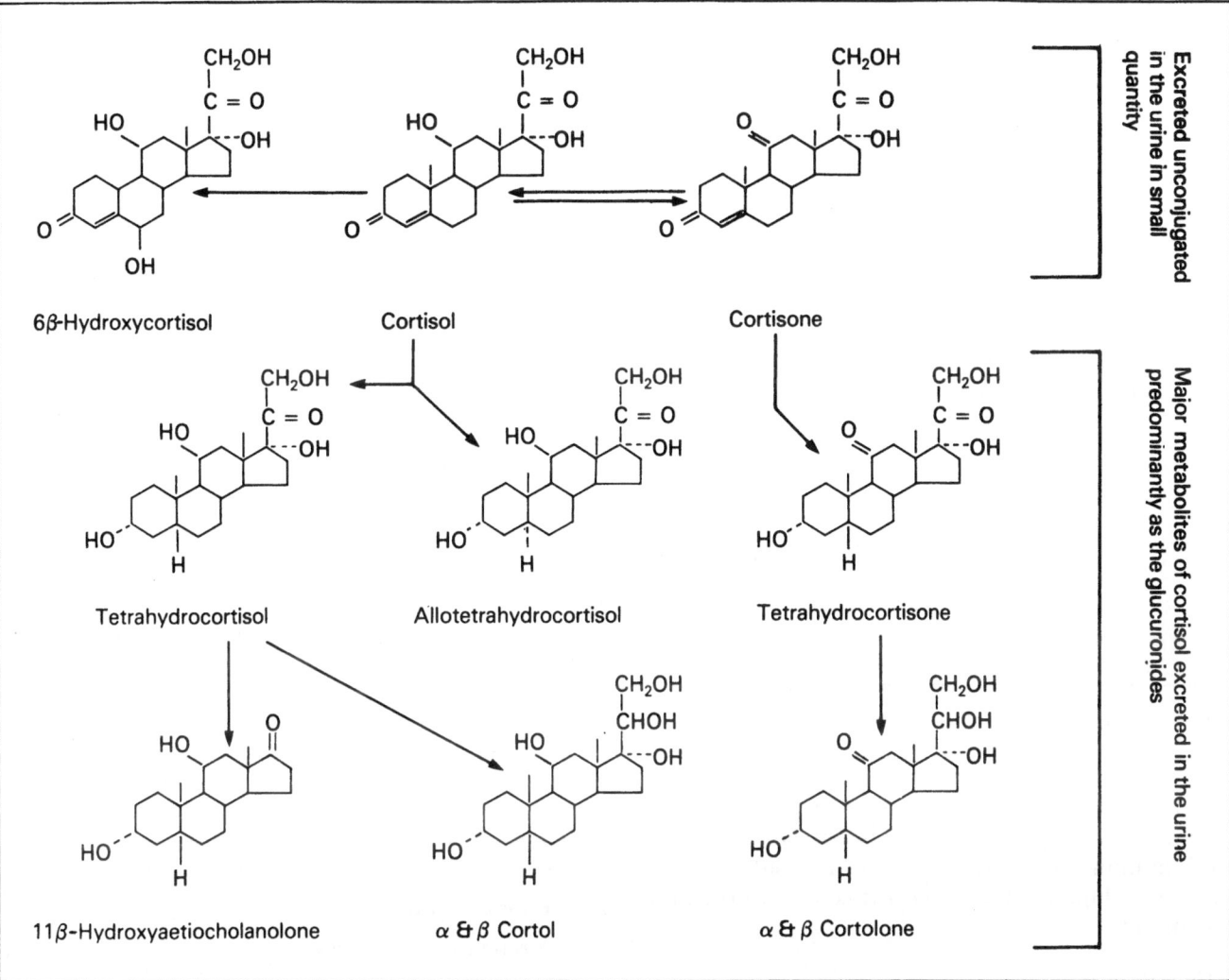

LD-14. Major steps in the metabolism of cortisol by the liver.

- glucocorticoid
- mineralocorticoid

Glucocorticoids

These are concerned with organic metabolism – carbohydrate, fat and protein – inflammation and wound healing, muscle and myocardial integrity.

Table 5. Actions of glucocorticoids

Actions	Effect of excess
Metabolic	
Gluconeogenesis	increased
Liver glycogen deposition	increased
Protein catabolism	increased
Protein synthesis	decreased
Amino acid uptake peripherally	decreased
Amino acid uptake by liver	increased
Fat deposition on face	increased
neck	increased
trunk	increased
Uric acid secretion	increased
Sodium retention	increased
Potassium excretion	increased
Anti-inflammatory	
Host response to infection	decreased
Lysosomes	stabilized
Lymphocyte transformation	inhibited
Granuloma formation	inhibited
Immunosuppresive	
Delayed hypersensitivity	decreased
Antibody production	decreased
Circulating lymphocytes	decreased
Size of lymph nodes – thymus	decreased
spleen	decreased
Eosinophils	decreased
Other	
Some enzymes	induced
Haematopoeisis	increased
Free water clearance	increased

CH₂OH shown as CH_2OH in structures.

LD-15. *Conjugation and excretion of corticosteroids.*

Cortisol is the most important glucocorticoid in man. Some of the more important actions of glucocorticoids are outlined in Table 5. The anti-inflammatory and immunosuppressive actions are only evident when very high levels of glucocorticoids are present in the blood.

Mineralocorticoids

These are concerned with electrolyte balance. Aldosterone is the most important mineralocorticoid.

The major actions of mineralocorticoids are on ion transport by epithelial cells. They result in sodium conservation and loss of potassium.

Absence of mineralocorticoid activity may result in lethal wastage of sodium and retention of potassium.

The action of aldosterone on the distal tubule is shown in LD-15.1. The same exchange takes place in the:

- sweat glands
- salivary glands
- intestinal mucosa

between intra- and extracellular water.

Aldosterone acts at cellular level by affecting the activity of the sodium pump. Aldosterone is bound in target cells to specific cytoplasmic G nuclear receptors. This nuclear binding induces the new protein synthesis, and this protein in turn either increases the efficiency of the sodium pump or increases the permeability of cell membranes to sodium.

Absence of mineralocorticoid activity results in:

- loss of sodium
- retention of potassium

Excess mineralocorticoid activity causes:

- potassium depletion
- excessive sodium retention, with consequent oedema
  hypertension and
  suppression of renin production

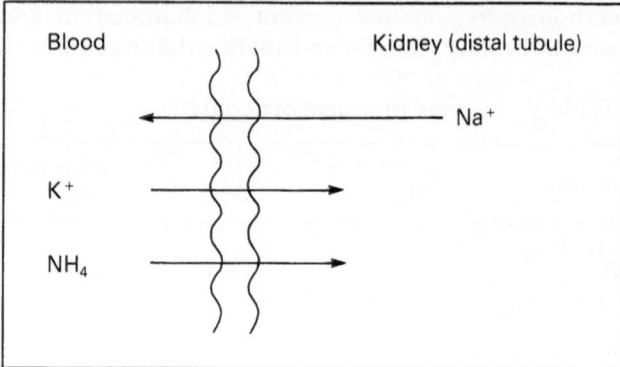

LD-15.1. *Effects of aldosterone on the kidney.*

Most steroids, both naturally occurring and synthetic, contain some aspects of both types of activity although either glucocorticoid or mineralocorticoid activity predominates. The relative glucocorticoid, mineralocorticoid and anti-inflammatory activity of various naturally occurring and synthetic corticosteroids is shown in Table 6.

Physiological actions of adrenal androgens and oestrogens

Under normal circumstances in adults, adrenal production of androgens and oestrogens is trivial compared with production by the gonads. A pathological increase in androgen production can, however, induce the development of masculine secondary sex characteristics. Feminization due to adrenal oestrogen production is rare.

In the adult male the effects of excess adrenal androgen may pass unnoticed but in the female or immature male they are conspicuous.

Excess production of oestrogen by the adrenal might go unnoticed in the adult female but would lead to loss of masculine features in the male and to enlargement of the breasts in both males or immature females.

Disease of the adrenal cortex

Diseases of the adrenal cortex are rare. Many more patients will be investigated to exclude the possibility of a diagnosis of adrenal disease than to confirm it.

As with most endocrine glands, disease of the adrenal cortex falls into two groups:

- increased production of hormones
- decreased production of hormones

Increased production of cortisol

The combination of symptoms found when there is increased production of cortisol by the adrenal cortex is called 'Cushing's syndrome'. It may result from increased stimulation of the adrenal cortex by increased production of ACTH by the anterior pituitary (Cushing's disease) or by the production of ACTH by a tumour, e.g. in the bronchus (ectopic ACTH production). Increased production of cortisol may also be the result of an autonomous tumour of the adrenal cortex.

Excess aldosterone production

This may result from:

- autonomous production by the adrenal
  This is primary hyperaldosteronism and may be due to hyperplasia or tumour (Conn's syndrome).
- increased activity in the renin–angiotensin system
  This can occur in unilateral renal artery obstruction.

Decreased hormone production

Failure of production of ACTH will lead to a reduction in the secretion of cortisol.

Destruction of the adrenal cortex will reduce the production of both cortisol and aldosterone (Addison's disease). Abnormalities in the biosynthesis pathways involved in steroid biosynthesis will result in a reduction in the secretion of both cortisol and aldosterone.

THE ADRENAL MEDULLA

The adrenal medulla is composed of cords of polyhedral or columnar chromaffin cells. These cells

Table 6. Relative potencies of natural and synthetic corticosteroids

	Sodium retention (mineralocorticoid activity)	Liver glycogen deposition (glucocorticoid activity)	Anti-inflammatory activity
Cortisol	1	1	1
Cortisone	0.8	0.8	0.8
Corticosterone	15	0.35	0.3
11-deoxycorticosterone	30	0	0
Aldosterone	3000	0.3	
Prednisolone	0.8	4	4
Dexamethasone	0	30	30
Betamethasone	0	30	30
Triamcinolone	0	5	5
9α-fluorocortisol	125	10	10

contain storage particles of chromaffin granules which resemble the granulated vesicles of sympathetic nerve endings. Individual chromaffin cells synthesize and store large amounts of either noradrenaline or adrenaline.

The adrenal medulla is innervated by preganglionic sympathetic neurons from the splanchnic nerves. The nerve supply of the adrenal chromaffin cell is therefore analogous to that of the postganglionic sympathetic neurone.

The blood supply of the adrenal medulla is derived from the adrenal cortex and is thus very rich in steroid hormones.

Physiology and biochemistry of catecholamines

In mammals adrenaline is located almost exclusively in the chromaffin cells of the adrenal medulla where it is stored in high concentration. Noradrenaline on the other hand is widely distributed.

Noradrenaline is also found in the adrenal medulla and also in the peripheral sympathetic nerves, the central nervous system and the extra-adrenal chromaffin cells.

Catecholamine biosynthesis

The steps are outlined in LD-15.2.

Corticosteroids stimulate the conversion of noradrenaline to adrenaline by inducing the synthesis of phenylethanolamine N-methyl transferase (PNMT).

As noradrenaline accumulates in tissues it inhibits the formation of tyrosine. Mild stimulation of the appropriate nerves increases tyrosine hydroxylase activity by releasing the most recently synthesized pool of catecholamines. This reduces the feedback inhibition on the enzyme (tyrosine hydroxylase).

Protracted nervous stimulation increases synthesis of both tyrosine hydroxylase and dopamine β-hydroxylase.

In prolonged stress the sensitivity of the effector organs to catecholamines increases. It requires less and less catecholamine to elicit any given effect.

Inactivation of catecholamines

Three mechanisms exist to inactivate catecholamines:

- reuptake of catecholamines
- enzymatic deactivation by catechol-o-methyl transferase
- enzymatic deactivation by monamine oxidase

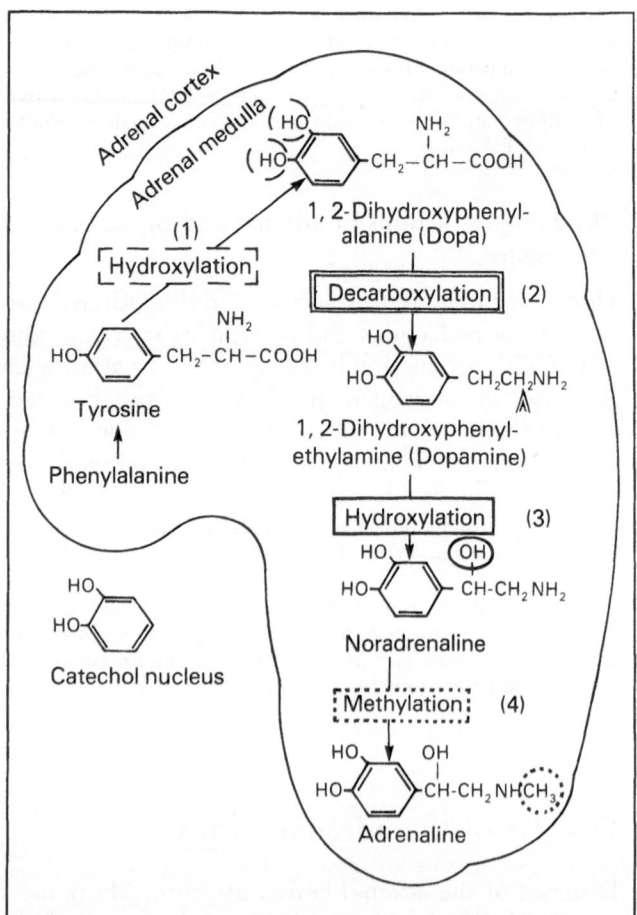

LD-15.2. *Catecholamine biosynthesis.*

Reuptake of catecholamines

This is achieved by postganglionic neuron storage granules. It is probably the most important method of inactivation. Reuptake is blocked by cocaine and by the tricyclic antidepressants. Inhibition of uptake is associated with supersensitivity to noradrenaline.

Enzymatic deactivation by catechol-o-methyl transferase (LD-16)

This enzyme is found in plasma, liver and kidney. Inhibition of catechol-o-methyl transferase results in a prolongation of the effect of sympathetic nerve stimulation or of administered catecholamines.

Enzymatic deactivation by monoamine oxidase

Monoamine oxidase is thought to be responsible for the rapid inactivation of catecholamines in the cytoplasm of the nerve endings.

Inhibition of monoamine oxidase results in an increase in stored catecholamines.

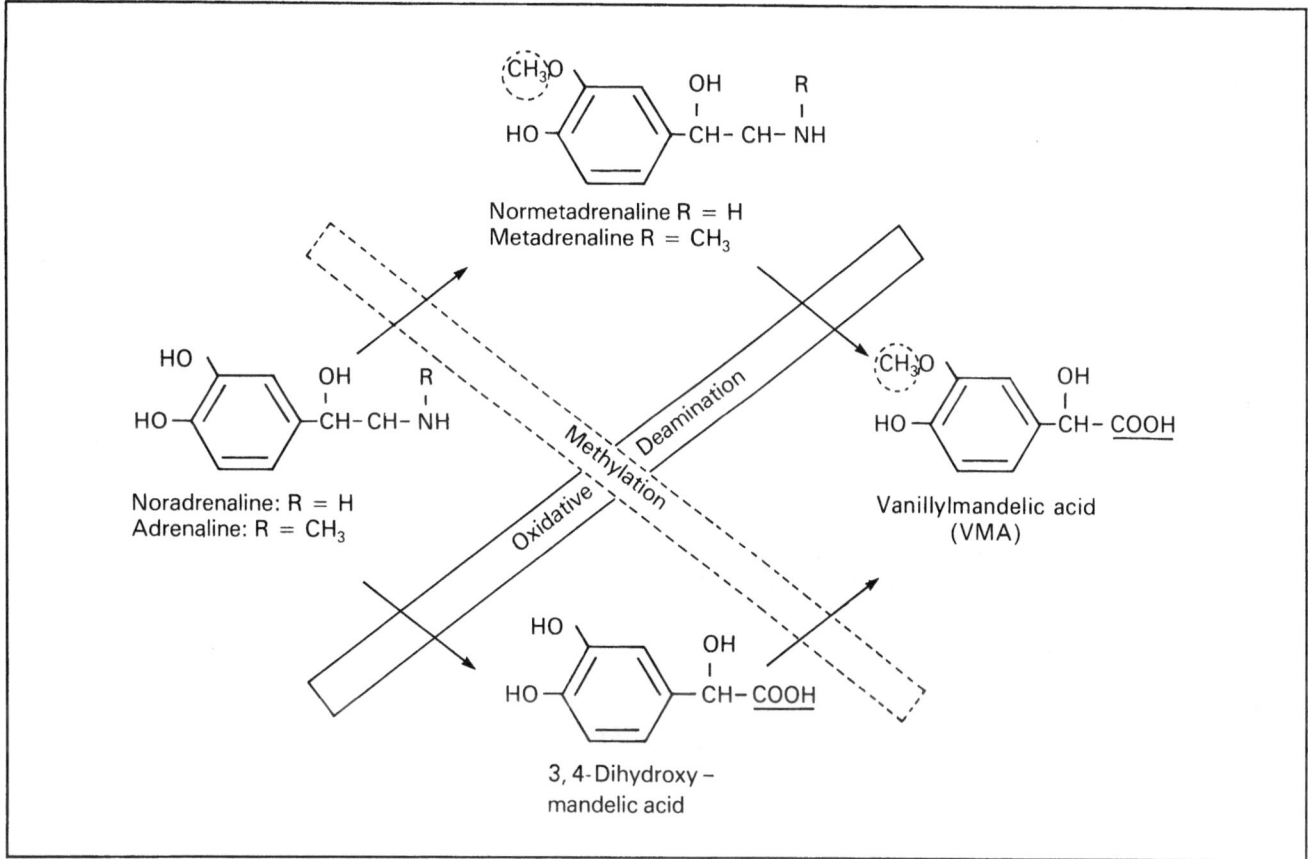

LD-16. *Inactivation of catecholamines.*

Regardless of which enzyme has the initial action the end product in both cases is vanilylmandelic acid (VMA).

A small proportion of the catecholamines produced is excreted unchanged in the urine, and a small proportion is excreted as sulphates or glucuronides.

Physiological actions of catecholamines

There are two types of receptor upon which sympathomimetic amines may act to elicit a response. These sites are classified on the basis of their responses to a series of sympathomimetic amines:

- α-receptor·
- β-receptor

The concept of α- and β-receptor sites simplifies the classification of both sympathomimetic drugs and adrenergic blocking agents.

An abbreviated list of the major α- and β-adrenergic responses elicited by the catecholamines is given in Table 7.

Table 7. α- and β-adrenergic responses

Effects	α	β
Vascular		
Constriction of veins and arteries	+	−
Dilatation, arteries	−	+
Cardiac		
Increase in heart rate	○	+
Increase in atrial contractability and conduction rate	○	+
Increase in conduction velocity and A–V node	○	+
Increase in ventricular contractability and in conduction velocity	○	+
Pulmonary		
Dilatation of bronchial musculature	○	+
Metabolic		
Increase in blood glucose and free fatty acid levels, inhibition of insulin release	○	+

Disease of the adrenal medulla

No known diseases are caused by adrenal medullary insufficiency. There are no indications that the adrenal medulla has an important function different from other sympathetic ganglia. The removal of the adrenal medulla does not in itself threaten life although the effects of acute stress, e.g. sudden hypoglycaemia, may be more profound in the absence of adrenal medullary hormone production.

There is no requirement for catecholamine replacement therapy after adrenalectomy or following destruction of the adrenal by disease, e.g. tuberculosis.

Phaeochromocytoma

Phaeochromocytomas are rare catecholamine-producing tumours that arise from chromaffin cells usually in the adrenal medulla but also from chromaffin cells in other sites.

The classical manifestations of phaeochromocytoma are the result of excessive catecholamine production and release. In some patients α-receptor stimulation predominates, in others β-receptor stimulation.

Phaeochromocytoma is a diagnosis often suspected but rarely confirmed.

THE THYROID GLAND

Anatomy

The thyroid gland is located in the anterior part of the neck. It consists of two lobes joined by a central isthmus. The gland is superficially placed but in the adult it varies in position (LD-17).

Its weight varies from 1.5 g in the newborn to 15–30 g in the adult. The weight depends on dietary iodine intake which varies with geographical location. The higher the intake of iodine, the smaller the gland and vice versa. Each lobe is conical and about 5 cm long, 3 cm wide and 2 cm thick.

The gland has a thin capsule from which connective tissue septa extend dividing the parenchyma into regular lobules. Each lobule consists of 20–40 follicles and is composed of a single layer of cuboid cells.

Within the follicle is the homogeneous colloid glycoprotein–thyroglobulin.

Blood supply

The thyroid is an exceptionally vascular organ. The arterial supply is mainly from the two superior

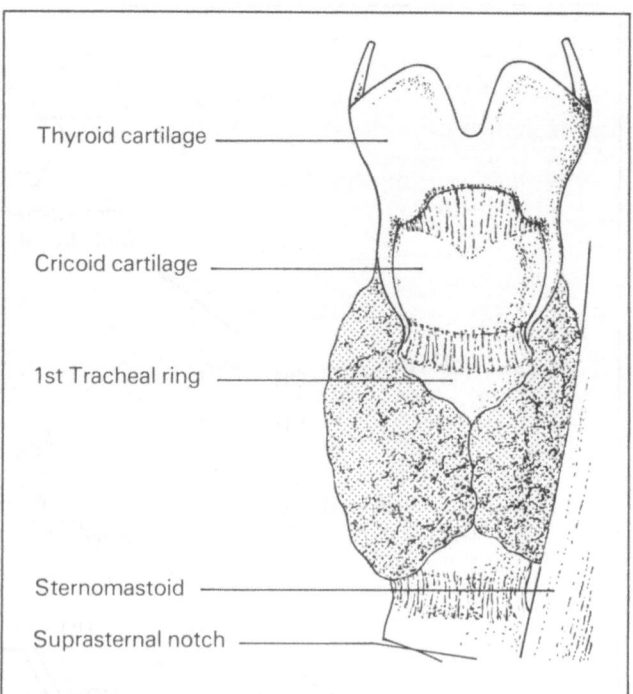

LD-17. *The thyroid gland.*

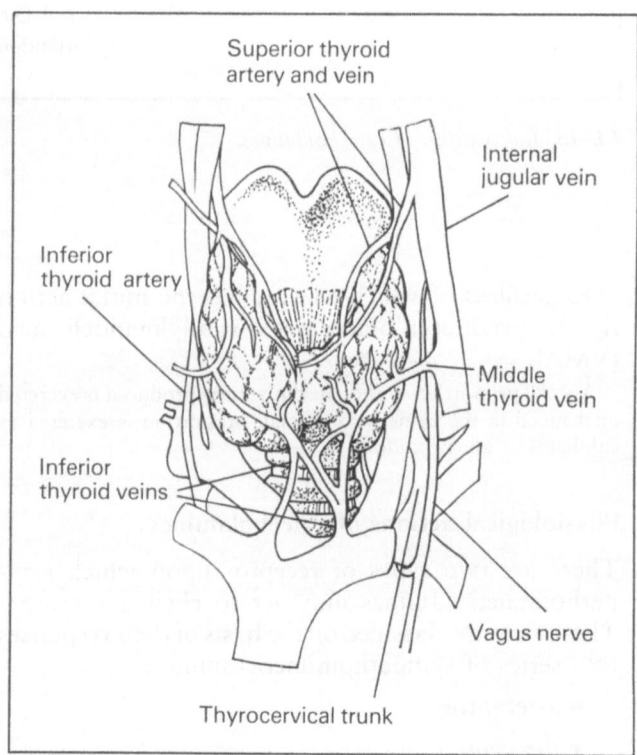

LD.18. *Diagrammatic representation of the thyroid gland showing its major blood supply.*

thyroid arteries and the two inferior thyroid arteries (LD-18).

The inferior thyroid arteries also supply the parathyroid glands and the inferior laryngeal nerves.

Lymph drainage

The individual follicles are close to a rich plexus of lymph vessels. The main lymph vessels leave the thyroid gland with the veins.

Nerves

The nerves can alter the flow of arterial blood within the gland but have no clear effect upon its function.

Normally the thyroid gland is situated opposite the fifth, sixth and seventh cervical vertebrae and is bound to the trachea, over the second, third and fourth tracheal rings, by the pretracheal fascia.

The superior thyroid arteries arise from the external carotids and the inferior from the thyrocervical trunk of the subclavian artery.

Stimulation of sympathetic nerves to the thyroid, from the superior cervical ganglion and carotid plexus, produces vaso-constriction while stimulation of the parasympathetic nerves, from the supralaryngeal branch of the vagus, causes vasodilation.

Embryology and ontogeny

In man, the thyroid arises from two structures:

- the central median thyroid diverticulum
- the ultimobranchial bodies

The central median thyroid diverticulum, derived from the floor of the pharynx, gives rise to the follicular cells in which the iodinated thyroid hormones are synthesized. The lateral ultimobranchial bodies produce the calcitonin-synthesizing 'C' or parafollicular cells.

The median thyroid mass grows forward and caudally to become a diverticulum lined with foregut epithelium. By 30 days two connected lateral lobes are present with the stalk forming the thyroglossal duct. Ectopic thyroid tissue can occur at any point along the line of the duct (LD-19).

Thyroglobulin synthesis and the ability to concentrate iodine and synthesize thyroid hormones appear early – after about 5 weeks. TSH does not appear in fetal plasma until about the 10th week and remains low until about 18 weeks when a rise in circulating TSH causes an abrupt rise in circulating thyroid hormones. There is no transfer or interchange across the placenta of either TSH or thyroid hormones.

The foramen caecum marks the origin of the thyroglossal duct. The ultimobranchial bodies contribute to the lateral lobes of the thyroid after 40 days.

By the 10th week the tubular arrangements are established and follicles present. These increase in size, but not in number, after the embryo is about 160 mm long.

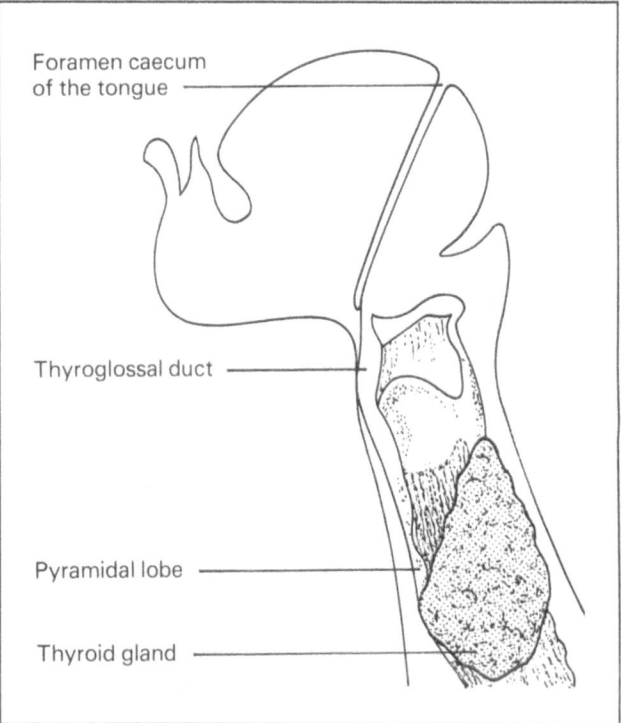

LD-19. *Development of the thyroid.*

The early development appears to be independent of TSH but subsequent growth and development require the trophic hormone.

Histology

The primary structural and secretory units of the gland are the follicles, separated by small quantities of connective tissue containing the parafollicular 'C' cells (CP-1).

The height of the cuboidal cells lining the follicles varies from 3 to 20 μm depending upon the stimulatory status. Microvilli protrude from the apical surfaces into the colloid and pores in the endothelial lining of the capillaries allow plasma to come into direct contact with the basement membrane of the cells so permitting free diffusion.

There are about three million follicles with diameters from 50 to 500 μm in adult man. The wide variation in size reflects environmental factors and the stimulatory state of the gland.

Physiology and biochemistry

Thyroid hormone synthesis and secretion

The thyroid cell has special functions relating to the synthesis of thyroid hormones but it also has func-

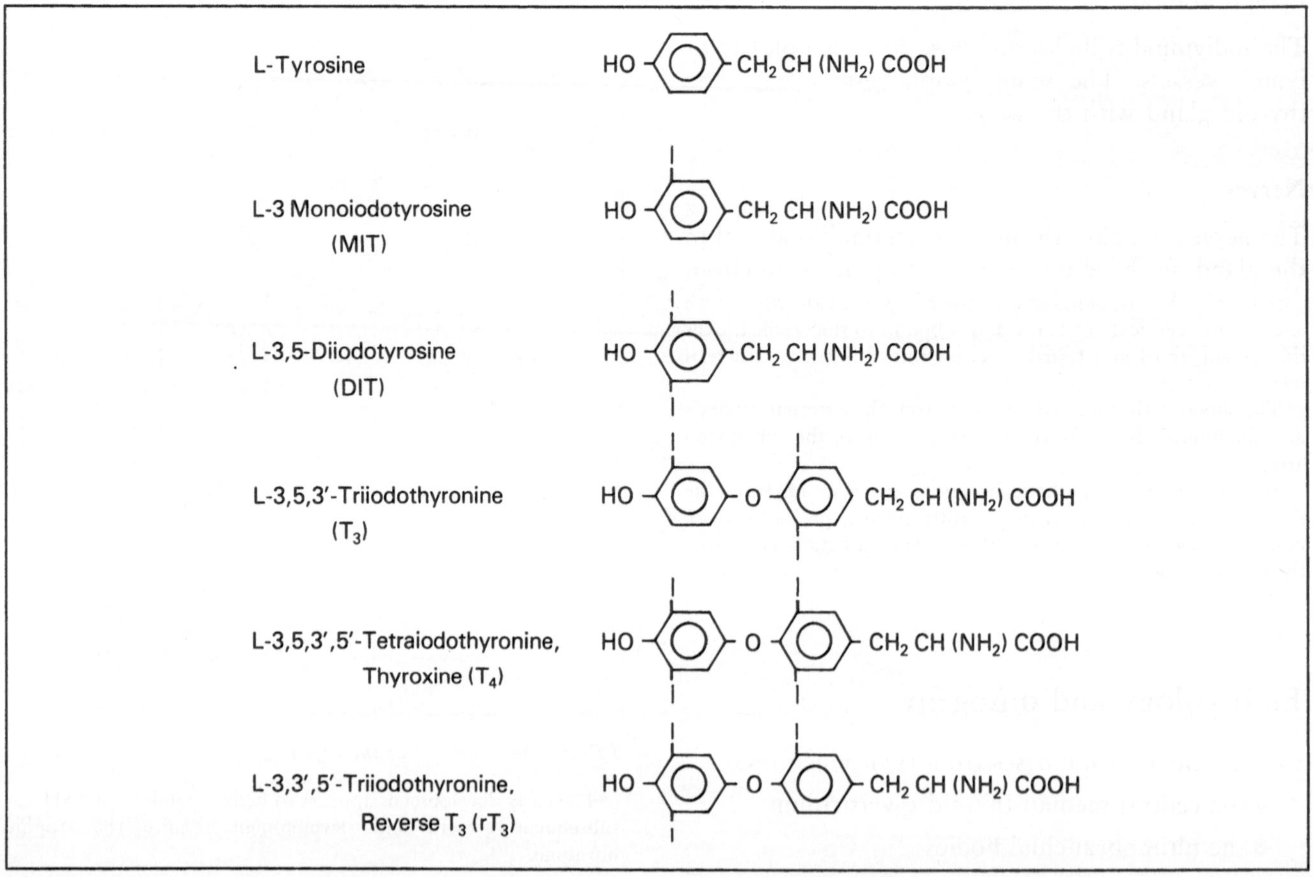

L-Tyrosine $\quad$ HO $-\langle\bigcirc\rangle- CH_2\,CH\,(NH_2)\,COOH$

L-3 Monoiodotyrosine
(MIT) $\quad$ HO $-\langle\bigcirc\rangle- CH_2\,CH\,(NH_2)\,COOH$

L-3,5-Diiodotyrosine
(DIT) $\quad$ HO $-\langle\bigcirc\rangle- CH_2\,CH\,(NH_2)\,COOH$

L-3,5,3'-Triiodothyronine
(T$_3$) $\quad$ HO $-\langle\bigcirc\rangle- O -\langle\bigcirc\rangle- CH_2\,CH\,(NH_2)\,COOH$

L-3,5,3',5'-Tetraiodothyronine,
Thyroxine (T$_4$) $\quad$ HO $-\langle\bigcirc\rangle- O -\langle\bigcirc\rangle- CH_2\,CH\,(NH_2)\,COOH$

L-3,3',5'-Triiodothyronine,
Reverse T$_3$ (rT$_3$) $\quad$ HO $-\langle\bigcirc\rangle- O -\langle\bigcirc\rangle- CH_2\,CH\,(NH_2)\,COOH$

LD-20. *Iodinated derivatives of tyrosine.*

tions common to most cells. Many of the functions of the thyroid cell are sensitive to TSH (Table 8).

Table 8. Thyroid processes stimulated by TSH

Iodoprotein T$_3$, T$_4$ synthesis
Colloid resorption
Iodide uptake by thyroid
Release of T$_3$, T$_4$
Oxygen consumption by thyroid
Glucose oxidation
Pyridine nucleotide coenzyme oxidation
Synthesis of phospholipid, proteins, nucleic acids
Cell membrane permeability

The thyroid gland synthesizes a number of iodinated derivatives of tyrosine (LD-20) and two of these, L-thyroxine (T$_4$) and L-tri-iodothyrone (T$_3$), have hormonal activity.

The synthesis and secretion involves discrete steps (LD-21):

• iodide trapping (1)

• iodination of tyrosine (2)

• coupling of iodotyrosines (3)

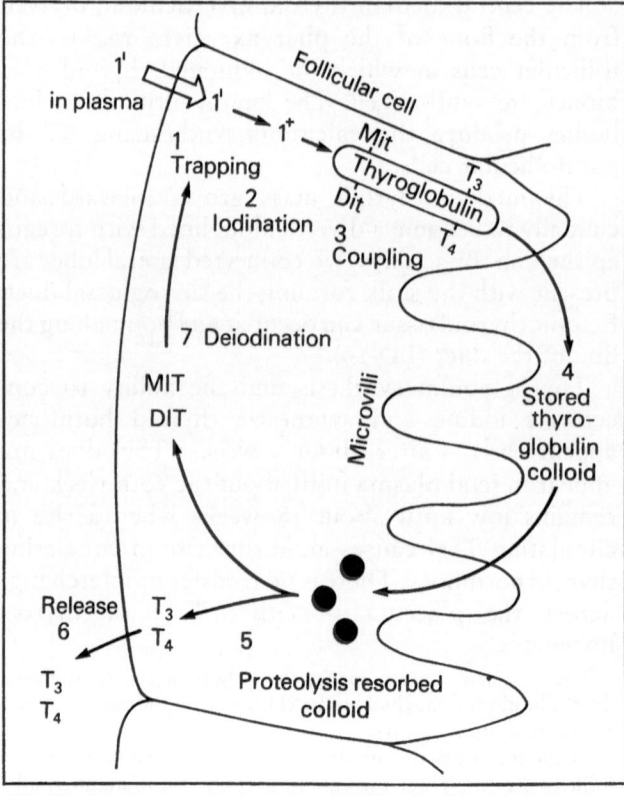

LD-21. *Synthesis storage and secretion of thyroid hormones.*

- storage of thyroglobulin (4)
- proteolysis of thyroglobulin (5)
- release of thyroid hormone to the circulation (6)
- deiodination of iodotyrosines (7)

Iodide trapping (1)

A daily intake of 50–100 μg iodide from drinking water, sea foods, milk and eggs is required for normal thyroid function. Too little iodine in the diet leads to hypertrophy of the gland or goitre.

Absorption of the ingested iodide is virtually complete and 90% of the iodide so acquired is found in the thyroid – sufficient for several months' thyroxine synthesis.

The iodide is concentrated in the thyroid gland by an active, energy-requiring process against a concentration gradient of about 1–40.

Other tissues, e.g. salivary glands, stomach, mammary glands and placenta, take up iodide but do not synthesize thyroxine.

The TSH-stimulated trapping of iodide can be blocked by perchlorate, thiocyanate, nitrate and pertechnetate ions.

Iodination of tyrosine (2)

The trapped iodide is activated by a peroxidase and H_2O_2 and incorporated into the tyrosine residues of the thyroglobulin forming monoiodotyrosine (MIT) and di-iodotyrosine (DIT).

The active form of iodide may be an iodinium ion or a free radical. The iodination may be blocked by reducing agents, e.g. thiouracil, or compounds with an aromatic ring, e.g. resorcinol.

Coupling of iodotyrosines (3)

T3 and T4 are formed by the coupling within the thyroglobulin of MIT and DIT and two DIT units respectively. Iodination and coupling take place in the microvilli.

Storage of thyroglobulin (4)

The glycoprotein thyroglobulin is the main constituent of the colloid stored in the follicular lumen. About 1% of the stored hormone is released each day.

Only about 15 of the 120 tyrosine residues in the thyroglobulin are iodinated. Minute amounts of thyroglobulin are found in the serum of normal subjects.

Proteolysis of thyroglobulin (5)

On TSH stimulation thyroglobulin is taken up by the microvilli of the cell as a colloid droplet. Lysosomal peptidases and acid proteases release T4 and T3 from the thyroglobulin.

Release of thyroid hormone to the circulation (6)

On release from the follicular cell the T4 and T3 are almost totally bound to plasma proteins.

A euthyroid man will produce 100 μg T4 and 20 μg T3 per day from the thyroid gland. Twenty-five per cent of this T4 is converted peripherally to T3. Reverse T3 (rT3) (LD-20) is produced intrathyroidally and peripherally. Reverse T3 has *no* hormonal action.

Deiodination of iodotyrosines (7)

MIT and DIT released from thyroglobulin are deiodinated thus conserving iodide within the follicular cell for new hormone synthesis.

Transport of thyroid hormones

In plasma, 75% of the T4 is bound by the glycoprotein thyroxine binding globulin (TBG); thyroxine binding albumin (TBPA) binds 15% and albumin the remaining 10%. T3 is bound less strongly to TBG and not to TBPA at all.

Small amounts of the hormones, (0.03% total T4 and 0.3% T3) remain free to act upon the tissues where they become bound to intracellular proteins. The small free hormone fraction actually determines the metabolic status of the individual. Thus major fluctuations in the amount of total thyroxine can occur due to alterations in the binding proteins without affecting the patient's true thyroid status.

The amount of protein-bound hormone in plasma is influenced by:

- increased TBG – pregnancy, oestrogen-containing oral contraceptives
- drugs binding to the same proteins – diphenylhydantoin
- decreased TBG – corticosteroid therapy, androgens
- hereditary deficiency state
- decreased proteins – liver dysfunction

Metabolism of the thyroid hormones

T4 and T3 are present in the plasma at average concentrations of 105 and 1.5 nmol/l respectively. T3 is some four times more potent than T4.

T4 is metabolized through monodeiodination to T3 and rT3 in liver, kidney and other tissues. Normally, some 25% of secreted T4 goes to T3, but

- starvation
- surgery
- corticosteroid therapy

- propranolol treatment
- other factors

decrease the amount of T_3 formed while increasing that of rT_3.

The plasma half-life of T_4 is 7 days, that of T_3 1.5 days and of rT_3 very short. Small amounts of T_4 and T_3 are excreted 'free' in the urine and as conjugates in bile.

As rT_3 has negligible metabolic effects, conversion of T_4 to rT_3 acts as a peripheral mechanism controlling the amount of active hormone available. The thyroxine nucleus of T_3 and T_4 is degraded to pyruvate, lactate and acetate after deiodination.

Regulation of thyroid hormone secretion

In addition to stimulating iodide trapping and thyroid hormone synthesis the pituitary trophic hormone, TSH, has a number of cyclic AMP-mediated actions (Table 8), including colloid resorption, which lead to hormone release.

The amount of TSH secreted depends upon a balance between stimulation of the pituitary by hypothalamic thyrotrophin releasing hormone (TRH) and the negative feedback inhibition of the 'free' thyroid hormones (see LD-4). The latter is the more important mechanism.

Minor elevation of T_4 or T_3, even within the normal reference ranges, will diminish TSH secretion. Conversely, lowering of T_3 or T_4 levels increases TSH release as the feedback inhibition is removed. There is some evidence of a positive feedback effect upon the hypothalamus (LD-5).

TSH secretion exhibits a peak in the early hours of the day with a nadir around 08:00. TRH may be modulated by circulating cortisol.

The inverse relationship with corticosteroids is demonstrated by the rebound TSH hypersecretion which occurs when corticosteroid therapy, which eliminates the TSH response to TRH, ceases.

Table 9. Processes affected by thyroid hormones

Oxygen consumption
Basal metabolic rate
Calorigenesis
Temperature regulation
Oxygen transport

Muscle contraction and cardiac output
Pre- and postnatal growth
Brain development

Protein and RNA synthesis
Lipid turnover
Glucose oxidation
Glycogen synthesis
Coenzyme synthesis and vitamin metabolism
Cell membrane enzyme activity
Secretion and turnover of other hormones

The actions of thyroid hormones

In the euthyroid state, optimal concentrations of thyroid hormones efficiently balance metabolic processes. The primary site of action of T_3 and T_4 is not known but fine control of the balance is achieved by regulation of enzyme activities while coarser control involves synthesis of RNA and proteins. Many synthetic and metabolic processes are influenced by thyroid hormone action (Table 9). In general increased thyroid hormone levels stimulate these processes.

The actions of thyroid hormones were thought to be mediated through catecholamines but while they may summate they are separate. Evidence indicates that thyroid hormones affect those processes mediated by β-adrenergic receptors.

No receptors for thyroid hormones have been detected on cell membrane, but cytosol and nuclear receptors have been identified.

Thyroid function and age

Fetal T_4 levels are low in early pregnancy partly because of the low TBG levels. These rise progressively to term.

Deficiency of thyroid hormone during the first 3 months of life retards myelination, neural and intellectual development. These effects may not be completely reversed by subsequent T_4 therapy.

Thyroid function reaches adult equilibrium around age 20 years and remains stable for the next 30 years. It then declines as the T_4 degradation rate decreases and thyroid secretion rate slows. Serum T_3 levels may decline with age while T_4 levels remain normal or slightly depressed relative to younger adults.

In the immediate neonatal period a burst of TSH secretion occurs. TSH levels may reach 100 mU/l but fall rapidly over the next 6 hours to reach adult levels within 48 hours. T_4 levels remain high during the first week of life but by 7–21 days attain normal adult values. Conversely neonatal T_3 levels are at first low but return rapidly to normal adult levels while the initially high rT_3 shows a concomitant fall.

Calcitonin

This polypeptide of 32 amino acids is secreted not only by the parafollicular (C) cells of the thyroid but also by the thymus and parathyroid glands. It is discussed in detail in the following section.

Recommended reading

Evered, D.C. (1976). *Diseases of the Thyroid*. (London: Pitman Medical)
Werner, S.C. and Ingbar, S.H. (eds.) (1978). *The Thyroid. A Fundamental and Clinical Text*. 4th Edn. (New York: Harper & Row.)

THE PARATHYROID GLAND AND CALCIUM METABOLISM

Bone metabolism

In an adult the skeleton consists almost entirely of bone, which is a living tissue with remarkable properties. It is both strong and light and is able to remodel itself to withstand normal or new stresses.

Bone serves as a store for calcium and phosphorus; the skeleton contains 99% of the body's calcium, 88% of its phosphorus, and 50% of its magnesium.

Bone is not inert but is continually being broken down and replaced. In a normal adult the exchange of calcium between bone and plasma is about 12 mmol (480 mg) daily.

Bone consists of two main components. About 70% is inorganic – largely hydroxyapatite, $Ca_{10}(PO_4)OH_2$, and the remainder is organic – principally collagen. The strength of bone depends on both components:

- the mineral to give rigidity
- the lattice work of collagen fibres to give resilience

This is similar to reinforced concrete which depends for its strength both on the concrete and on the wire framework. The importance of the collagen of bone is illustrated by the fragility of the skeleton in osteogenesis imperfecta, a condition in which collagen is abnormal.

Bone deposition is largely carried out by osteoblasts, while bone resorption is carried out by osteoclasts. The osteocytes within the bone are capable of both functions to a limited extent.

Calcium metabolism

Dietary calcium

In Western countries adults consume 20–25 mmol (800–1000 mg) of calcium daily, but in many developing countries the calcium intake is usually much less – of the order of 5–10 mmol (200–400 mg) daily.

The minimum daily requirement for calcium is not known and no definite calcium deficiency disease has ever been convincingly described. Normal people can adapt to a calcium intake as low as 4 mmol (160 mg) daily, for the proportion of the dietary calcium which is absorbed increases and the calcium balance is maintained.

This adaptation was initially demonstrated in rats but was also shown to be true of man in long-term studies of the inmates in the Oslo gaol on low calcium diets. The adaptive process involves the hormones parathyroid hormone and 1,25-dihydroxycholecalciferol, which are described later.

Calcium absorption

Calcium may be absorbed in all parts of the small intestine by an active transport mechanism which depends on the hormone 1,25-dihydroxycholecalciferol. Calcium absorption also takes place by diffusion. In a normal adult on a Western diet containing 25 mmol (1000 mg) of calcium per day, the faeces contain about 21 mmol (840 mg) per day and the net absorption is about 4 mmol (160 mg) per day. (LD-22, Table 10).

Table 10. Factors influencing calcium absorption

Increased	Hyperparathyroidism
	Vitamin D excess
Decreased	Old age
	Vitamin D deficiency
	Renal failure
	Malabsorption

Urinary excretion of calcium

About 175 mmol (7.9 g) of calcium is filtered through the glomeruli daily and most is reabsorbed in the

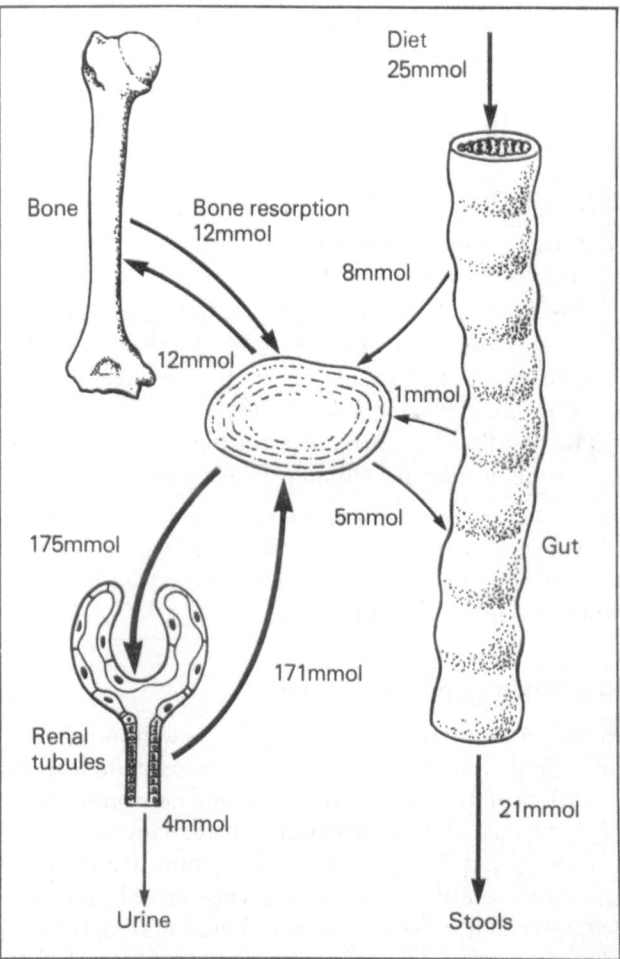

LD-22. Calcium balance in the body.

tubules. The urinary excretion of calcium in normal people is between 2 and 10 mmol daily (80–400 mg/day) (LD-22).

Tubular reabsorption of calcium is increased by parathyroid hormone and excessive renal calcium reabsorption is one of the reasons for the hypercalcaemia of hyperparathyroidism.

Plasma calcium

The total plasma calcium is normally 2.2–2.6 mmol/l (8.8–10.4 mg/dl) but the precise normal range depends upon several factors, particularly the analytical method used. Just under half of the total plasma calcium is ionized calcium and physiologically active; most of the rest is bound to albumin.

One important consequence of the protein-binding of calcium in plasma is that the plasma total calcium level depends in part on the plasma albumin. A patient with a low plasma albumin (because of chronic liver disease or the nephrotic syndrome, for example) inevitably has a low total plasma calcium even if his ionized calcium level is normal. Conversely if the plasma albumin is raised, (for example, after the prolonged application of a tourniquet) the total calcium is also raised.

One careful study showed that the mean proportions of the various calcium fractions in normal subjects were:

ionized calcium 1.16 mmol/l (44.6%)
protein bound calcium 1.07 mmol/l (41.2%)
complexed calcium (as CaHPO$_4$ and calcium
salts of organic acids) 0.37 mmol/l (14.2%).

The ionized and complexed fractions are together known as the 'diffusible calcium'. Methods for the measurement of plasma levels of ionic or ultrafiltrable calcium are not widely available.

One method for 'correcting' the plasma calcium for variations in plasma albumin is as follows:

for every 1 g/l by which the plasma albumin falls short of 40 g/l, 0.02 mmol/l is added to the value for the total plasma calcium. A similar subtraction is made when the plasma albumin exceeds 40 g/l.

Regulation of plasma calcium

In health the plasma calcium remains within narrow limits and abnormalities in the plasma calcium lead to disorders of muscles and to neural and neuromuscular problems as well as psychiatric disturbances.

Low blood levels of ionized calcium are characterized by tetanic spasms or convulsions. High levels are particularly associated with cardiac arrhythmias and, if prolonged, renal failure due to calcium deposition in the kidneys.

The principal factors responsible for the maintenance of the plasma calcium are physicochemical factors and three hormones:

- parathyroid hormone
- calcitonin
- 1,25-dihydroxycholecalciferol, a metabolite of vitamin D$_3$ (cholecalciferol)

The chemical equilibrium between bone mineral and interstitial fluid is thought to maintain the plasma calcium around 1.7–2.0 mmol/l. In intact animals hormones, particularly parathyroid hormone, maintain the plasma calcium at about 2.5 mmol/l.

Parathyroid hormone (PTH)

Source

There are generally four parathyroid glands but a larger or smaller number may be found. They develop from the third and fourth branchial pouches and usually lie in the neck just posterior to the thyroid gland. Accessory parathyroid tissue is occasionally found in the mediastinum.

The secretory activity of parathyroid glands is not controlled by the nervous systems and the glands function satisfactorily after transplantation to another part of the body.

Parathyroid glands are very small; few measure more than 8 mm in any dimension. Histologically two cell types are recognized: chief cells which produce parathyroid hormone and oxyntic cells whose function is unknown. Normally parathyroid glands also contain an appreciable number of fat cells.

Chemistry

Parathyroid hormone (parathormone or PTH) is a polypeptide with a single chain containing some 84 amino acids. The full sequence of the human hormone has not yet been determined but the biological activity is held by the 34 amino acids at the amino-terminal (n-terminal).

Like insulin, parathyroid hormone is derived from a larger precursor molecule. This 'proparathyroid hormone' has 90 amino acids. It is normally present in parathyroid cells but not in blood.

Secretion

The secretion of PTH increases linearly as the plasma ionized calcium level falls and is suppressed when the plasma calcium is above the normal range (LD-23).

Actions

PTH acts directly on the kidney to decrease the

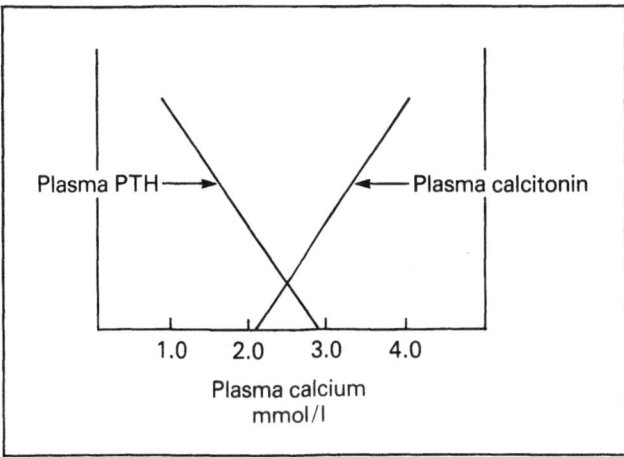

LD-23. Secretion of parathyroid hormone.

excretion of calcium and increase that of phosphate. In addition it increases the production of 1,25-dihydroxycholecalciferol by the kidney and so indirectly increases active transport absorption of calcium in the gut. PTH also stimulates the resorption of bone both by osteoclasts and by osteocytes. These actions together serve to raise the plasma calcium. The action of parathyroid hormone, at least in the kidney, is mediated by adenylcyclase and cyclic AMP.

Calcitonin

Calcitonin (sometimes called thyrocalcitonin) is produced by the parafollicular cells (C cells) of the thyroid gland.

It is a single chain polypeptide of 32 amino acids, and is secreted when the plasma ionized calcium level rises. Its main action is the lowering of plasma calcium by inhibiting bone resorption. It also increases phosphate excretion in the urine.

Parafollicular cells are larger than the ordinary follicular cells and contain granules which in some species can be stained with special stains.

Calcitonin release is stimulated by the hormones gastrin, cholecystokinin, pancreozymin and glucagon. It is thought that this mechanism minimizes any change in plasma calcium level while a meal is being absorbed.

1,25-dihydroxycholecalciferol (DHCC)

This hormone is a metabolite of vitamin D_3 (cholecalciferol).

Source

Vitamin D_3 can be obtained in the diet from foodstuffs such as fatty fish (herring, salmon, sardines) and to a small extent eggs, or can be synthesized in the skin from 7-dehydrocholesterol under the influence of ultraviolet light.

Vitamin D_2 (ergocalciferol) is a chemically similar compound which can be made industrially from a plant sterol, ergosterol. It is used for the fortification of margarine (LD-24).

Vitamin D is absorbed in the upper part of the small intestine and bile salts are needed for this. Vitamin D and its metabolites are stored in adipose tissue and muscles and, in man, these stores may contain enough vitamin D for several years.

Vitamin D_3 is hydroxylated in the liver to 25-hydroxycholecalciferol which is the principal form in which vitamin D is carried in the plasma. In turn 25-hydroxycholecalciferol is further hydroxylated in the kidney either to 1,25-dihydroxycholecalciferol or to a physiologically inactive metabolite, 24,25-dihydroxycholecalciferol (LD-25). It seems likely that vitamin D_2 is treated in the same way but this has not yet been confirmed.

Vitamin D deficiency occurs only when the diet lacks vitamin D and the patient has inadequate exposure to ultraviolet light. Deficiency of vitamin D causes rickets in children and osteomalacia in adults.

Secretion

The production of 1,25-dihydroxycholecalciferol varies with the plasma calcium. It is increased when the plasma calcium is low and negligible when the plasma calcium is raised at which time the kidney switches over to the production of 24,25-dihydroxycholecalciferol.

It is probable that parathyroid hormone and calcitonin are involved in this regulation – parathyroid hormone promoting the production of 1,25-dihydroxycholecalciferol while calcitonin increases that of 24,25-dihydroxycholecalciferol.

Actions

1,25-dihydroxycholecalciferol is essential for the active transport of calcium in the small intestine.

It probably operates by promoting the synthesis of a calcium binding protein. The other actions of 1,25-dihydroxycholecalciferol are not yet fully understood. It has a direct action on bone cells in culture to increase bone resorption. Its physiological action in allowing bone mineralization (for example when rickets is being treated) is not yet understood.

Other hormones affecting bone

An adequate supply of growth hormone is needed for

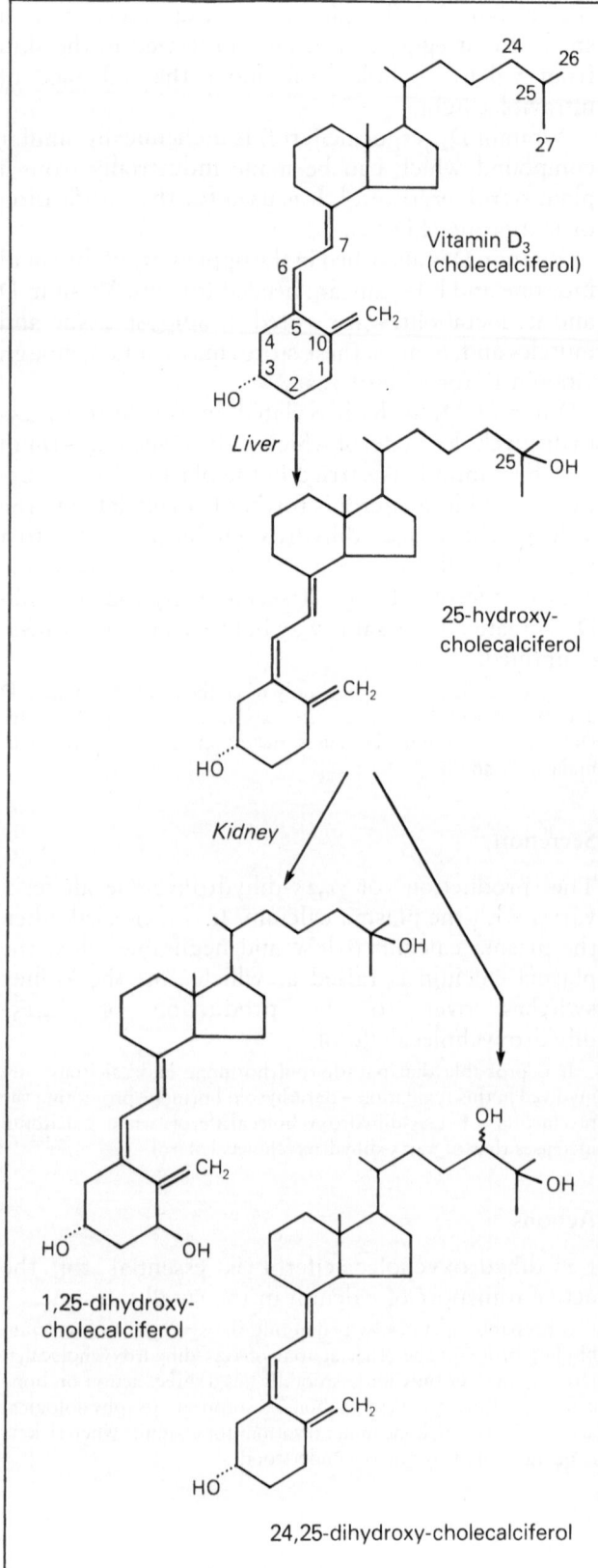

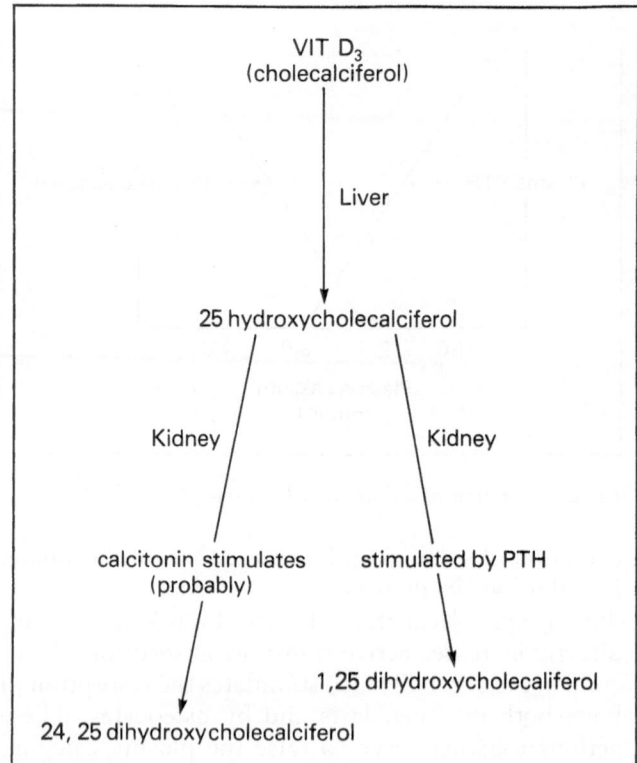

LD-25. *Hydroxylation of vitamin D₃.*

the proliferation of the cells of the epiphyseal cartilage and therefore for growth in the length of bone in children and adolescents.

Excessive circulating levels of the thyroid hormones cause bone loss as do the excessive circulating levels of glucocorticoids which occur in Cushing's syndrome or as a result of steroid therapy. Low levels of thyroid hormone in childhood retard longitudinal bone growth and in infancy lead to epiphyseal dysgenesis.

Phosphorus metabolism

An adult contains about 25 mmol (770 g) of phosphorus and about four fifths of that is in the bone. Much of the remainder plays an essential role in the composition of all cells as organic phosphates, phospholipids and nucleic acids.

The plasma inorganic phosphate level is normally between 0.8 and 1.4 mmol/l (2.5–4.5 mg/dl). Higher values are found in children. Almost all of the inorganic phosphate in the plasma is diffusible.

Phosphate passes into the glomerular filtrate and about 90% of the filtered phosphate is reabsorbed. This reabsorption is decreased by parathyroid hor-

LD-24. *The metabolism of vitamin D₃ (cholecalciferol). It is likely that vitamin D₂ is metabolized a similar way.*

mone and by calcitonin and increased when parathyroid hormone is absent.

Dietary deficiency of phosphate does not occur since all foods contain phosphate. Phosphate depletion with a low plasma phosphate level occurs in patients with excessive phosphate losses in the urine as a result of hyperparathyroidism or inherited disorders of renal tubular function.

A high plasma phosphate level is found in hypoparathyroidism or renal failure.

Magnesium metabolism

An adult has about 1000 mmol (25 g) of magnesium in his body. About half is found in bone and half in the cells where it plays an essential role in the function of many enzymes. The plasma magnesium level is usually between 0.7 and 1.0 mmol/l.

Low values occur in a few patients with intestinal resections; high values are found in renal failure.

The factors affecting plasma magnesium homeostasis are not fully understood but magnesium deficiency can cause hypocalcaemia by interfering with the action of parathyroid hormone.

Recommended for further reading

Paterson, C.R. (1975). *Metabolic Disorders of Bone*. (Oxford: Blackwell).
Vaughan, J. (1975). *Physiology of Bone*. 2nd Edn. (Oxford: Oxford UP).

THE GONADS

Gonadal differentiation and development

During the 5th fetal week, the coelomic epithelium positioned along the posterior abdominal wall just medial to the mesonephros (the primitive kidney), thickens bilaterally to form the genital ridges. These ridges become colonized with primitive germ cells with migrate from the wall of the yolk sac.

Until the 17 mm stage these ridges in both sexes share a common development and at this indifferent stage the gonad is composed of a

- medulla – potentially a testis
- cortex – potentially an ovary

Lateral to the genital ridges lie the mesonephric (Wolffian) ducts and lateral to these the paramesonephric (Müllerian) ducts.

In the presence of a Y chromosome the testis develops from the medulla whilst the cortex regresses. The androgens produced by the primitive fetal testes cause differentiation of the mesonephric ducts into the male genital ducts and regression of the paramesonephric ducts.

Lack of a Y chromosome leads to further development of the cortex into an ovary and the paramesonephric ducts into

- uterus
- upper part of the vagina

These changes are outlined in LD-26 and will be further described under Disorders of Gender Differentiation (Section IV, p. 160 and LD-49, 50).

Regression of the paramesonephric ducts is now regarded as an endocrine-induced phenomenon. Josso has demonstrated that this regression or de-differentiation of the Müllerian ducts is induced by a substance, probably of amino-acid structure, elaborated by the fetal Sertoli cells.

Müllerian inhibiting activity in the human testis persists until the 24th fetal week, but is only important around the 9th fetal week. Lack of this substance results in persistence of Müllerian ductal tissue in otherwise normal males.

The testis

At birth the volume of the testis varies from 0.5 to 1.0 ml. In general, average testicular growth is approximately 1.0 ml per year till puberty.

The increase in testicular volume is due to the development of the seminiferous tubules in preparation for the onset of spermatogenesis at puberty.

Spermatogenesis

Spermatozoa develop from a series of synchronized mitotic divisions of the type A spermatogonia (germinal cells) to form intermediate spermatogonia and type B spermatogonia. These latter divide to form primary and secondary spermatocytes.

By the process of meiosis, haploid spermatids are formed prior to the final transformation to spermatozoa. The spermatozoa are shed into the lumen of the tubules and migrate via the rete testis to the epididymides and thence to storage in the seminal vesicles.

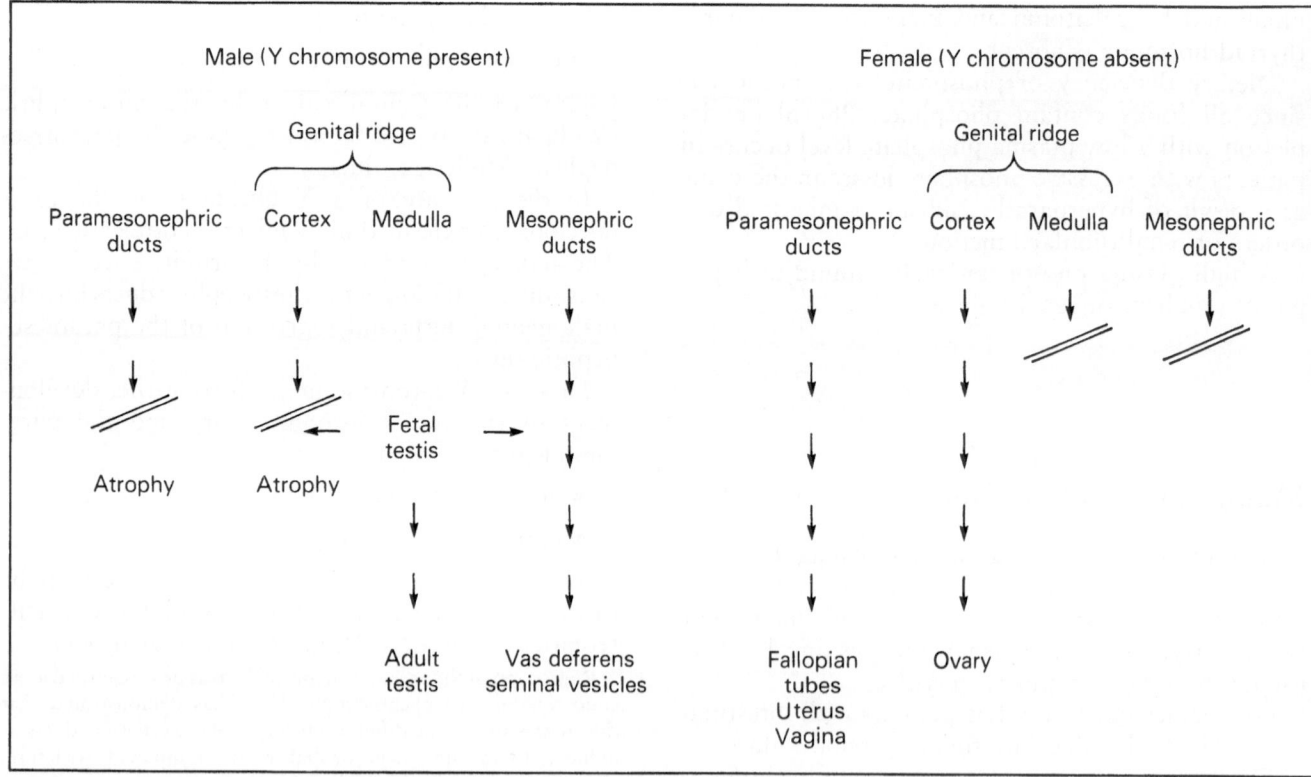

LD-26. *Gonadal differentiation.*

The process may be schematized as:

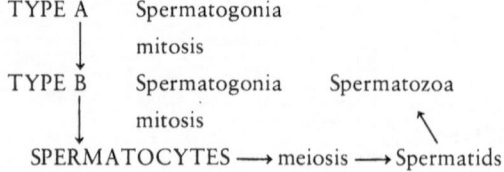

Testicular hormonal function

The Leydig cells are the hormonally active cells of the testes.

Testosterone is the principal androgen produced in the testis and is synthesized from cholesterol mainly via 17α-hydroxy pregnenolone although some is derived via 17α-hydroxy progesterone.

Other androgens produced by the testis include:

- dihydrotestosterone (DHT)
- $\Delta 4$-androstenedione
- dehydroepiandrosterone (DHEA)

Control of testicular function

The major control of testosterone secretion is by LH on the Leydig cells.

Spermatogenesis is dependant on the actions of both LH and FSH.

LH stimulates the mitotic divisions of type A spermatogonia. FSH has a major anabolic effect on the seminiferous tubules thus enhancing spermatogenesis. This effect requires the presence of testosterone and is thus indirectly dependant on LH which is the main stimulator of Leydig cell testosterone production (LD-27).

It would appear that type A spermatogonia are under the control of testosterone while progression to primary and secondary spermatocytes requires neither gonadotrophins nor testosterone. Spermatid formation and their maturation to spermatozoa are controlled by FSH.

The ovary

The definitive ovary lies at the upper and outer angle of the broad ligament and nestles under the fimbriated extremity of the fallopian tube.

The germ cells or oogonia arise from the sex cords which form islands in the gonadal cortex. They rapidly increase in number by mitosis and by the 8th fetal week number about 500 000. Oocytes are formed by the haploid division of oogonia. The follicular cell or granulosa cell arises from the coelomic epithelium

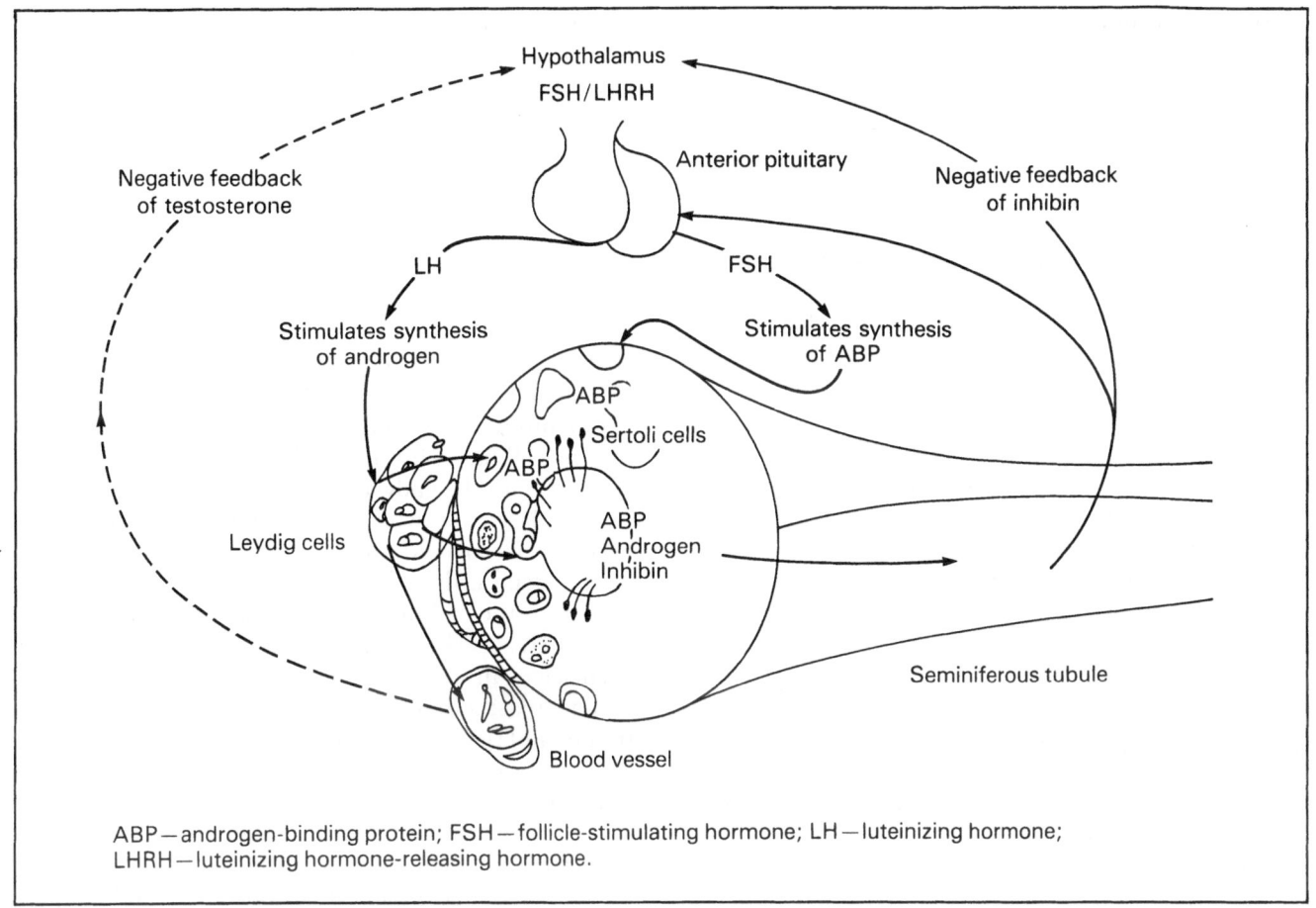

ABP—androgen-binding protein; FSH—follicle-stimulating hormone; LH—luteinizing hormone; LHRH—luteinizing hormone-releasing hormone.

LD-27. *Testis – anatomy and pituitary control.*

which forms deep folds between the clusters of oogonia. Finally, the mature oocyte is surrounded by a layer of granulosa cells but separated from them by a zona pellucida (a layer of mucopolysaccharides). These are called 'graafian follicles'.

Follicular development

At birth, there are several hundreds of thousands of primordial follicles in the ovary.

During infancy and childhood, these follicles undergo varying degrees of maturation and secrete small amounts of oestrogens.

About the age of 8 years appreciable plasma oestrogens levels are detected but not till several years after the first menstrual period do the plasma oestrogen levels reach the high cyclic concentrations of the fertile female.

Of the many primordial follicles, only a relatively small number mature. The smaller follicles, destined to become atretic, do not develop the capacity to elaborate oestrogens and the steroidogenic activity is confined to the preovulatory follicle.

It is thought that ovarian oestrogens stimulate the development of increased binding sites for FSH on the granulosa cells and the action of the FSH in turn is to increase binding sites for LH. LH then induces increased secretion of oestrogens as a herald of puberty.

Corpus luteum

Following rupture of the graafian follicle with extrusion of the ovum, the follicle collapses and beneath a small amount of blood lie the hyperplastic thecal cells. Within 48 hours, the site is vascularized by vessels from the theca and luteinization begins. By 7–10 days post ovulation, the corpus luteum is a bright yellow convoluted structure about 1 cm in diameter and persists until a few days before the onset of menstruation.

If pregnancy supervenes, the corpus luteum enlarges, and becomes highly vascularized elaborating progesterone. If there is no pregnancy the retrogression phase is characterized by hyalinization followed by fibrous replacement. These scars on the ovary are termed 'corpora albicanta'.

Hormonal production and control

Control of development is less clearly defined for the ovary than it is for the testis. Development of the oocytes and granulosa cells into mature graafian follicles is controlled by gonadotrophins and by the 25th week of fetal life mature follicles are present.

Once the trigger mechanism has been set off, no further stimulus is required for continuing maturation of newly formed follicles.

Ovarian tissue has a complete complement of enzymes for the elaboration of progesterone and the active oestrogens from cholesterol. The various steps in this pathway are shown in LD-28.

This synthesis is controlled by the pituitary gonadotrophins LH and FSH. In the normal postpubertal female, there is an ordered sequence of events throughout the menstrual cycle which is reflected in the plasma hormone levels of the various hormones both trophic and steroidal (LD-8b).

THE PLACENTA

The placenta consists primarily of the continuous layers of the endometrium and the chorion. When functionally active, it serves as a

- resorption
- excretory
- exchange

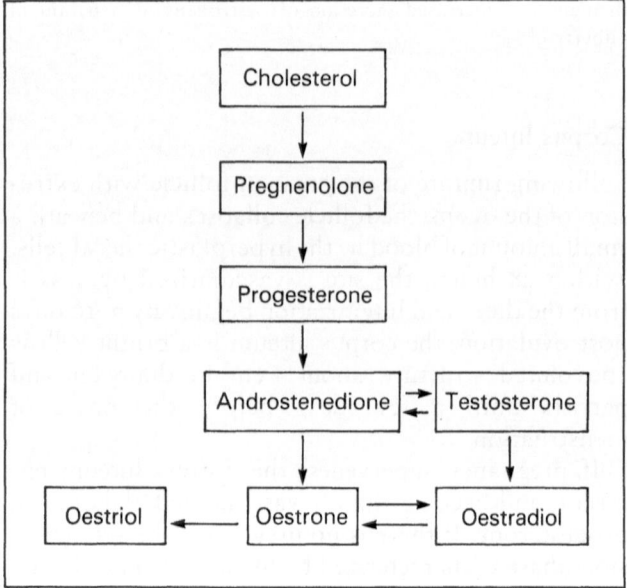

LD-28. *Ovarian hormonal production.*

organ for the fetus, providing the transport facilities for food and waste products.

In addition, it forms a screen between mother and fetus and acts as a barrier to allow separate development of the fetus.

Development

By the 4th day after conception, the fertilized ovum, or blastula, is in the uterus. The ovum orientates itself (by an unknown mechanism) so that the pole containing the cell mass destined to become the embryo is in contact with the endometrial surface. Thereafter, the whole blastocyst penetrates the mucosa until its area of entry is covered by an overgrowth of maternal endometrial cells.

Within the endometrium, the stromal cells enlarge and increase their cytoplasmic content arranging themselves in a mosaic pattern.

The plasmodial trophoblast invades the uterine tissues and the uterine blood vessels and forms a loose spongework containing anastomosing spaces filled with maternal blood. The invading processes of trophoblast are termed the primary villi and the spaces the intervillous spaces. In these spaces maternal blood vessels maintain a constant blood supply.

Later the primary villi are invaded by a core of mesenchyme of the chorion and start to branch within the intervillous space. These branches are the secondary villi. There then develop, within these secondary villi, vascular spaces which link with radicles of the umbilical vessels. Now known as tertiary villi, these elements are the medium by which exchange of nourishment from mother to fetus and of waste products from fetus to mother is effected.

Endocrine function

While the placenta has been demonstrated to have many specific enzyme activities relative to steroidogenesis, it is more proper to regard the fetus and placenta as a single unit in steroid biosynthesis. The steps of principal interest are:

- cholesterol production
  The fetus (adrenal) but not the placenta, can elaborate cholesterol from acetate radicals, but the further transformation to progesterone and cortisol is small.

- dehydroepiandrosterone and 17β-oestradiol

  The placenta can synthesize progesterone from cholesterol only, but will transform it further to dehydroepiandrosterone and 17β-oestradiol

- progesterone

  The fetus receives its principal supply of progesterone from the placenta and from it elaborates cortisol.

- aldosterone

  This is synthesized by the fetus via deoxycorticosterone and corticosterone.

- oestriol pool

  Both the fetus and placenta contribute to the oestriol pool utilizing dehydroepiandrosterone as precursor.

The importance of this fetoplacental unit is that its efficiency determines to a large extent fetal viability. Fetal well-being correlates well with maternal oestriol excretion since this steroid arises largely in the fetal compartment.

During pregnancy, maternal urinary levels of oestriol rise rapidly between the 20th and 40th weeks of gestation (4–30 mg per 24 h). A fall to around one half of these levels indicates retardation of fetal development, while near term low levels indicate impending fetal death.

- human placental lactogen (HPL)

  This polypeptide hormone which has radioimmunological characteristics similar to human growth hormone is detectable in plasma 5–6 weeks after the last menstrual period and gradually rises during pregnancy. It has lactogenic properties but the importance of HPL in preparing the breasts for lactation remains to be proven in humans.

- chorionic gonadotrophin (HCG)

  This is a glycoprotein, of molecular weight 36 000–40 000, bearing structural similarities to luteinizing hormone (LH) from the pituitary. It is thought that this hormone is responsible for at least the early oestrogen and progesterone secretion by the corpus luteum of pregnancy. Therapeutically it has been used in the induction of ovulation but also in hypogonadotrophic states in males and females.

THE BREAST

Embryology

It is the possession of mammary glands for suckling the young that gives the mammalian order its name.

In the early embryo, mammary ridges develop as ingrowths of ectoderm extending from the fore limb bud to the level of the hind limb bud. In species with characteristically large litters of young, e.g. dog, cat, pig, this situation persists, so that several pairs of breasts come to develop. In some, e.g. cattle, a composite mammary gland or udder persists in the caudal area, and the cephalad primitive mammary glands undergo regression.

In humans all but the most cephalad parts of the mammary ridges regress leading to development of a single pair of mammary glands. Rarely, regression of the caudal parts of the mammary ridges is incomplete, resulting in accessory mammary glands or nipples.

Anatomy

The mammary glands lie within the superficial fascia and are more extensive than is apparent, a point of importance if surgical excision has to be undertaken.

The breast extends from the level of the second to the sixth rib and from virtually the midline anteriorly to the midaxillary line. It overlies the pectoralis major muscle and also parts of the external oblique and serratus anterior muscles. There is an axillary tail extending up behind the anterior fold of the axilla. It is composed of glandular tissue supported by fibrous septa which divide the gland into 12–20 lobes each roughly conical in shape. The apex of each lobe opens as a duct at the nipple. Lobes are divided into a number of lobules and these are supported by the fibrous and fatty tissue which constitutes the stroma of the breast.

The nipple surmounts the breast and onto it open the lactiferous ducts, 12–20 in number. It is surrounded by the areola. Both the nipples and the areola contain erectile tissue. In fair-skinned people the nipple and areola are pink in colour, darkening permanently to brown during the first pregnancy.

In the young child the mammary glands are small and there is little difference between those of the male and female. At puberty the female mammary gland increases rapidly in size. During pregnancy the breasts enlarge and are at their greatest size during lactation.

Vascular supply

The arterial supply to the breast comes from perforating branches of the internal mammary (internal thoracic) artery and from the lateral thoracic artery. The intercostal arteries also supply additional branches.

Venous drainage is to the internal mammary and axillary veins and to a lesser extent the external jugular vein.

Lymphatic drainage

Lymphatic drainage is important because of the lymphatic spread in breast cancer. Lymphatics form a perilobular plexus from which vessels emerge. They follow the lactiferous ducts to the areola where they form a subareolar lymphatic plexus. From this plexus efferent lymphatics flow to the anterior (pectoral) axillary lymph glands and thence to the other groups of axillary lymph glands and the glands along the internal mammary artery. There are some lymphatics which cross the midline to the opposite breast and others which drain to the upper part of the rectus sheath.

Nerve supply

The nerve supply is from the fourth, fifth and sixth intercostal nerves. Sympathetic nerves are distributed along these nerves especially to the nipple and areola.

Physiology

In the male, and in the female before puberty, the breast is rudimentary, consisting of ducts within a fibroalveolar stroma. There is no significant alveolar development.

At puberty the female breast develops mainly by branching of the duct system and an increase in the fibroalveolar stroma. Alveolar development occurs particularly during pregnancy in preparation for lactation when the epithelial cells become cuboidal and filled with fat globules. When lactation ceases the alveolar element regresses.

Development at puberty is oestrogen stimulated, but the progestogenic hormones of the corpus luteum stimulate some minor alveolar development.

In pregnancy alveolar development is under the influence of prolactin and placental lactogen, acting upon the breast but primed by oestrogen and progesterone.

Lactation is initiated and sustained by prolactin. It is released from the pituitary under the control of a hypothalamic inhibitory factor – prolactin inhibitory factor (PIF). Ejection of the milk is a function of oxytocin which is derived from the posterior pituitary. The secretion of oxytocin is initiated by a suckling reflex.

The endocrine control of the breast is outlined in LD-29.

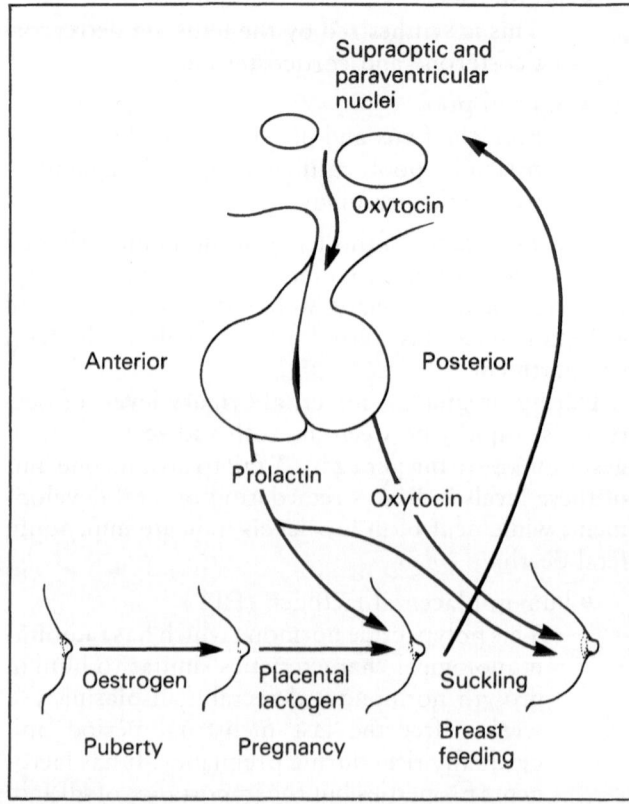

LD-29. *The endocrine control of the breast.*

CARBOHYDRATE AND ENERGY METABOLISM

Carbohydrate metabolism

Digestion of carbohydrate (LD-30)

Starch consists of long chains of glucose molecules joined together in:

- a single line (amylose), or
- a branched line (amylopectin)

The enzyme ptyalin in the saliva starts the de-

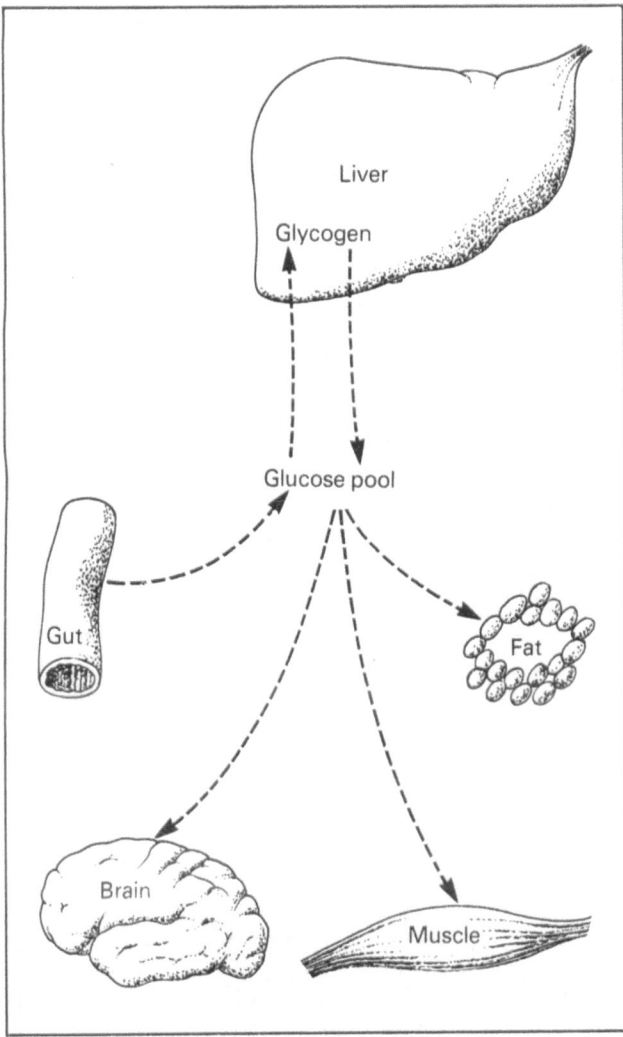

LD-30. Carbohydrate metabolism.

polymerization of the glucose units and the process is completed by maltase in the duodenum. Further enzymes break down the other naturally occurring disaccharides such as:

- sucrose, producing glucose and fructose
- lactose, producing glucose and galactose

Glycogen formation (glycogenesis)

Glycogen, the storage carbohydrate of the liver, is synthesized and broken down by different enzyme systems. Synthesis starts with the phosphorylation of glucose to glucose-6-phosphate by the enzyme hexokinase. This is rearranged by phosphoglucomutase to glucose-1-phosphate which is built into glycogen.

Galactose, fructose and a number of other monosaccharides less commonly present in the diet are converted into glucose or glycogen in the liver.

Glycogen breakdown (glycogenolysis)

Glycogenolysis is started by a phosphorylase to produce glucose-1-phosphate which is changed to glucose-6-phosphate before glucose-6-phosphatase releases free glucose.

Glycolysis

Glucose is metabolized in all cells by the same system – glycolysis. This breaks the 6 carbon units down into 3 carbon molecules (LD-31). Normally this produces pyruvic acid although, in a state of oxygen lack, it is reduced to lactic acid. Pyruvic acid is further metabolized to give the 2 carbon compound acetyl coenzyme-A which is then oxidized to carbon dioxide and water via the tricarboxylic acid cycle. Thus, this cycle and the earlier process of glycolysis change glucose into carbon dioxide and water with the provision of energy (LD-32).

The speed of glycolysis depends initially on the rate of transport of glucose into cells and then on the balance of the two phosphorylation steps.

The further metabolism of glucose from the pyruvic acid stage follows the so-called tricarboxylic acid cycle first described by Krebs.

Gluconeogenesis

This is the process which generates glucose from non-carbohydrate sources, mainly protein. It is active in starvation and even more so in uncontrolled diabetes with rapid loss of weight. Insulin inhibits and the adrenal corticosteroids stimulate gluconeogenesis in

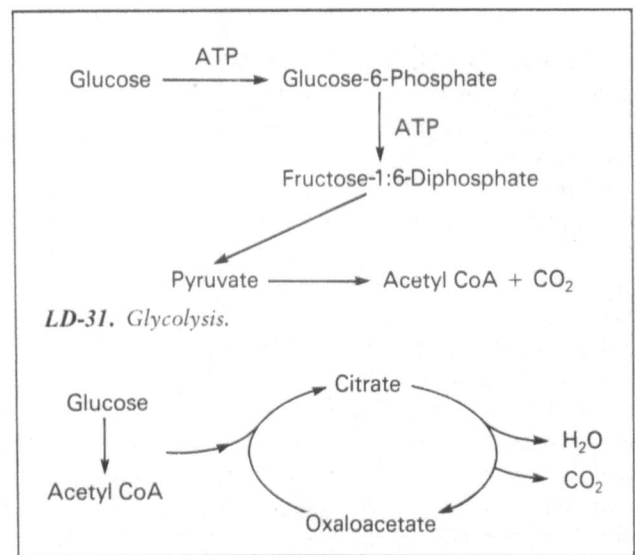

LD-31. Glycolysis.

LD-32. Tricarboxylic acid cycle.

the liver and kidneys. This cycle is localized in the mitochondria of cells where most oxidative energy is produced.

In anaerobic conditions, glycolysis is accelerated – the so-called 'Paskar effect' – for Paskar noticed a similar phenomenon in fermentation by yeast.

Protein metabolism (LD-33)

Proteins are split in the course of digestion into single amino acids which are absorbed into the portal system and taken to the liver. Here, and in the kidneys, they can be used as building-blocks for protein, and in the liver many, but not all, amino acids can be metabolized to glucose.

In the course of gluconeogenesis, complex proteins break down into their constituent amino acids and

then into 3 carbon intermediates such as pyruvic acid. From this stage on, the process is really glycolysis in reverse. Although glucose can be formed from most amino acids, three, namely leucine, phenylalanine and tyrosine produce ketones rather than glucose.

Fat metabolism (LD-34)

Fat consists of a combination of three molecules of fatty acid with one molecule of glycerol. Fat is situated mainly in the adipose tissue under the skin and around the viscera, constituting the organ which supplies most of the energy in conditions of starvation.

The first stage of the breakdown of fat is the hydrolysis of neutral fat or triglyceride to free fatty acids and glycerol, although di- and monoglyce-

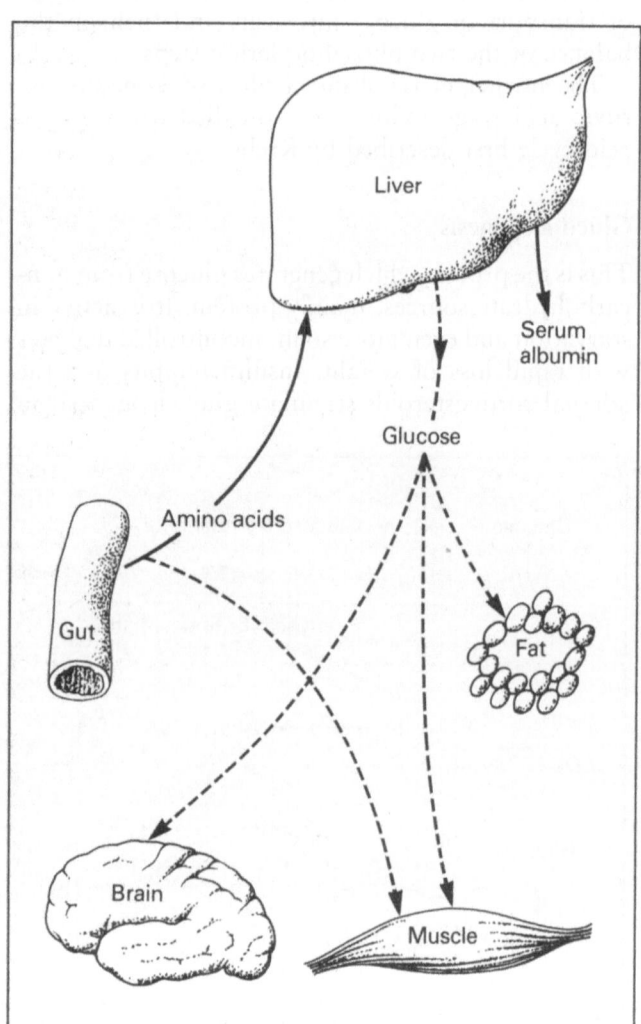

LD-33. Protein metabolism.

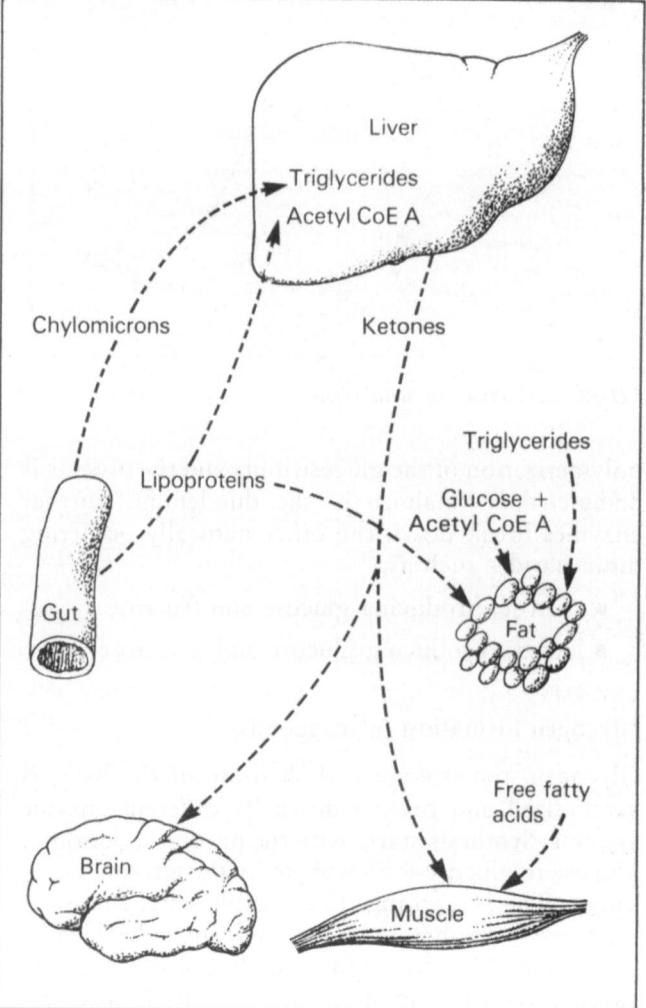

LD-34. Fat metabolism.

LD-35. *Structure of triglycerides.*

rides may delay the release of glycerol from which glucose can be formed (LD-35).

Fat synthesis is stimulated by insulin and its breakdown (lipolysis) by a number of hormones such as growth hormones, glucagon and ACTH.

Lipogenesis does not follow the path of lipolysis in reverse. The long chain fatty acids are formed from acetyl coenzyme-A, a product of glycolysis. This occurs usually in the cell cytoplasm outside the mitochondria.

In uncontrolled diabetes the synthesis of fatty acids is much decreased. This may be because of inadequate formation of glycerol from glucose and from the impaired oxidation of glucose by the pentose phosphate shunt. This in turn leads to a low level of NADP (TPM) which is necessary for the synthesis of fatty acids. Thirdly, in diabetes the high intracellular levels of fatty acids probably greatly inhibit the synthesis of more fatty acids.

Homeostasis of blood glucose

Glucose is the main carbohydrate fuel of the body and its concentration in the plasma varies little in the normal individual in spite of irregular meals of differing size or of starvation. The slight rise in plasma glucose which follows meals is derived mainly from starches which are hydrolysed in the upper alimentary tract and are absorbed into the portal venous system which takes the glucose to the liver. Here, any glucose in excess of intermediate requirements is laid down as glycogen (glycogenesis). On the other hand, when dietary carbohydrate is not available, glycogen is broken down to release glucose into the blood (glycogenolysis).

In starvation, the glycogen store in the liver would suffice to maintain normal levels of plasma glucose for only a few hours unless glucose could be produced from other sources. The main such source is protein in the muscles of the body (gluconeogenesis).

The normal range of plasma glucose is about 3.3–6.6 mmol/l (60–120 mg/dl). Lower values would deprive the brain of its main fuel and, conversely, high concentrations of glucose cause an osmotic diuresis. Any plasma glucose level represents the balance between inflow of glucose and its uptake by the tissues. The inflow may be from the gut and/or the liver by the breakdown of glycogen or protein.

Hormonal control of metabolism

Insulin

Source and chemistry

Insulin is formed in the beta cells of the pancreatic islets of Langerhans. It is stored in these cells as granules which are then moved through a system of microtubules to the periphery by a process requiring energy, calcium ions and cyclic AMP. Most commercial insulin is extracted in acid alcohol from porcine or bovine pancreas and can now be highly purified.

The structure of the insulin molecule was unravelled by X-ray crystallography and it was the first protein to be synthesized (by Sanger). It consists of A and B chains which are linked by disulphide bridges. The sequence of amino acids varies from one animal species to another and these differences produce antibodies when the insulin from one species is injected into another. Pig insulin differs from human insulin in only one amino acid while beef insulin differs in three and, as can be expected, is more antigenic in man.

Insulin in the granules is stored in units of 6 molecules, in association with 2 atoms of zinc, and in a larger precursor form called proinsulin (LD-36).

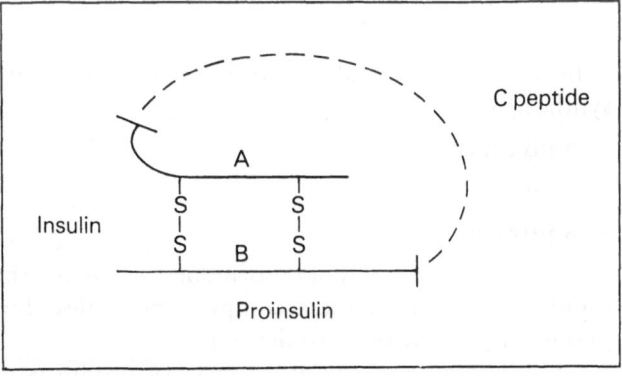

LD-36. *Proinsulin.*

When insulin is secreted, there is an enzymic cleavage which separates the connecting peptide (C peptide) joining the end of the A and B chains, releasing the active hormone and an equimolar proportion of C peptide.

Various substances can stimulate the secretion of insulin. The most important is glucose but certain amino acids share this property, especially arginine and leucine, and indeed seem to be more important than glucose in fetal life. Therapeutically, it is important to note that the sulphonylurea oral hypoglycaemic agents act in this way.

Insulin and C peptide can be measured very specifically in plasma by radioimmunoassay. The amount of C peptide considerably exceeds that of insulin because it is not destroyed so quickly in the body. C peptide does not occur in significant amounts in commercial insulins so that its level in plasma can be used as a measure of endogenous insulin secretion.

Actions

Insulin is the only hormone which both stimulates storage of glucose and its uptake as a fuel in muscle. It also has important anabolic effects on protein and fat. The main actions are summarized in Table 11 and are considered under the headings of early and late effects.

Table 11. Mode of action of insulin

Action	Effect
Early	Increases glucose uptake and metabolism in muscle and adipose tissue Reduces fatty acid release from adipose tissue Increases glycogen formation in liver Increases protein synthesis
Late	Increases activity of some enzymes concerned with glucose metabolism, e.g. liver glucokinase (phosphorylation of glucose) Reduces activity of enzymes concerned in gluconeogenesis

Insulin is thus an anabolic hormone concerned with synthesis of:

- glycogen
- fat
- protein

In its absence, catabolic hormones stimulate the rapid breakdown of these complex molecules thus permitting ketoacidosis to develop.

At a cellular level the main effect of insulin is to increase the rate of glucose transport across cell membranes of insulin-sensitive tissues. It also lowers the level of the intracellular messenger cyclic AMP in fat, thus inhibiting lipase and fat breakdown. Insulin appears to stimulate protein synthesis by an effect on the intracellular ribosomes.

There is a considerable range of sensitivity to insulin in the various tissues of the body, muscle and adipose tissue being highly sensitive. The liver also responds to even lower concentrations of insulin in the plasma although the effect is not seen for about a quarter of an hour. The brain is an example of a tissue that is relatively unresponsive to insulin.

Metabolic consequences of insulin deficiency

In the absence of insulin, less glucose enters muscle and fat cells so that its level rises in the plasma. There is increased breakdown of triglyceride to free fatty acids and glycerol, the levels of both rising in the plasma. Increased fat breakdown in the liver causes increased production of ketones, the levels of which also rise in the plasma. Some ketones are oxidized in muscle but the capacity for this in an uncontrolled diabetic is less than normal so the ketonaemia progresses and produces a metabolic acidosis.

Protein breakdown is rapid and the activity of enzymes concerned with gluconeogenesis is increased, thus further raising the plasma glucose mainly from protein in the liver. Synthesis of protein and triglyceride virtually ceases.

Effect of hormones other than insulin on plasma glucose

The main hormones which oppose the action of insulin on the plasma glucose are:

- growth hormone
- glucagon
- adrenaline
- ACTH
- adrenal corticosteroids

Growth hormone (GH)

Normally GH stimulates increased insulin production. When insulin is reduced or absent in a diabetic, it is a powerful antagonist to the action of any remaining insulin. It also stimulates lipolysis, sparing the oxidation of glucose and causing its level in the plasma to rise still further.

Active acromegaly is often associated with diabetes and with extreme insulin resistance. In hypopituitarism there is increased sensitivity to insulin.

Glucagon

This is produced by the alpha cells of the pancreatic islets of Langerhans and also in the small intestine, but it is pancreatic glucagon which classically raises plasma glucose levels by increasing hepatic glycogen breakdown. This it does by raising intracellular cyclic AMP levels and also by increasing gluconeogenesis.

Glucagon activates hepatic phosphorylase which produces glucose-1-phosphate and this is then quickly hydrolysed to glucose. Plasma glucagon levels increase in the first week or so of starvation, thus maintaining a normal plasma glucose level.

Adrenaline

Like glucagon, this raises the level of glucose in the plasma by increasing the hepatic cyclic AMP and thus activating phosphorylase leading to glucose release from glycogen. Unlike glucagon, however, it inhibits the secretion of insulin.

Glucocorticoids

These are steroids with an oxygen on C_{11} which raise plasma glucose levels and increase hepatic glycogen.

ACTH has, indirectly, the same effect. The plasma glucose rises because of an increase in gluconeogenesis from protein in the liver and an inhibition of glucose uptake by cells.

Gut hormones

Enormously improved chemical purification techniques and radioimmunoassays have recently allowed the identification of a number of simple peptide hormones which occur in parts of the gut and the nervous system.

The gut hormones can be divided into two families, the one chemically related to gastrin and the other to secretin. The many chemical similarities presumably stem from a common evolutionary origin.

These hormones can act in three different ways as:

- peptidergic neurotransmitters
- locally acting humoral agents
- circulatory hormones.

The gut hormones include:

- gastrin
- secretin
- cholecystokinin–pancreozymin
- motilin
- enteroglucagon

They differ in their distribution in the gut.

- Somatostatin

  This is an example of a brain–gut hormone with such a variety of actions that these must be called in to play locally as the occasion demands. In the pancreas, for instance, it is secreted by D cells which occur among the glucagon secreting cells and which lie close to the beta cells which secrete insulin.

- Enteroglucagon

  This is a powerful stimulus for insulin release by the beta cells.

- Gastric inhibitory peptide (GIP)

  This is one of the secretin family and not only inhibits gastric acid secretion but also stimulates insulin release if glucose is present. In this context, GIP may be regarded as a 'glucose-dependent insulin-releasing peptide'. Its action may largely explain why the same amount of glucose given orally produces a larger insulin response than when administered intravenously. Plasma levels of GIP have been reported to be increased in obesity and in diabetes mellitus, possibly due to deficient feedback control.

- Vasoactive inhibitory peptide (VIP)

  This is another member of the secretin family which, among other actions, stimulates insulin release.

Many of the gut hormones are concerned with the process of digestion but they are still only partly understood and the significance of their integrating action in the brain, gut and its associated organs is still to be recognized.

NORMAL GROWTH AND DEVELOPMENT

Growth in the fetus

The principal events of organ and tissue differentiation occur during the first 3 months of fetal life but the fetus of 3 months is only about 8 cm in length and 14 g in weight. Fetal growth then proceeds rapidly in the 2nd and 3rd trimesters of pregnancy. The peak velocity of growth in length is reached during the 2nd trimester but that for weight is reached during the 3rd trimester at about 34 weeks. Brain growth involving neurons is very active about 26 weeks whereas glial tissue is increasing most rapidly at term. Adverse factors early in pregnancy may lead to congenital defects in various organs whereas, when they occur later in pregnancy, the fetus may suffer from intra-uterine growth retardation.

Monitoring fetal growth

The most accurate clinical guide to fetal growth from the 13th week of pregnancy is by consecutive measurements of the biparietal diameter of the fetal head made by ultrasound. By this method, retardation of brain growth, as reflected by head growth, has been detected as early as 26 weeks. Crown–rump lengths and abdominal circumference measurements by ultrasound are useful adjuncts in the study of fetal growth.

Poor fetal growth is usually associated with insufficiency of the fetoplacental unit. The placental function may be monitored by serial estimations of the plasma levels of progesterone and human placental lactogen and the 24 hour quantitative excretion of pregnanediol. As the fetus plays an important part in the synthesis of oestriol, serial estimations of plasma oestriol and the quantitative 24 hour excretion of total oestrogens provide a better indication of the function of the fetoplacental unit as a whole. When intrauterine growth retardation is detected, the obstetrician may decide to induce labour or perform a caesarean section.

Birth weight and gestational age

Normal fetal growth during the last weeks of pregnancy, as reflected by body weight, is shown on Charts 1 and 2. They give the birth weight for boys and girls, based on data from infants born at various times during pregnancy, according to the gestational age calculated from the first day of the last menstrual period.

Light-for-dates and preterm infants

Newborn infants are termed light-for-dates when the birth weight falls below the 10th centile for their appropriate gestational age and sex. Infants born prematurely, that is, before the 37th completed week of pregnancy, are known as preterm infants and they have a low birth weight. However, the birth weight is usually normal for the gestational age when the pregnancy has been developing satisfactorily until the occurrence of an acute event leading to early delivery.

Where there has been a more chronic intrauterine abnormality then these babies are likely to be light-for-dates.

Both preterm infants and light-for-dates infants are liable to present clinical problems in the neonatal period and possibly later. Many show catch-up growth in infancy and early childhood but some, particularly the light-for-dates group, may continue to be of short stature in adult life.

Heavy-for-dates infants

Some infants have a birth weight greater than the 90th centile and are 'heavy-for-dates'. Hereditary factors are the usual cause but maternal diabetes mellitus must be excluded if it is not already known to be present.

Growth in infancy, childhood and adolescence

Length and weight

The growth rate is very rapid after birth and slows down markedly during the 1st and 2nd years of life and more gradually throughout childhood until the growth spurt associated with puberty takes place. Thereafter, growth ceases when the adult height is reached.

An infant-measuring table is required for the accurate measurement of the length of infants and a stadiometer for the height of children and adolescents.

Charts 3 and 4 show the height plotted against age for boys and girls.

The centre line, or 50th centile, on the height chart is the mean and represents the growth curve of the average boy or girl. The 97th and 3rd centiles are ($\pm$) 1.88 standard deviations from the mean. A boy or girl whose height lies above the 97th centile is taller than 97% of the population of that sex, whereas a boy or girl whose height lies on the 3rd centile is taller than only 3%. In addition, the charts show a -3 standard deviation line: a boy or girl whose height lies on the -3 standard deviation line is taller than only 0.13% of the normal population and therefore has a severe degree of short stature, requiring full investigation. In addition, these charts show the stages of puberty for age, according to Tanner, and these will be explained later.

Weight

Charts 5 and 6 illustrate the weight for age for boys and girls.

Height velocity

Charts 7 and 8 show the height velocity for boys and girls. It should be noted that the peak of the growth spurt related to puberty in boys occurs at 14 years on average whereas in girls it occurs 2 years earlier around 12 years of age.

Bone age

The bone age is a further parameter of growth and development which, though it shows considerable variation at adolescence, is used extensively in the clinical study of growth disorders. Tanner and his colleagues have described a detailed method of assessment, based on an X-ray of the left wrist and hand, which may be used from the age of 1 year. Under 1 year of age, a lateral X-ray of the knee and the ankle provides a better estimate of skeletal maturity.

Stages in the development of the lower end of the radius and ulna, the carpal bones and the metacarpal bones and the phalanges of the first, third and fifth digits are used to compile a maturity score on which the bone age is based. In the United States, Gruelich and Pyle have published an atlas of the stages of development on X-ray of the left wrist and hand in boys and girls. From these, the bone age may be determined by visual comparison. Using this atlas, the average normal British child's bone age is 6 months retarded in comparison with the Tanner method.

Development of trunk and limb fat

In both sexes the thickness of the skin and subcutaneous fat, in the subscapular region (representing trunk fat) and over the triceps (representing limb fat) increases rapidly during the first year of life and then decreases during the middle years of childhood.

In boys, from the age of about 8 years, trunk fat increases but there is very little concomitant increase of fat in the limbs. In girls, an increase of both trunk and limb fat takes place from about the age of 7 years and continues throughout puberty. A girl who has tended to be obese during childhood may thus become markedly so as an adolescent.

Pubertal growth

During puberty, there is a spurt in linear growth, the sexual organs mature, the secondary sexual characteristics develop and the individual becomes fertile (LD-37.1, 37.2, 37.3).

Tanner has described stages of puberty numbered 1 to 5, based on the growth of the penis and pubic hair in boys and the growth of the breasts and pubic hair in girls. Genital maturity ratings in boys are shown in LD-37.1, breast development ratings in girls in LD-37.2 and pubic hair ratings in boys and girls in LD-37.3. The ages of appearance of the various features of puberty are shown for boys on Chart 3 and girls on Chart 4 with centiles indicating the range of normal for each.

In boys

In the average boy, the penis begins to enlarge at the age of 12 years with a normal variation of ± 2 years and pubic hair appears between the ages of 12 and 13 years with a normal variation of ± 2 years.

According to British standards, a boy is showing precocious puberty if growth of the penis occurs before the age of 10 years, or delayed puberty if growth of the penis does not begin until later than 14 years.

In girls

In the average girl, breast development begins at the age of 11 years with a normal variation of ± 2 years and pubic hair appears between the ages of 11 and 12 years, with a normal variation of ± 2 years. The average age at menarche is 13 years and 95% of girls first menstruate between their 11th and 15th birthdays. According to British standards, a girl is showing precocious puberty if the breasts enlarge before the age of 9 years and delayed puberty if they show no enlargement after the age of 13 years.

In normal individuals, as well as differences in the age of onset of puberty, there is variation in the time taken to advance through the various stages. Although breast development in girls usually takes place only 1 year earlier than penile development in boys, the growth spurt occurs about 2 years earlier in girls so that the adult height is reached at a correspondingly younger age. Characteristic psychological and behavioural changes during puberty accompany the physical features outlined.

Hormonal control of growth

In the fetus

The rate of growth of the fetus initially depends upon the drive to growth within the cells themselves. This is genetically determined. Thereafter, suboptimal environmental influences may impede growth whereas various hormones are thought to be stimulatory.

The hypothalamic–pituitary axis is functional in the 2nd trimester of pregnancy and the feedback mechanisms between the hormones of the target glands and the axis gradually mature. Growth hormone is produced by the fetal anterior pituitary gland and stimulates the production of somatomedins – polypeptide growth factors chiefly in the liver. However, growth hormone lacks the prominent role in fetal growth that it has after birth. This is cor-

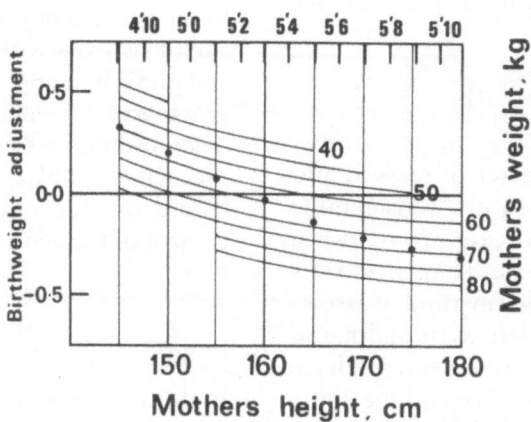

Adjustment for mother's height and mid-pregnancy weight to be added
or subtracted from birthweight. Scale as in chart below. If weight
unknown, use ● points.

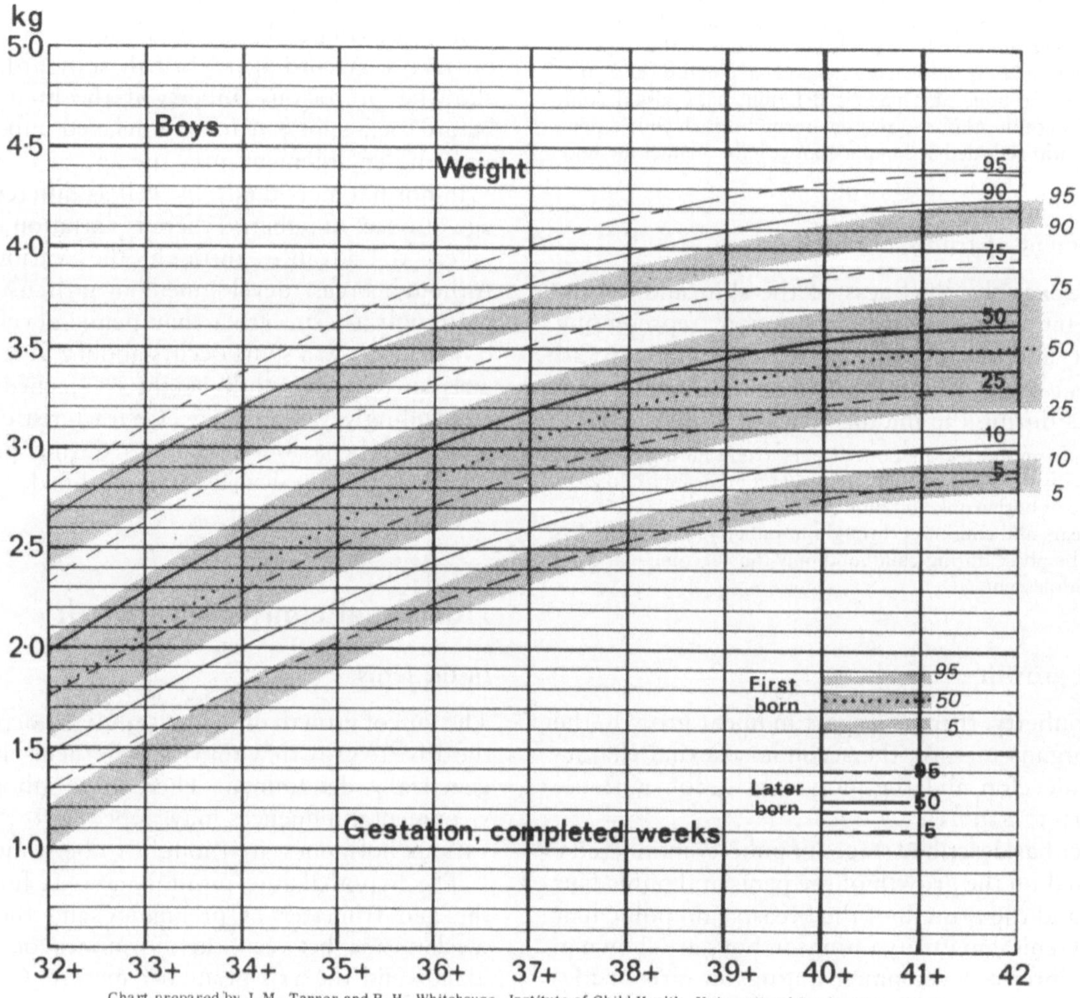

Chart prepared by J. M. Tanner and R.H. Whitehouse, Institute of Child Health, University of London, from data by
A.M. Thomson, W.Z. Billewicz and F.E. Hytten in J. Obstet. Gynaec. Brit. Cwlth., 75, 9O3, 1968.

Chart 1

Charts 1 and 2 *Standards for birth weight at gestation
periods from 32 to 42 weeks allowing for maternal height and
weight for boys and girls. (After Tanner, J.M. and Whitehouse,
R.H. (1976) Archives of Diseases of Childhood, 45, 566)*

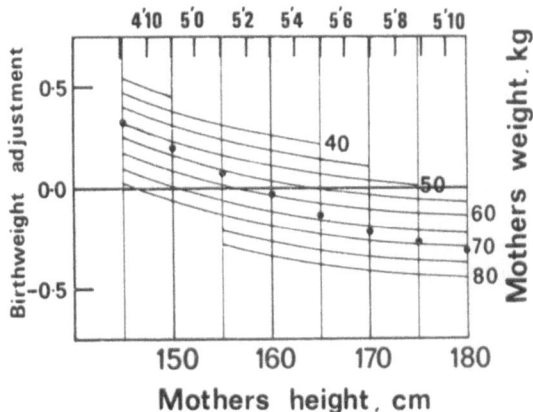

Adjustment for mother's height and mid-pregnancy weight to be added or subtracted from birthweight. Scale as in chart below. If weight unknown, use ● points.

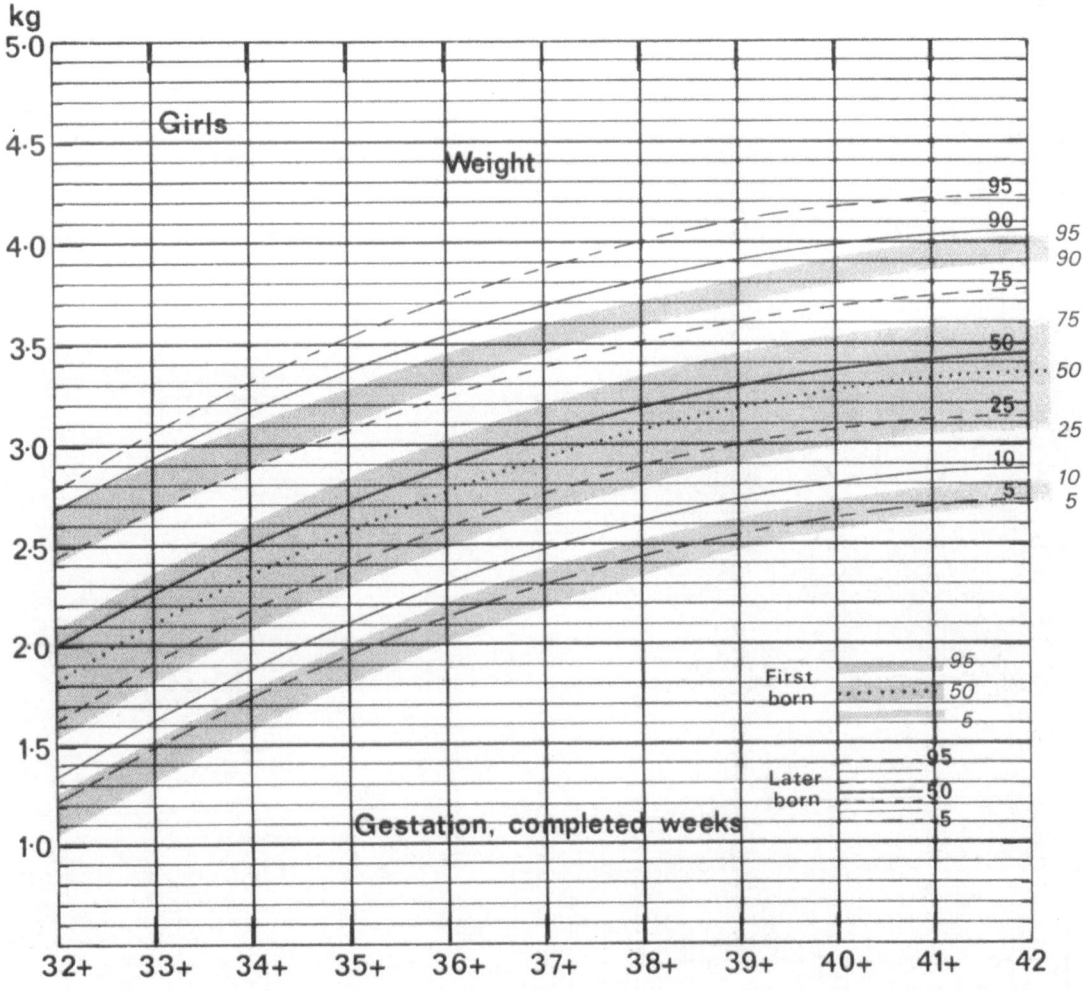

Chart prepared by J. M. Tanner and R. H. Whitehouse, Institute of Child Health, University of London, from data by A. M. Thomson, W. Z. Billewicz and F. E. Hytten in J. Obstet. Gynaec. Brit. Cwlth., 75, 903, 1968.

Chart 2

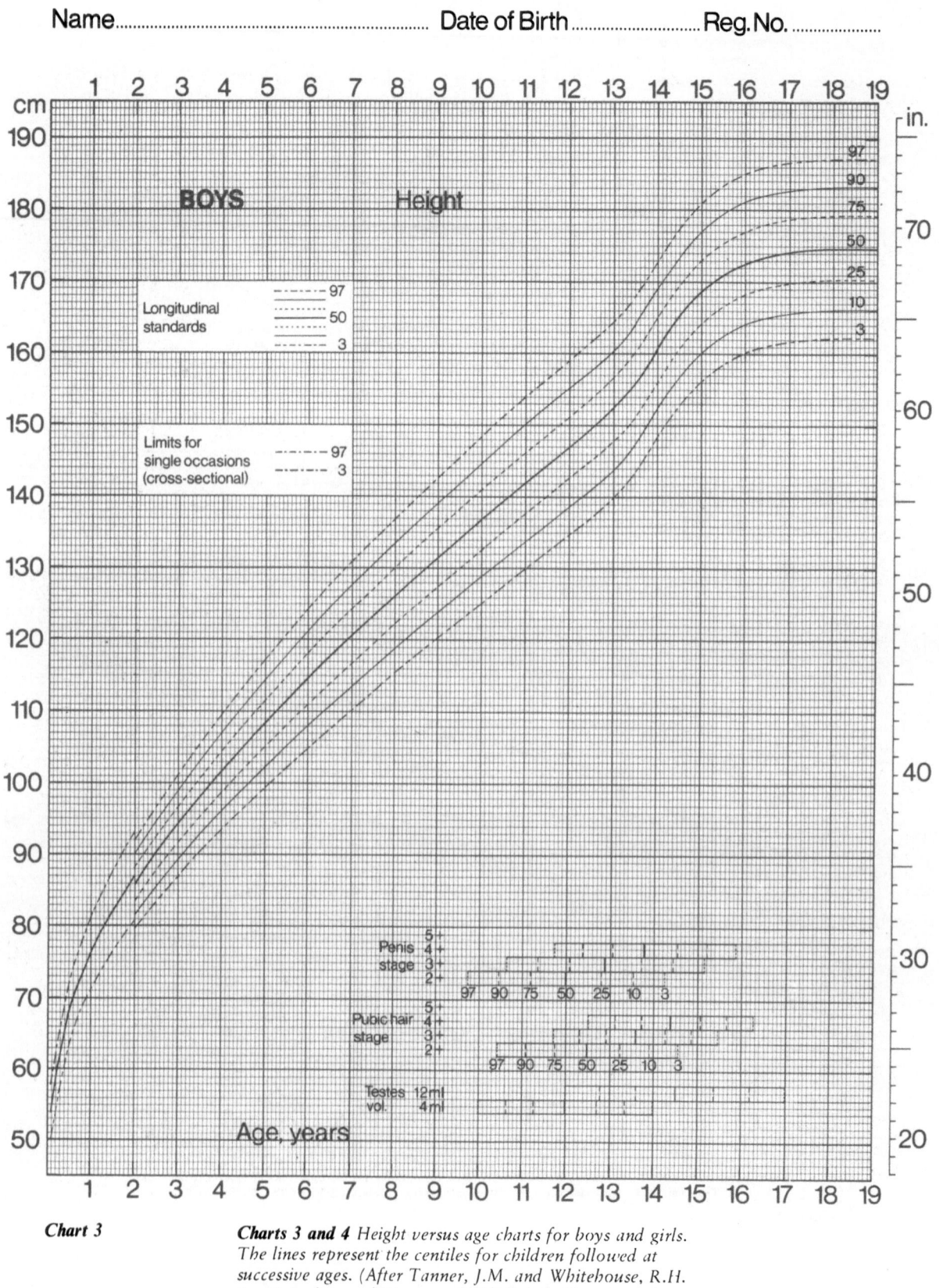

Chart 3

Charts 3 and 4 *Height versus age charts for boys and girls. The lines represent the centiles for children followed at successive ages. (After Tanner, J.M. and Whitehouse, R.H. (1976) Archives of Diseases of Childhood,* **51**, *170–179)*

Name.. Date of Birth .... Reg. No. ....

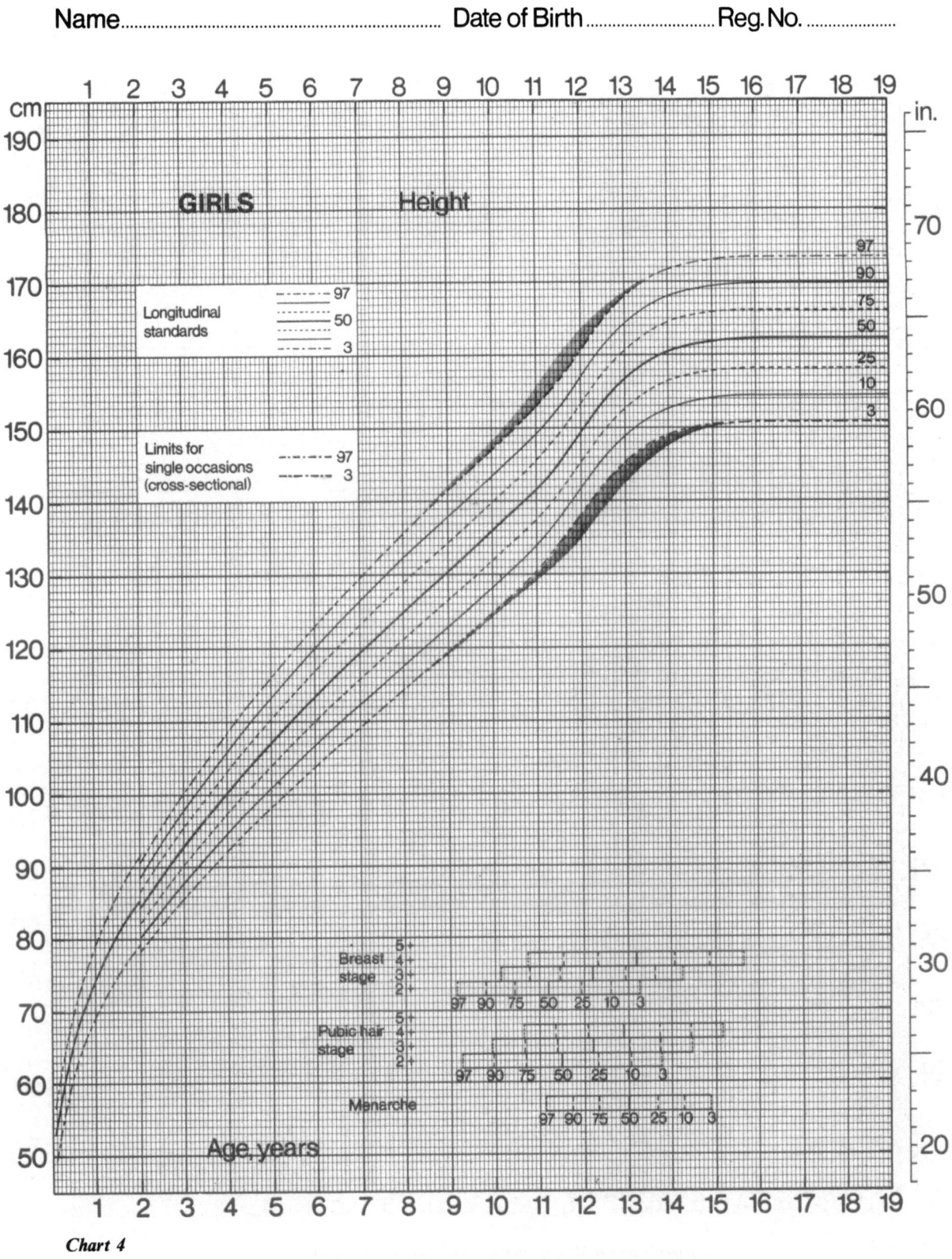

Chart 4

Name .... Date of Birth .... Reg. No. ....

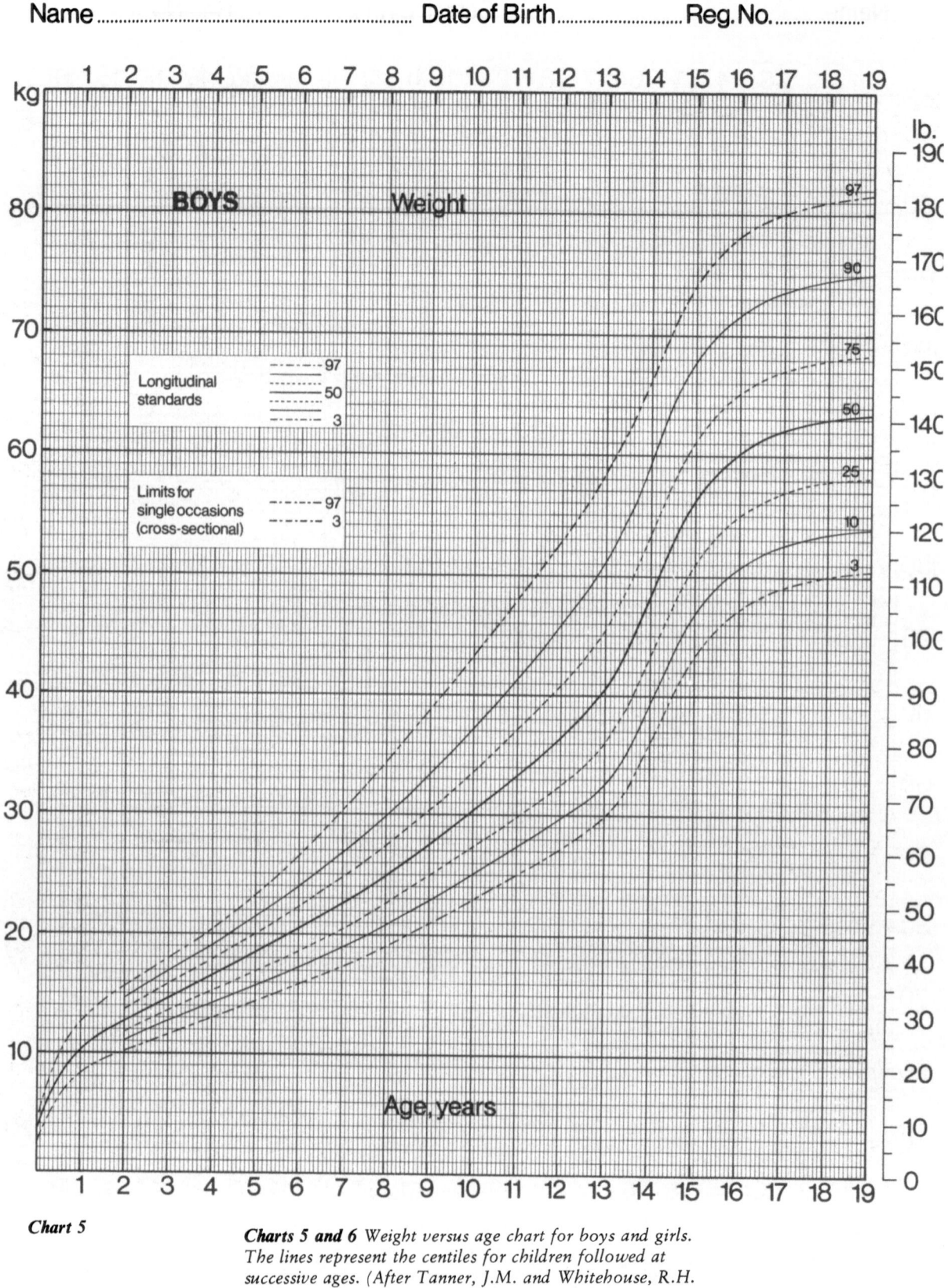

Chart 5

Charts 5 and 6 *Weight versus age chart for boys and girls.*
The lines represent the centiles for children followed at
successive ages. (After Tanner, J.M. and Whitehouse, R.H.
(1976) Archives of Diseases of Childhood, **51**,*170–179*)

Name .... Date of Birth .... Reg. No. ....

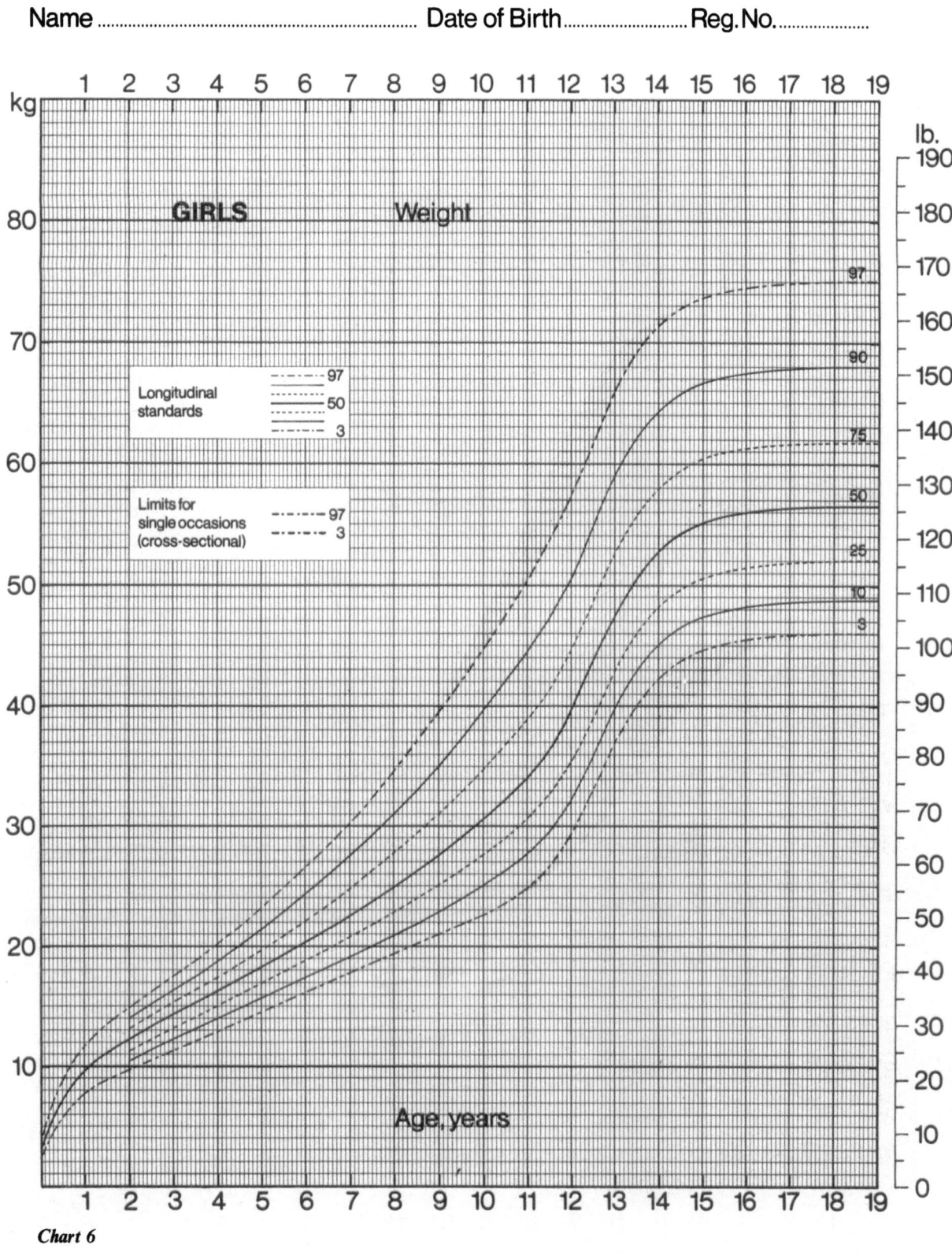

Chart 6

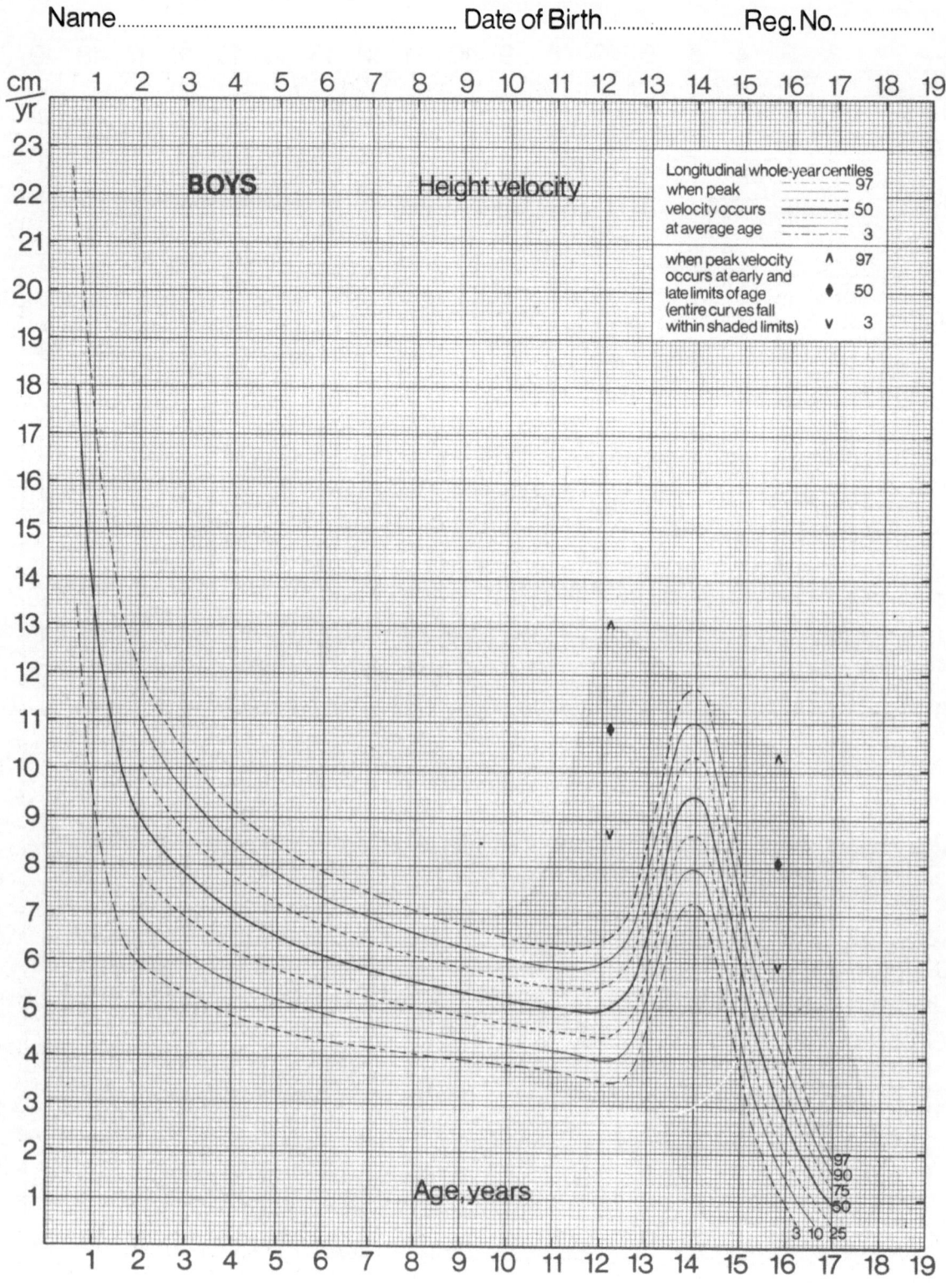

Name.. Date of Birth......................... Reg. No. ....

BOYS　　　　　Height velocity

Longitudinal whole-year centiles
when peak
velocity occurs
at average age

when peak velocity
occurs at early and
late limits of age
(entire curves fall
within shaded limits)

Age, years

Chart 7　　　　**Charts 7 and 8** *Height velocity chart for boys and girls. The
centiles are appropriate to children who have their peak velocity
at the average age. (After Tanner, J.M. and Whitehouse, R.H.
(1976)* Archives of Diseases of Childhood, **51**, *170–179)*

Name.. Date of Birth........................ Reg. No....................

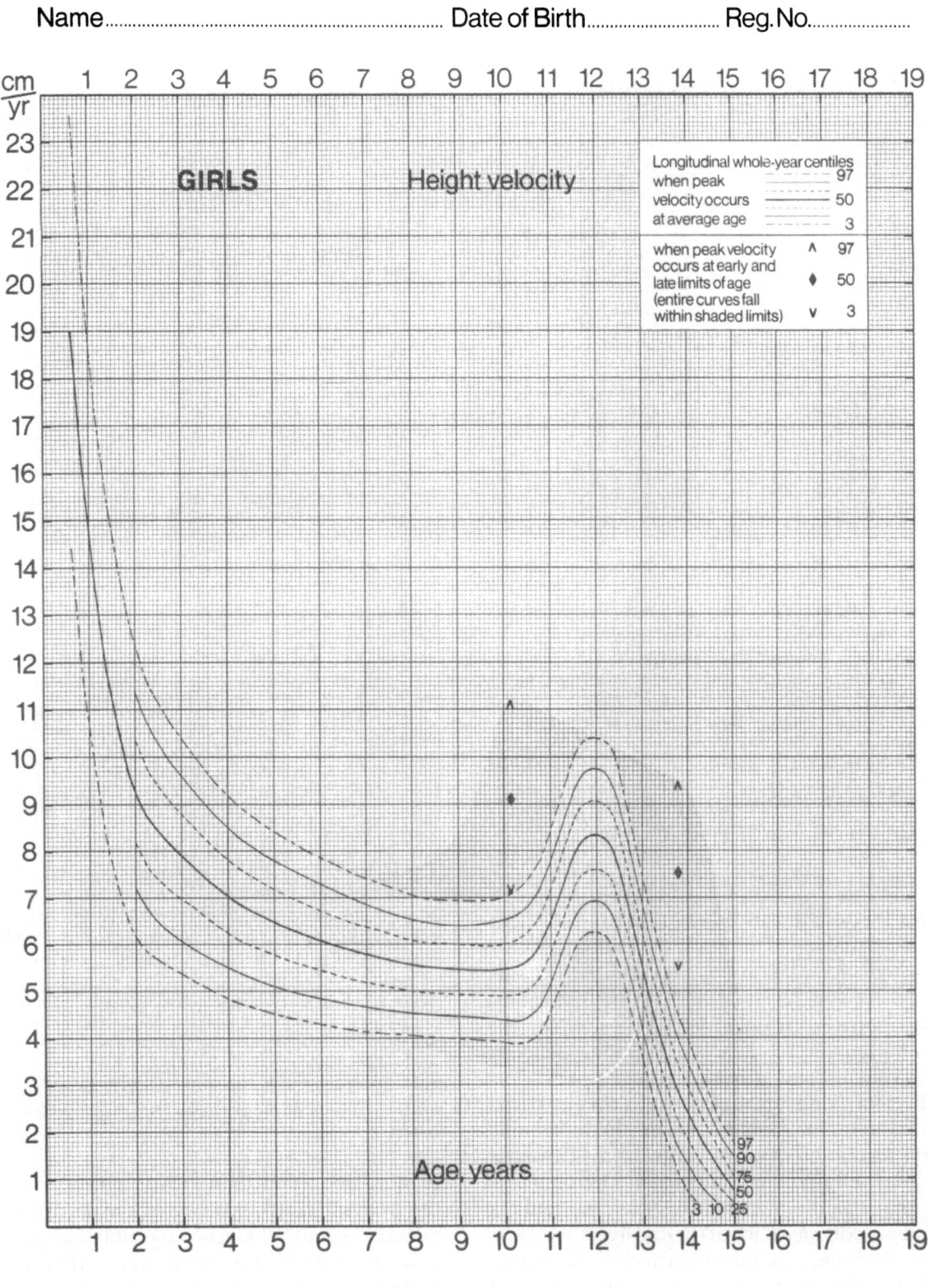

Chart 8

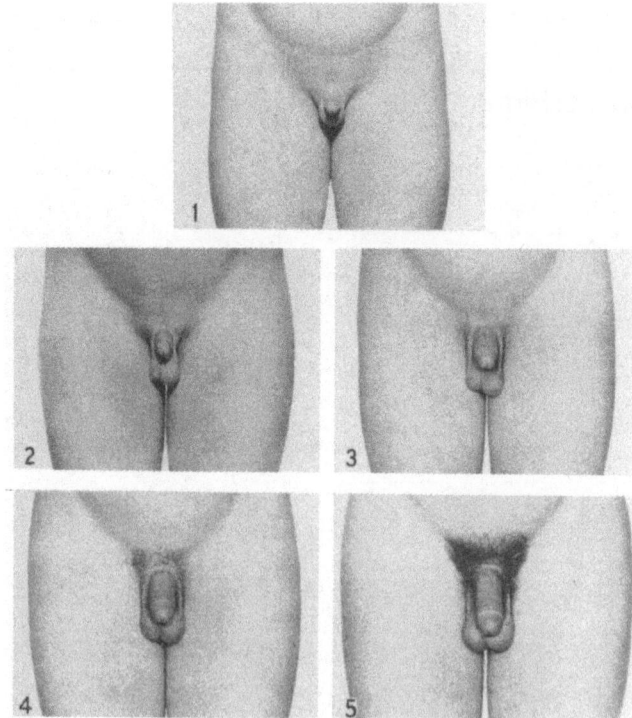

LD-37.1. Standards for genital maturity ratings *(by courtesy of J.M. Tanner).*
Stage 1: Pre-adolescent. The testes, scrotum and penis are about the same size and shape as in early childhood.
Stage 2: The testes and scrotum are enlarged slightly. The skin of the scrotum is reddened and changed in texture. There is little or no enlargement of the penis at this stage.
Stage 3: The penis is enlarged slightly, mainly in length. There is further growth of the testes and scrotum.
Stage 4: The penis is enlarged further in length and breadth and there is development of the glans. The testes and scrotum are further increased in size and the scrotal skin is darker.
Stage 5: The genitalia are of adult size and shape.

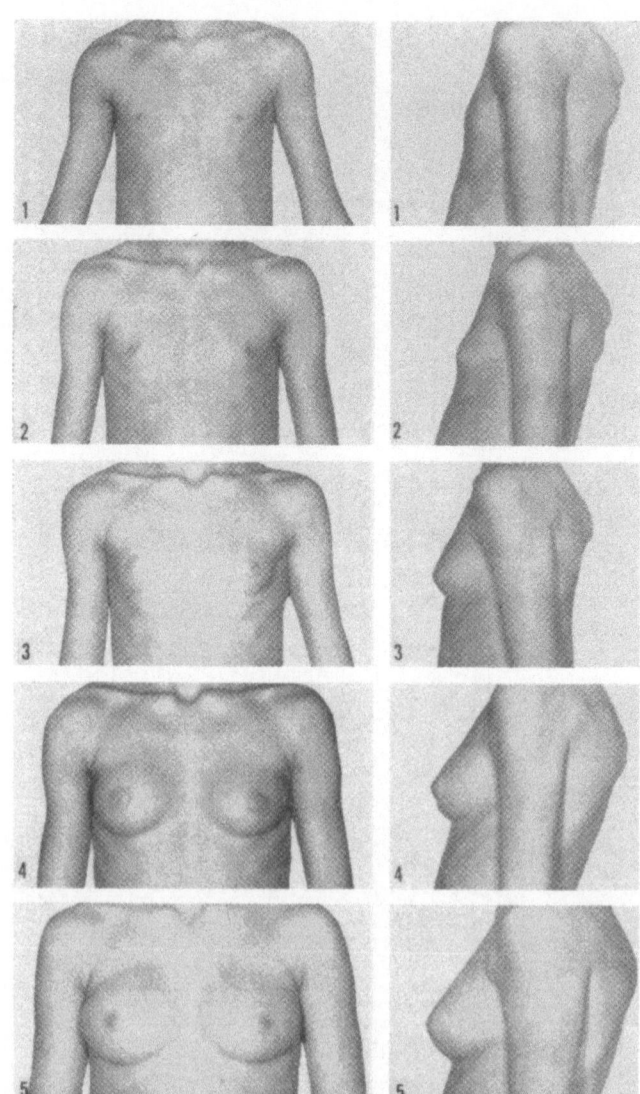

LD-37.2. Standards for breast development ratings *(by courtesy of J.M. Tanner).*
Stage 1: Pre-adolescent. There is elevation of the papilla only.
Stage 2: Breast bud stage: elevation of the breast and papilla as a small mound. There is enlargement of the areola diameter.
Stage 3: There is further enlargement and elevation of the breast and areola with no separation of their contours.
Stage 4: The areola and papilla form a secondary mound projecting above the contour of the breast.
Stage 5: Mature stage: there is projection of the papilla only, due to recession of the areola to the general contour of the breast.

roborated by experiments on rats, rabbits and sheep and, in the human situation, by the finding of only mild growth retardation in anencephalic infants known to have a severe deficit in hypothalamic–pituitary function. In addition, it is usual for the length of even severely growth hormone deficient children to be normal at birth.

The thyroid gland functions towards the end of the first trimester and comes under hypothalamic pituitary regulation a few weeks later. Placental permeability of thyroxine and triiodothyroinine is low so that the thyroid physiology of the fetus is virtually autonomous. Lack of thyroid hormones does not have a strikingly detrimental effect on the length at birth but the infants show delay in brain and skeletal maturation.

In terms of fetal growth, it seems likely that insulin is more important than growth hormone or thyroid hormones. It is strongly anabolic in the presence of an adequate supply of glucose and amino acids and it has been recognized for many years that infants of diabetic mothers, who produce more insulin in response to an abnormally high glucose load, are excessively long and heavy at birth.

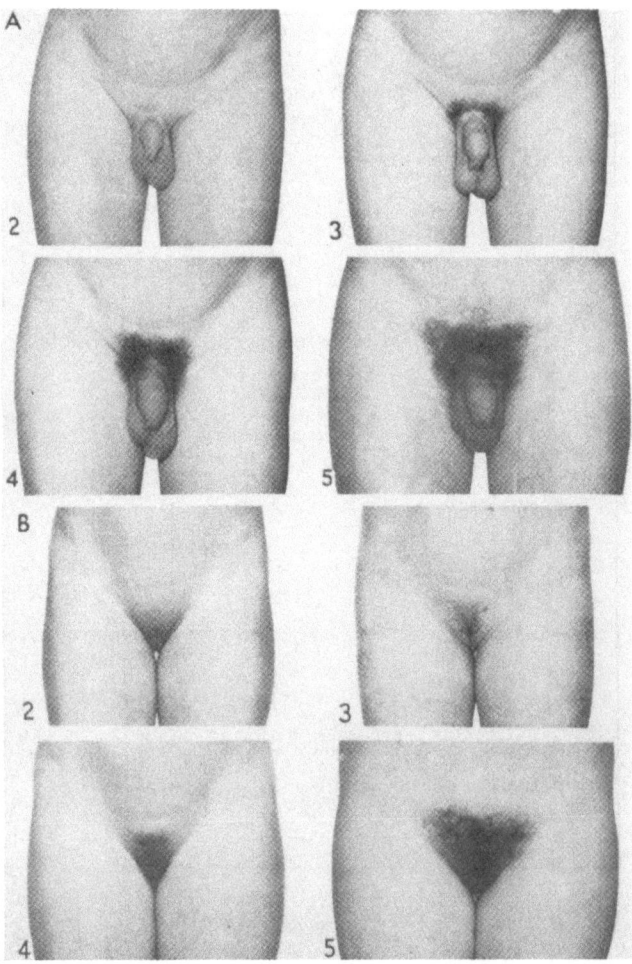

LD-37.3. Standards for pubic hair ratings in boys (A) and girls (B) *(by courtesy of J.M. Tanner).*
Stage 1: *Pre-adolescent. No pubic hair is present.*
Stage 2: *There is a sparse growth of long, slightly pigmented, downy hair, straight or slightly curled, chiefly at the base of the penis or along the labia.*
Stage 3: *The hair is considerably darker, coarser and more curled. It spreads sparsely over the junction of the pubes.*
Stage 4: *The hair is now adult in type, but the area covered is still considerably smaller than in the adult. There is no spread to the medial surface of the thighs.*
Stage 5: *The hair is adult in quality and type with a distribution of the horizontal (or classically 'feminine') pattern. Spread to the medial surface of the thighs is present (In males, spread up the linea alba occurs late and is rated as stage 6).*

Little is known about the part other hormones play in fetal growth. Cortisol may antagonize some growth promoting factors. The roles of the adrenal androgens and sex hormones are unknown but they might exert an anabolic effect. Human placental lactogen, which is secreted on the maternal side of the fetoplacental unit, probably promotes sparing of glucose and amino acids which are thus made available for transfer to the fetus. The place of prolactin has yet to be clarified.

In infancy, childhood and adolescence

In infancy, childhood and adolescence, many hormones interact in achieving normal skeletal growth. Limb growth takes place at the ends of the long bones in the growth plates. These consist of multiplying cartilage cells which gradually calcify to form bone.

The hypothalamus controls the output of growth hormone from the anterior pituitary gland and is itself influenced by stimuli from the higher centres. The action of growth hormone on the growth plates is mediated through that of the somatomedins. Stimulation of growth is accompanied by an advance in bone age.

The thyroid hormones, thyroxine and triiodothyronine, act directly on the growth plates to promote growth and have a more marked effect in advancing the bone age than growth hormone.

The sex steroids, testosterone and the oestrogens, are responsible for the growth spurt which occurs during puberty and act in addition to the basal growth due to other hormones. Sex steroids have a marked effect in advancing bone age and, when the epiphyses fuse, growth ceases altogether.

It is likely that adrenal androgens also play a part in the growth spurt. It is not known whether cortisol secreted by the adrenal glands in normal physiological amounts has any effect on growth one way or the other. It has been known for many years that abnormally high levels of cortisol or corticosteroids inhibit growth and this is a major problem in the use of prolonged corticosteroid drug therapy during childhood.

Insulin and certain insulin-like peptides, parathormone, calcitonin, and possibly prolactin may also contribute to the normal growth process.

During puberty

It has been postulated that puberty begins when the hypothalamus becomes less sensitive to the feedback mechanism of the small quantities of sex hormones produced during childhood by the gonads. The plasma levels of luteinizing hormone and follicle stimulating hormone rise, followed by an increase in testosterone or oestrogen. In both sexes the plasma levels of androgens of adrenal origin rise before there is clinical evidence of puberty and before there is a rise of LH, FSH, testosterone or oestrogen. Thereafter, the adrenal androgens continue to rise throughout puberty.

The large quantity of adrenal androgens produced may play a part in the anabolic processes associated with skeletal maturation and the growth spurt.

Recommended for further reading

Tanner, J.M. (1978) *Foetus into Man: Physical Growth from Conception to Maturity*. (London: Open Books).

Tanner, J.M., Whitehouse, R.H., Marshall, W.A., Healy, M.J.R. and Goldstein, H. (1976). *Assessment of Skeletal Maturity and Prediction of Adult Height*. (London. Academic Press).

Gruelich, W.W. and Pyle, S.I. (1959). *Radiographic Atlas of Skeletal Development of the Hand and Wrist*. 2nd Edn. (Stanford: Stanford UP; London: Oxford UP).

Grunewald, P. (ed.), (1975). *The Placenta and its Maternal Supply Line. Effects of Insufficiency on the Fetus*. (Lancaster: MTP).

Campbell, S. and Newman, G.B. (1971). Growth of the fetal biparietal diameter during normal pregnancy. *J. Obstet. Gynaecol. Br. Commonw.*, **78**, 513.

Fancourt, R., Campbell, S., Harvey, D. and Norman, A.P. (1976). Follow-up study of small-for-dates babies. *Br. Med. J.*, **1**, 1435.

Apparatus

An accurate apparatus for measuring the length of infants, the height of children and the skinfold thickness may be obtained from Holtain Ltd, Croswell, Crymmich, Pembrokeshire.

IMMUNOLOGY

Endocrinology and immunology relate to one another in a number of ways. These are discussed under the following headings:

- basic concepts of immunology
- disordered immunity and the endocrine system
- hormonal effects on the immune system

Basic concepts of immunology

Development of the immune response has made an enormous contribution to the survival of the species. Impairment of the immune response is associated with increased susceptibility to infection, for example:

- in immune deficiency syndromes
- with high doses of steroids
- with immunosuppressive drugs
- in autoimmune diseases

As shown in LD-38, the immune reaction is de-

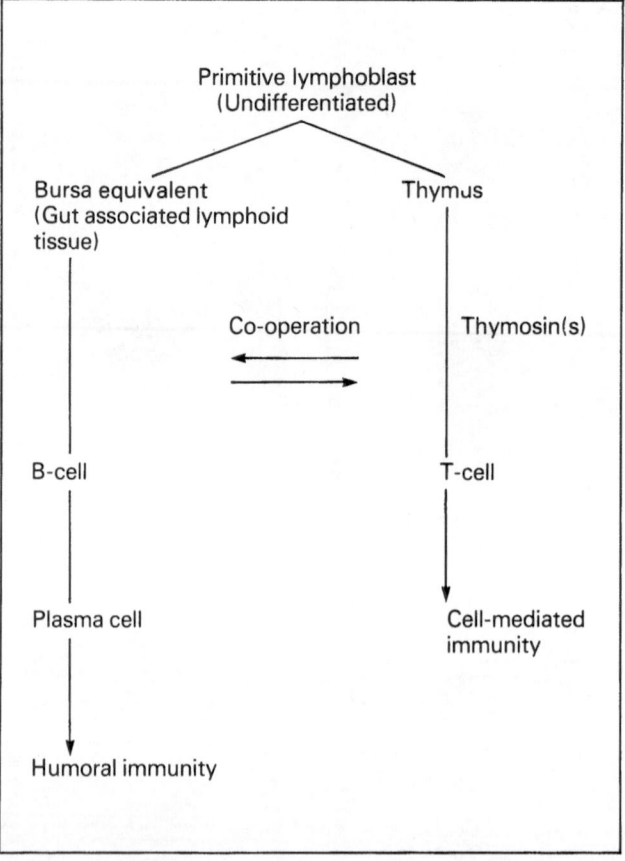

LD-38. *Basic immunology.*

pendent on two cell lines derived from a common progenitor cell – the primitive lymphoblast; they function in a co-operative manner.

These two strains of cells are

- T- (or thymus-derived) cells
- B- (or bursa-equivalent) cells

They are, between them, capable of reacting in a highly specific way with certain molecular configurations – *antigens*. Although co-operating in both functions the

- T-cell is mainly responsible for *cell-mediated immunity*
- B-cell is mainly responsible for *humoral immunity* or the production of antibodies.

T-cells

T-cells or 'thymus-derived cells' are mainly responsible for cell-mediated immunity. They are 'educated' – possibly under the influence of thymosin, a thymic hormone – to perform their later functions.

The mature T-cell recirculates:

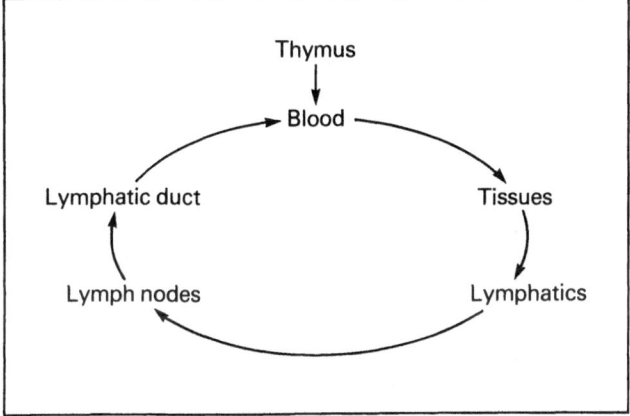

This ability to recirculate allows the T-cells to 'police' the frontiers of the body where contact with antigens – micro-organisms and malignant cells – may be made.

Although not fixed in tissues, at any given time fairly large numbers of T-cells may be found in the

- spleen – in the thymus-dependent areas of the perarteriolar cuff
- lymph nodes – in the paracortical areas

Different types of T-cells are now recognized but are morphologically indistinguishable:

- effector T-cells – with a subtype called 'Killer' cells or K-cells
- suppressor T-cells

Effector T-cells

These are made up of families or clones each of which responds to a specific antigen. When contact is made between such cells and the appropriate antigen the response is proliferation of the cells of that clone. These T-cells are responsible for such cell-mediated immune reactions as

- the tuberculin reaction and
- allograft rejection

Suppressor T-cells

These perform a very important role in suppressing other cells involved in the immune reaction and thus in the fine regulation of the immune response.

B-cells

B-cells or 'bursa-equivalent cells' mature in mammals in the gut lymphoid tissue. This consists of the tonsil, Peyer's patches and condensations of lymphoid tissue scattered throughout the gastrointestinal tract.

The characteristic organized element of gut-associated lymphoid tissue is the *lymphoid follicle* with its germinal centre. This is typically seen in lymph nodes. The B-cells reside mainly in the germinal centres and do not recirculate to the same extent as T-cells. It is there that the B-cell makes contact with the antigen.

B-cells react to antigenic stimulation by transforming into plasma cells which produce antibodies.

At least five types of antibody are known. All are immunoglobulins:

- IgM (a pentamer)
  This is characteristic of the early antibody response.

- IgG (a monomer)
  This is usually responsible for the definitive circulating antibody response.

- IgA (a dimer)
  This is responsible for the antibody defence of mucous membranes.

- IgE
  This is fixed in tissues and is responsible for immediate hypersensitivity and allergic reactions.

- IgD
  No precise role has yet been defined.

The two arms of the immune response are more separate in birds than in mammals. The term 'bursa-equivalent', and thus 'B-cells' derives from the fact that in birds most of the gut lymphoid tissue is gathered in a lymphoid organ in the distal gastrointestinal tract – the bursa of Fabricius.

Disordered immunity and the endocrine system

In normal circumstances the immune response does not react with the body's own constituents. In autoimmune disorders, somehow the mechanisms providing this protection break down and the immune system attacks the body's own tissues.

Autoimmune disorders may attack the body either by damaging a specific organ or by causing damage to many tissues which apparently share a common antigen. Thus autoimmune diseases may be:

- organ-specific
  Hashimoto's thyroiditis
  Graves' disease
  some cases of Addison's disease

- non-organ-specific
  systemic lupus erythematosus
  rheumatoid arthritis
  polyarteritis nodosa

Although endocrine glands may be damaged indirectly in the non-organ-specific autoimmune diseases most of the autoimmune endocrine diseases are of the organ-specific type. These are summarized:

- Autoimmune aetiology established
  Hashimoto's thyroiditis
  Graves' disease
    hyperthyroidism
    exophthalmos
    pretibial myxoedema
  Addison's disease (some cases)
  premature ovarian failure (some cases)

- Autoimmune aetiology – possible in some cases
  primary hypothyroidism
  diabetes mellitus
  idiopathic hypoparathyroidism
  idiopathic hypopituitarism
  infertility

Organ failure is the usual end result of organ-specific autoimmune disease of the endocrine system. A unique exception is the hyperthyroidism of Graves's disease where an antibody to the TSH-receptor on the thyroid cell actually stimulates the gland.

Mechanisms of autoimmunity

There are currently many theories about the mechanism for the breakdown of the immune system's recognition of self. These theories are not all mutually exclusive. They include:

- the privileged site theory
  Some tissues – the eye and the central nervous system – contain no lymphatic pathways and only when damaged are they exposed to the immune system thus eliciting a response. Sympathetic ophthalmia is an example.

- alteration of some normal tissue component by physical or chemical injury
  This could render the tissue antigenic. An example of this is the ability of some drugs to link with platelets causing antibody response which destroys the platelets.

- increased concentration of the body's own antigens
  Above a certain threshold this could elicit an

antibody response. This mechanism has been proposed to explain the antibodies to thyroglobulin which arise transiently in subacute thyroiditis when much larger amounts of thyroglobulin than usual reach the circulation.

- antigenic similarity
  Some body component may be similar to an invading micro-organism. This can give rise to cross-reactions between the antibodies directed against the micro-organism and the body tissue. An example is the cross-reactivity of antibodies to some β-haemolytic streptococci and the endocardium.

- modification of cell antigenicity
  A virus may be incorporated into the cell genome with the production of new antigens in the cell.

- failure of suppressor T-cell function
  This can result in the emergence of a 'forbidden clone' reactive against the bodies tissues.

There is currently much interest in the last of these theories. It may be responsible for a number of organ-specific autoimmune disorders – the hyperthyroidism of Graves's disease.

In practice, in autoimmune disease in humans, much more is known about the role of antibodies than about the role of cell-mediated immunity in disease. This reflects the relative ease of studying an antibody and the relative difficulty in studying the cellular element.

Antibodies may cause disease when directed against such tissues as renal glomerular basement membrane, or against the TSH receptor of the thyroid cell as in Graves's disease.

In other situations, the cause of autoimmune disease appears to be due to cell-mediated immunity, the antibody resulting as an epiphenomenon rather than as a pathogenetic force. Such antibodies may nonetheless be important diagnostically:

- antibodies to thyroglobulin
- antibodies to thyroid microsomes

These can aid diagnosis in Hashimoto's disease though there is no evidence that they are important in the pathogenesis of the disease.

Management of autoimmune endocrine disease

Since most of the autoimmune disorders involving the endocrine system lead to destruction of the specific gland, hormonal replacement therapy is all that is usually required.

The exception, of course, is hyperthyroid Graves's disease where antithyroid drug therapy or partial ablation or removal on the gland is necessary.

It is currently not feasible, nor indeed necessary, to treat the underlying disorder of the immune system.

Hormonal effects on the immune system

What little is known of the factors which control the immune response points to a critical role being exerted by hormones.

In general, immunological activity of females is greater than that of males. This difference appears to depend on the immune response being influenced by hormonal control

- oestrogen stimulating
- androgens depressing

The individual differences in the response between males and females may seem small enough but on a population basis they may explain the greater susceptibility of females to autoimmune disease.

Thyroid hormones and growth hormone also appear to play a special part in stimulating the immune response. In hyperthyroidism there is marked hyperplasia of all lymphoid tissue which in part is related to the degree of hyperthyroidism.

Steroids and other immunosuppressive drugs are commonly used in the management of immunologically mediated diseases, indeed a branch of pathology has developed as a result of the complications that may arise from such therapy. This important topic is considered in Section V (p. 195).

Thymosin

Probably the most important hormone controlling the immune response is thymosin. This hormone – conceivably more than one – is responsible for the maturation of lymphocytes within the thymus. The exact nature of thymosin is unknown and little is known of its physiology.

PRINCIPLES OF INVESTIGATION OF FUNCTIONAL ENDOCRINE DISEASE

The presentation of an endocrine disorder may be such that the clinician does not immediately suspect an endocrine basis to the problem. It is a fundamental concept of modern investigative procedures that early detection of an endocrine disorder should be achieved before clinical features of that disorder have become advanced. The presenting problems are dealt with at length in Section III.

When an endocrine problem is considered the logical laboratory investigation of the condition is made easier by following some basic rules.

For practical purposes there are two types of functional endocrine disorder:

- hormone overproduction
- hormone underproduction

Abnormalities of endocrine gland structure are usually, but not always, associated with disordered function. Techniques for investigating structural abnormalities are outlined later.

Although clinical examination frequently reveals the end organ effects of hormonal excess or deficiency these significant findings are often non-specific, e.g. an abnormal pulse rate in thyroid disorders.

General principles

Basal hormone levels in plasma and urine

Specific assessment of glandular function usually includes measurement of an appropriate basal plasma hormone level. In endocrine dysfunction these may differ significantly from the reference range values found in normal individuals. It should be remembered that a reference range for hospital patients may be different from that for normal healthy individuals. As most, reference ranges relate to 95% confidence limits; 5% of any population may be outside the range and yet have no abnormality. With advances in laboratory techniques hormone assays are widely available and direct measurement of clinically important hormones in blood can be made. Measurements in urine are becoming less common but where the hormone is measured in a 24 hour urine sample this may be a better index of hormonal output. This eliminates sampling problems due to the short-term fluctuations which sometimes occur in blood levels.

Tables 12 and 13 illustrate some of the hormone measurements available to clinicians.

Table 12. Hormones commonly measured in blood serum and urine

Blood serum	Urine
Thyroxine	Pregnancy oestrogens
Triiodothyronine	Chorionic gonadotrophin
TSH	
LH	
FSH	
Growth hormone	
Prolactin	
Placental lactogen	
Cortisol	
Oestradiol	

Table 13. Hormones less commonly measured in blood serum

ACTH	Progesterone
Insulin	Testosterone
Glucagon	Oestriol
Gastrin	17α-hydroxy progesterone
Renin	
Angiotensin	
Aldosterone	

Stimulation and suppression tests

Basal hormone levels may not be discriminatory. Therefore dynamic tests of endocrine function are often applied.

Suspected *hyperfunction* of an endocrine gland is tested by a procedure designed to *suppress* the gland's activity while the appropriate test for *hypofunction* is a *stimulation* test. Tables 14 and 15 give examples.

It is important to realize that a gland's function reflects its previous state of stimulation which may have been abnormal for a considerable time. To produce a response, a dynamic stimulatory test may have to be repeated or modified, e.g. prolonged synacthen test.

Hormonal measurements in relation to dynamic tests may indicate that the disorder is:

- *Primary* – originating in the endocrine gland

- *Secondary* – to some dysfunction of the controlling mechanisms (LD-39, LD-40).

Measurement of more than one hormone may reveal endocrine insufficiency before gross abnormality develops, e.g. T4 may be borderline low but TSH will be high in subclinical hypothyroidism.

Factors other than disease may affect hormone levels (Table 16). There are few absolutes in endocrine measurements and comparison with previously determined ranges is essential. Further it is essential that tests are carried out under conditions comparable to those which applied to the measurement of the laboratory reference values.

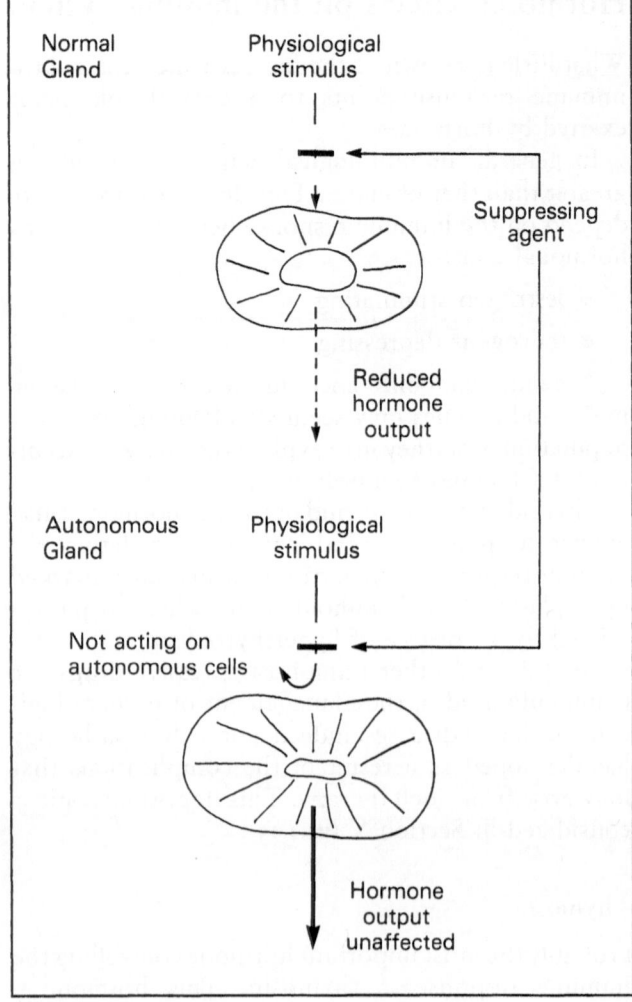

LD-39. Suppression tests.

Methods of hormone measurement

Radioimmunoassays introduced in 1960 (Yalow and Berson) have been largely responsible for the explosive growth in endocrinological knowledge. Before then, measurements depended upon relatively insensitive and non-specific chemical assays and bio-

Table 14. Suppression tests

Suspected hormonal disorder	Suppressing agent	Measured by
Adrenocortical hyperfunction	Betamethasone	Cortisol
Acromegaly	Glucose	Growth hormone
Insulinoma	Prolonged fasting	Glucose
Thyrotoxicosis	Tri-iodothyronine	Thyroidal iodine uptake
Hyperprolactinaemia	Bromocriptine	prolactin

Table 15. Stimulation tests

Suspected disorder	Stimulant	Measured by
Adrenocortical hypofunction	Synacthen	Cortisol
Growth hormone deficiency	Insulin	GH
Hypothyroidism	TRH	TSH
Hypothalamic–pituitary dysfunction	LHRH	LH + FSH
Testicular dysfunction	HCG	Testosterone

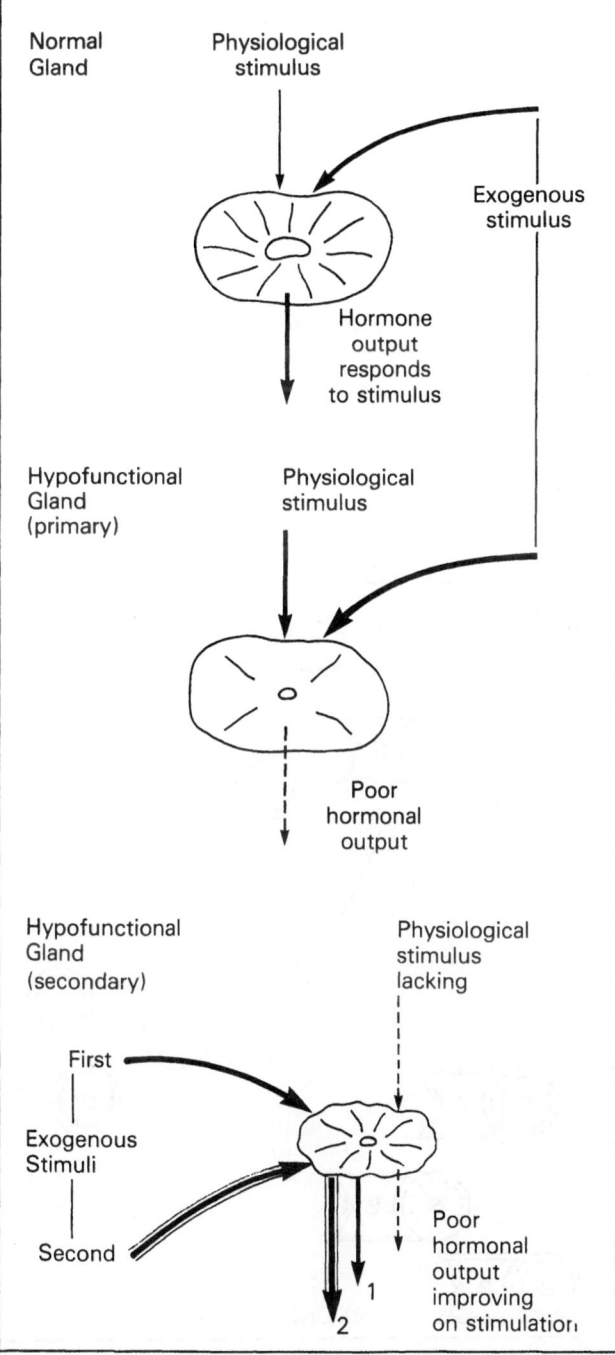

LD-40. Stimulation tests.

assays utilizing an appropriate physiological response in an animal or organ preparation.

A few hormones (e.g. some steroids) are still measured by chemical assays but currently most are measured by sensitive radioassays or cytochemical techniques and newer methods are continually being developed (Table 17).

The relative sensitivities of the various techniques are illustrated by the change in methods for measuring ACTH developed over the past 20 years:

	Detection Level (g/ml)
Bioassay (1962)	10^{-10}
Radioimmunoassay (1970)	10^{-12}
Cytochemical bioassay (1972)	10^{-15}

Radioassays

The principle of this type of assay is indicated in LD-41. The substance(s) to be measured reacts with a limited capacity specific binding reagent (B) which may be a specific antibody produced in another animal (radioimmunoassay), a naturally occurring binding protein (competitive protein binding assay) or a naturally occurring cell receptor (radioreceptor assay).

The binding reagent is saturated and 'S' is partitioned into *bound* and *free* phases, the ratio between the phases being determined by the amount of 'S' initially present. This distribution is quantified by adding a small amount of radioactively-labelled 'S' (S*) to the mixture of S and B and then measuring the radioactivity in the '*bound*' and '*free*' fractions. The values obtained are compared with those obtained in the same assay using known concentrations of standard hormone preparations in place of S.

Radioassays are advantageous in terms of sensitivity, molecular specificity and technical simplicity. Radioimmunoassays, however, have the inherent weakness that they detect immunologically reactive molecular structures which may not be the biologically active hormones.

Enzyme linked immunoassays

These assays, which utilize the activity of an enzyme instead of the radioactively labelled marker, are replacing some radioimmunoassays.

Bioassays

Bioassays of even greater sensitivity than radioimmunoassay and, more importantly, able to detect

Table 16. Factors affecting hormone levels

Factors	Hormone	Example	Value
Age	Thyroxine	At birth	200 nmol/l
		At 1 month	120 nmol/l
Sex	Testosterone	Females	below 2.78 nmol/l
		Males	over 8.68 nmol/l
Time of day	Cortisol	08:00	550 nmol/l
		Midnight	270 nmol/l
Time in menstrual cycle	LH	Week 1	7 miu/ml
		Week 2	40 miu/ml
		Week 3	10 miu/ml
Stress: e.g. admission to hospital	Cortisol	08:00	700 nmol/l
		Midnight	700 nmol/l
Drug therapy: e.g. diphenylhydantoin	Thyroxine	Before drug	120 nmol/l
		During therapy	70 nmol/l
Diet	Insulin release	(a) Delayed after carbohydrate deprivation	
		(b) Potentiated after carbohydrate excess	
Environment	Cortisol	Reversed diurnal variation on night work	
		Jet travel	

Table 17. Hormone assay methods

Chemical methods

Radioassays
(a) Radioimmunoassay
(b) Competitive protein binding assay
(c) Radioreceptor assay

Enzyme linked immunoassays

Bioassays
(a) Whole animals
(b) Organ preparations
(c) Cytochemical assays

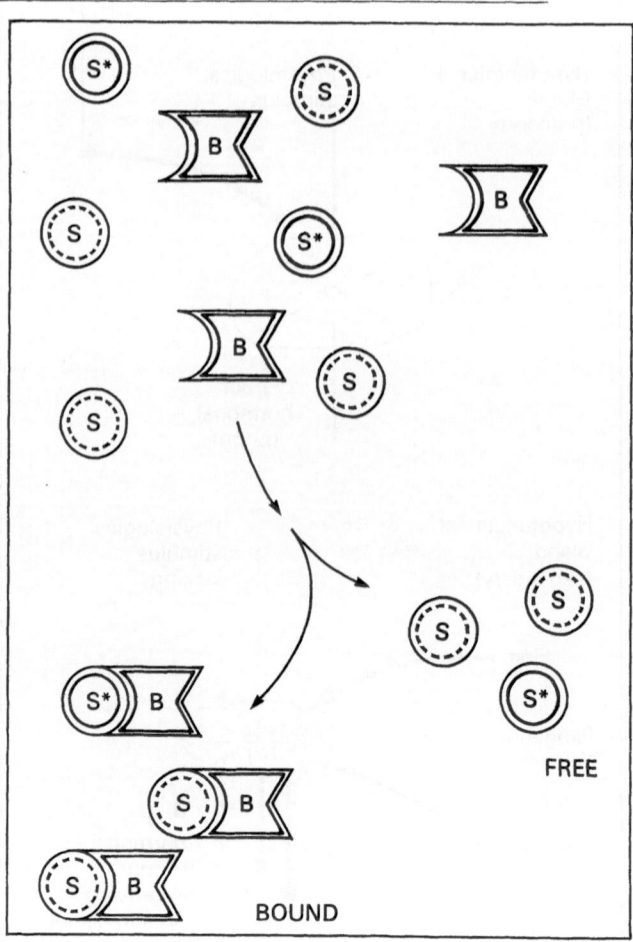

LD-41. *The principles of radioimmunoassay.*

biological activity have been developed using cytochemical techniques. That for ACTH utilizes the reduction in ascorbate levels of sections of adrenal glands *in vitro* following ACTH stimulation. The lowered ascorbate level results in the decreased deposition of an histochemical stain within the cell. Measurements of the stain intensity are made with a microdensitometer. The cytochemical assays are not as yet widely used for clinical purposes.

Bioassays using whole animals and organ preparations are very insensitive and are being used less and less.

Whatever the method used, direct measurement of hormonal levels in basal and dynamic states is the key to modern investigation of endocrine function.

ANCILLARY INVESTIGATIONS IN ENDOCRINE DISEASE

Nuclear medicine

Theory

All matter is composed of atoms. Each atom consists of a nucleus and an outer system of electrons.

The number of electrons in the atom is referred to as the atomic number (Z). The chemical properties of an element are determined by the number of electrons.

The nucleus consists of protons and neutrons. Protons carry a positive electric charge equal to the negative electric charge carried by the electron. Elements are characterized by the number of protons in their nuclei.

Neutrons have no electric charge.

Since the atom as a whole is electrically stable the number of protons in the nucleus must equal the number of electrons outside the nucleus. The mass number (A) of the atom is the total number of protons and neutrons in the nucleus.

Most pure elements consist of a mixture of atoms with the same number of electrons and therefore protons, but different mass numbers – i.e. they have different numbers of neutrons. Atoms of an element with the same number of protons but different numbers of neutrons are called *isotopes*. Since the number of protons and electrons is constant the chemical properties of the various isotopes of an element are the same.

Iodine atoms have 53 protons in their nuclei. The most stable isotope of iodine is ^{127}I which contains a total of 127 protons and neutrons (53 protons and 74 neutrons) in the nucleus. ^{131}I is the most commonly used radioactive isotope of iodine in medicine. Its nucleus consists of 53 protons and 78 neutrons.

Radiation

Isotopes may be stable or unstable. In unstable atoms there is an imbalance between the number of protons and neutrons and the atom disintegrates until the stable state is reached. This disintegration results in the emission of radiation and such atoms are called radioactive isotopes.

The radiation emitted may be of three types:

- alpha radiation
  This consists of nuclear particles. It does not penetrate human tissue and is of no medical value.
- beta radiation
  This consists of electrons and is of moderate penetrating power. Beta radiation is of limited medical use.
- gamma radiation
  This is not particular like alpha and beta radiation, but is an electromagnetic radiation similar to X-rays. It is of great medical value.

Radioactive isotopes may emit alpha, beta and gamma radiation but most emit only beta and gamma radiation.

A few isotopes are pure beta emitters:

- 18 fluorine
- 32 phosphorus

or pure gamma emitters:

- 99 technetium
- 51 chromium

A radioactive isotope is characterized, not only by the type(s) of radiation emitted but also by the rate at which it disintegrates. The time taken for 50% of the atoms to disintegrate is known as the half-life of the isotope.

The unit of radioactivity is the curie. 3.7×10^{10} disintegrations per second is one curie of activity.

The use of radioisotopes involves some absorption of energy by the tissue. The rad is the unit of absorbed dose and is the measure of the absorbed energy per gram of tissue.

Detection and measurement

The most important effects following the interactions between beta or gamma radiation and matter which can be used in detection are:

- ionization of gases, e.g. ionization chambers and Geiger counters.
- scintillation effects
- photographic film – used in scanning techniques

Diagnostic uses of radioisotopes

Tracer studies

Since isotopes are chemically identical, the body cannot distinguish between the radioactive and the stable element, thus enabling the radioactive isotope to be used in measuring the physiological processes.

This may be done *in vivo* by administering the isotope to the patient and following what happens to it over a period of time, or *in vitro* in radioimmunoassay tests.

The most commonly used *in vivo* isotope is ^{131}I for assessing the function of the thyroid gland. For example, in assessing the thyroid gland's functional ability to concentrate iodine and incorporate it into thyroid hormone, a small dose of ^{131}I is administered orally to the patient. The amount incorporated into the thyroid gland at any given time (usually $2\frac{1}{2}$–4 hours or 24 hours afterwards) is measured and expressed as a percentage of the original dose given.

Scanning

Radioisotopes may also be used to visualize organs either as the element itself or by binding it to substances which are specifically taken up by an organ. The distributed radioactivity can then be mapped out by an external radiation detector. These are of two types:

- rectilinear scanners

    Here a small motorized crystalline detector is driven back and forth in parallel sweeps above the patient in the area of the organ being examined. This enables a map of the isotope distribution to be gradually built up. It can provide a detailed and large picture of the organ being scrutinized but takes some time to perform and cannot be used for dynamic function studies (CP-2).

- gamma cameras

    Here one large or several smaller stationary crystalline detectors capable of visualizing the whole of an organ simultaneously take pictures very quickly. They can be used for dynamic studies of organ function.

Two kinds of clinical information are sought from organ imaging: first the demonstration of the presence, size, shape and position of an organ, and an assessment of its function, and second the detection of abnormalities within an organ and their number, size, shape and position.

The illustrations show some examples of the use of the radioisotope imaging technique (X-1 (T2602), X-2 (T2539), X-3 (822), X-4 (1325) and CP-3) and Table 18 gives a list of the more commonly used organ imaging techniques.

Therapeutic uses of radioisotopes

The solid isotopes such as radium, gold, cobalt and yttrium can be implanted into tumours (e.g. carcinoma of the tongue), inserted into body cavities (e.g. the uterine cavity in patients with carcinoma of the cervix) or applied to the surface of the body (e.g. in epithelioma of the skin).

Other radioisotopes may be given systemically; the most important example is the use of ^{131}I in the treatment of hyperthyroidism and in some forms of thyroid cancer.

Table 18. Organ imaging

Organ	Isotopes	Detects or visualizes
Thyroid	^{131}I ^{99}Tc	Primary and metastatic thyroid carcinomas. Ectopic thyroid tissue Functioning nodules
Adrenal	^{131}I ^{99}Tc	Adrenal adenoma or carcinoma. Adrenal size or function
Liver	^{198}Au ^{131}I ^{99}Tc	Tumours Abscesses Diffuse malfunctions – e.g. cirrhosis
Pancreas	^{75}Se	Cysts Tumours Pancreatitis – not reliable
Brain	^{99}Tc	Space occupying lesions Cerebrovascular abnormalities
Lung	^{131}I ^{99}Tc	Decreased perfusion due to tumours or emboli
Kidney	^{197}Hg ^{99}Tc	Renal position and function Tumours Cysts
Spleen	^{51}Cr ^{99}Tc	Splenic size, shape and position and may detect intrinsic abnormalities – not reliable
Heart	Various	Increasingly used for dynamic function studies

Radiology

Conventional radiology can be useful in confirming the presence of endocrine disease. Endocrine disorders may manifest themselves by the local destructive effect of an expanding lesion, e.g. destruction of the sella turcica by an expanding pituitary tumour, or by secondary hormonal effects on radio-opaque tissues such as bony changes in hyperparathyroidism.

Techniques using contrast media may be used to confirm or extend endocrine diagnoses, e.g. pneumoencephalography to outline the upward expansion of a pituitary tumour.

Computerized axial tomography (CAT) is a new non-invasive technique for giving cross-sectional images of the body. It is used in pituitary imaging and adrenal imaging, and will in future be extended to other endocrine organs.

Ultrasound

Diagnostic medical ultrasound is a rapidly expanding method of organ imaging, using a high frequency (2–8 MHz) sound beam and based on the analysis of echoes reflected back by body tissues (LD-42). The method:

- is non-invasive
- causes no patient discomfort
- is free from harmful biological effects.

Theory

Electrical excitation of a transducer containing a disc of piezo-electric material, usually quartz or synthetic ceramic, produces a pulse echo of sound equivalent to 1 wavelength and lasting 1 microsecond. The same transducer then acts as a receiving device for reflected echoes over the next 999 microseconds. This allows a pulse frequency of up to 1000 per second. A layer of vegetable oil acts as a conducting medium between the transducer and the skin.

Reflections occur at junctions between tissues of differing acoustic impedance and give information about organ anatomy. Backscatter reflections arise within organs, are dependent mainly on the amount of collagen present and reveal details of organ structure.

Electromechanical analysis of reflected echoes as organs are scanned by the transducer results in a visual display of cross-sectional anatomy and organ consistency. This image is presented on a television monitor in shades of black and white.

Applications

Diagnostic ultrasound was first applied to obstetrics but is now widely used for imaging liver, pancreas, kidney, spleen and most abdominal and pelvic masses. Specific endocrinological applications are outlined below.

Thyroid

Ultrasound allows evaluation of the cystic or solid nature of thyroid nodules and permits accurate measurement of size. Larger thyroid masses can

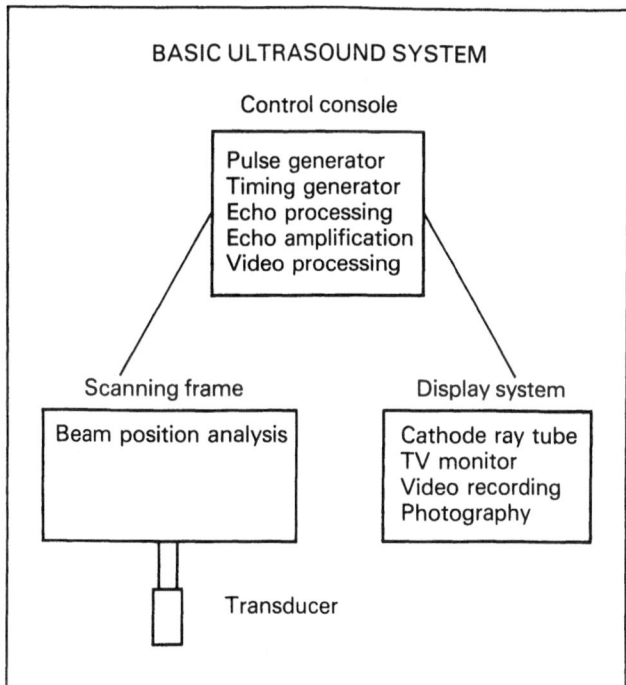

LD-42. Basic ultrasound system.

sometimes be differentiated into malignant and non-malignant lesions (X-5, X-6, 6a).

Pancreas

Ultrasound has been shown to be more reliable than isotope studies or angiography in the investigation of pancreatic lesions. An overall accuracy of 84% has been reported and tumours as small as 2 cm have been detected.

Adrenal

Most normal adrenal glands can now be visualized. Apart from showing enlargement, the solid or cystic nature of a lesion can be demonstrated.

Malignant structures are less echogenic than normal tissue, chronic inflammatory or fibrotic tissue is highly reflective, while cystic structures are echo-free. Bony structures and gas-containing organs cannot be investigated as they completely reflect the sound beam.

Laparotomy scars, obesity or intervening bowel gas all significantly interfere with resolution, making many patients unsuitable for ultrasonic investigation.

SECTION II

History and Physical Examination

INTRODUCTION

Endocrine symptoms and signs are rarely specific. This fact renders the art of recognizing an endocrine disorder in a patient more often one of suspecting rather than 'diagnosing' it. Unlike most other body 'systems' the very nature of the endocrine system ensures that the effects of disordered function in a gland affect the body as a whole and cause diffuse symptoms and signs.

Most patients with a cardiac problem present with one or more specific symptoms such as chest pain, palpitations and oedema. The problem then becomes one of resolving whether a cardiac problem exists or not and, if so, what is the nature of the disease process. Similarly, most patients with a gastro-enterological and neurological problem will present with a specific symptom – diarrhoea, or tremor – which will direct attention to that organ system.

In endocrine disease the patient may present with symptoms and signs involving almost any organ system. The diagnosis is then only made when the history and physical examination reveal a recognizable pattern of findings and provided the physician recognizes it.

A further problem is that often the features of the history and examination of an endocrine disease are but extensions of normality rather than clearcut abnormalities such as a chest pain, an abdominal mass, or paralysis of a leg. Thus the pale, wrinkled, fine-haired person (hypopituitarism), the woman with the fat neck (goitre), the mildly staring-eyed person (exophthalmos) and the obese, ruddy, plump lady (Cushing's syndrome) may well be subconsciously dismissed as being within the wide limits of normal human appearance.

An endocrine disorder must be suspected to be diagnosed.

The clinical features of endocrine disease can be divided into:

- local structural effects (Table 19)
- general hormonal or functional effects.

The general hormonal or functional effects of endocrine disease include:

- abnormalities of growth and development
- abnormalities of specific metabolic functions
  basal metabolic rate
  carbohydrate, protein and fat
  water and electrolytes
  calcium metabolism
  miscellaneous

Table 19. Local structural effects of endocrine diseases

Structure affected	Feature
Hypothalamus	Headaches
Pituitary	Headaches, visual defect
Thyroid	Pressure symptoms, dysphagia, goitre
Gonads	Testicular lumps, ovarian masses

HISTORY

Present history

Although patients with endocrine disease often present with complaints which are non-specific, there are a number of presenting complaints which are almost specifically endocrine (Table 20). Even if not the initial complaint, these symptoms or groups of symptoms (Table 21) may emerge on taking the history of the illness, and they should always be specifically asked about when endocrine disease is suspected.

Table 20. Endocrine presenting symptoms

	Symptom
General	Heat or cold intolerance
	Polyphagia
	Increased appetite
	Polydipsia
	Puffiness
	Salt craving
Structural	Increase in or excessive size (height or weight)
	Decrease in or diminished size (height or weight)
	Neck swelling
Skin and appendages	Hair loss or excess
	Increased or decreased sweating
	Pigmentation
Musculoskeletal	Bone pains or tenderness
	Periodic paralysis
Reproductive	Loss of libido
	Amenorrhoea
	Lack of or change in sexual development
	Galactorrhoea
	Failure of lactation
	Gynaecomastia
Eyes	Protrusion

Table 21. Symptom patterns of endocrine disease

Non-specific symptoms group	Especially if also	Think of
Bitemporal headaches and bitemporal vision disturbance	Increased size of body parts	Acromegaly Gigantism
Weakness, shakiness, and/or fainting spells	Related to food, fasting or exposure	Hypoglycaemia
Weight loss	With increased appetite	Diabetes mellitus Hyperthyroidism
Tiredness	Polyuria, polydipsia Cold intolerance, dry skin and weight gain Pigmentation	Diabetes mellitus Hypothyroidism Hypoadrenalism (Addison's disease)
Nervousness, shakiness	Weight loss, palpitations, heat intolerance and sweating, eye changes	Hyperthyroidism
Puffiness and fluid retention	Cold intolerance, dry skin, tiredness, hair loss, constipation	Hypothyroidism
Back pain	Loss in height or change in shape of spine	Osteoporosis
Palpitations	Nervousness, tremulousness, heat intolerance and sweating	Hyperthyroidism
Carpal tunnel syndrome	Weight loss, puffiness Increased size of body parts	Hypothyroidism Acromegaly

Past history

In addition to the usual questions on past history regarding medical, surgical, obstetrical and traumatic events, it is important to obtain details regarding:

- birth weight
- growth rate
- weight pattern
- stages of development

Where a disorder of growth or development is suspected family doctors and school records are often invaluable sources of information.

In patients presenting with a thyroid nodule, a past history of radiation to the head and neck in childhood must be sought.

Family history

A family history of thyroid disease and diabetes mellitus should be part of routine questioning since these conditions have a familial tendency. When patients present with:

- growth problems
- weight problems
- developmental problems
- hirsutism

questioning about a family tendency to these same problems is important in distinguishing an abnormality from a normal familial tendency.

In a few cases, some of the less common endocrine disorders such as phaeochromocytoma, hypoparathyroidism, medullary carcinoma of the thyroid gland and inborn errors of metabolism have distinct familial patterns.

Social and personal history

This seldom contributes towards diagnosis but, since many endocrine conditions require lifelong management and medication (e.g. diabetes, hypothyroidism, hypoadrenalinism), information about the patient's home and occupational background is essential for proper counselling.

Medication history

This includes, where appropriate, a history of the intake of illicit drugs some of which cause sympathomimetic features.

Medication history is significant because some drugs:

- can mimic endocrine diseases
- can cause endocrine diseases
- can interfere with the diagnosis of endocrine disease
- are given to control endocrine disease
- are given as replacement therapy

Further information is given in Table 22.

Table 22. Some examples of the endocrine effects of drugs

	Drug effects
Drugs mimicking endocrine disease	Chlorpromazine causing galactorrhoea Appetite suppressants mimicking hyperthyroidism
Drugs causing endocrine disease	Iodide-induced goitre Lithium-induced goitre and hypothyroidism Steroid-induced diabetes
Drugs interfering with endocrine diagnosis	Iodide invalidating ^{131}I uptake studies Oestrogens raising thyroid binding globulins Various medications increase urinary vanilmandelic acid (VMA)
Drugs used to treat endocrine disease	Propylthiouracil, carbimazole Bromocryptine Clomiphene
Drugs used in replacement therapy	Insulin Thyroxine Cortisol Calciferol

The drugs the patient is taking must be identified and the following information obtained about each drug.

- What dosage is it taken in?
- How often is it being taken?
- How long has it been taken for?

Patients who take replacement therapy which, when the dose is altered or omitted, may cause the rapid development of illness or even death, should always carry on their person details of the drug and dose. Diabetics, especially those receiving insulin, and patients receiving steroids are particularly at risk. This information can be carried on a card or bracelet, e.g. Medic Alert.

Specific questions

A number of questions relating to endocrine symptoms are part of the core enquiry which should be made of all patients. However, where an endocrine condition is suspected a number of specific questions must be asked. In the following sections both the core endocrine questions and the specific endocrine questions are listed under the following topics:

- general questions
- head and neck
- skin and appendages
- cardiorespiratory system
- gastrointestinal system
- urinary system
- reproductive system
- central nervous system
- musculoskeletal system

General

In addition to asking about:

- weakness
- syncope
- lassitude

ask about the features listed below.

Feature	Alteration	Significance
Growth rate	decreased	pituitary insufficiency hypothyroidism
	increased	gigantism eunochoidism
Temperature intolerance	heat intolerance	hyperthyroidism
	cold intolerance	hypothyroidism
	hot flushes	menopause ovarian failure
Weight change	decreased	hyperthyroidism Addison's disease hypopituitarism
	increased	obesity hypothyroidism Cushing's syndrome
	puffiness and fluid retention	hypothyroidism

Head and neck

In addition to asking about:

- headaches

enquire about the following features.

Feature	Alteration	Significance
Change in appearance	proptosis lid retraction puffiness	hyperthyroidism

hypothyroidism acromegaly |
| Diplopia | opthalmoplegia | thyroid eye disease |
| Voice change | gruff and hoarse deepening

dysarthria due to large tongue | hypothyroidism acromegaly virilization

hypothyroidism cretinism acromegaly |
| Swelling or pressure in the neck or dysphagia | | goitre |
| Pain in neck | especially if radiating to ear | thyroiditis |

Skin and appendages

In addition to enquiring into

- texture
  dry and coarse – hypothyroidism

ask about the following.

Feature	Alteration	Significance
Pigmentation	decreased increased	hypopituitarism Addison's disease
Sweating	decreased increased	hypothyroidism hyperthyroidism
Healing	impaired	diabetes mellitus
Bruising	increased	Cushing's syndrome
Flushing		carcinoid syndrome menopause
Scalp hair	decreased	hyperthyroidism hypothyroidism
Facial hair	decreased

increased | male hypogonadism hypopituitarism

female hirsutism virilization |
| Body hair | decreased excess or abnormal distribution | hypopituitarism hirsutism virilization |

Cardiorespiratory system

Ask specifically about palpitations which occur with hyperthyroidism or phaeochromocytoma.

Most of the other classical symptoms of cardiac or respiratory failure can occur if they arise from the underlying endocrine disorder, for example hyperaldosteronism causing hypertension and cardiac failure.

Gastrointestinal system

Ask first about the following.

Feature	Alteration	Significance
Appetite	decreased

increased | Addison's disease hypercalcaemia

hyperthyroidism diabetes mellitus |
| Altered bowel habit | diarrhoea

constipation | hyperthyroidism Addison's disease hypothyroidism |

Enquire in more detail about:

- eating habits and patterns
  This is particularly relevant to obesity and anorexia nervosa.
- salt craving
  This occurs in Addison's disease.

Urinary system

Patients must be specifically asked about polyuria and polydipsia, which can be symptoms of:

- diabetes mellitus
- diabetes insipidus
- hypercalcaemia
- hypokalaemia.

Reproductive system

Functional enquiries often only cover details of this system in a superficial way. It is, however, very important to take a detailed history relating to the reproductive system whenever endocrine disease is suspected. Many of the 'clue' endocrine symptoms relate to this system (see Table 20).

In addition to asking about:

- libido
- obstetrical history
- menstrual history

it may be necessary to obtain more detailed information of:

- fertility
- potency
- sexual intercourse as a measure of libido
- development of secondary sexual characteristics
- lactation

This is particularly relevant when one suspects a disorder of the pituitary or hypothalamus, the gonads or sexual differentiation.

Central nervous system

Specific parts of the enquiry should be gone into in more than usual detail as listed below.

Feature	Alteration	Significance
Nervousness	decreased	hypothyroidism
	increased	hyperthyroidism
Mental activity	decreased	hypothyroidism
Vision		diabetes mellitus
		pituitary tumours
Tremor		hyperthyroidism
Motor weakness	proximal paralysis	hyperthyroidism
	cranial nerve palsies	diabetes mellitus
		pituitary tumours
Sensory	paraesthesia	diabetes mellitus
	carpal tunnel	hypothyroidism
	syndrome	acromegaly

Musculoskeletal system

In addition to asking about joint pains which occur in acromegaly and hypothyroidism one may have to enquire about:

- increase in size of hands and feet
  Changes in shoe or glove sizes occur in acromegaly and gigantism.
- muscle cramps
  These occur in hypothyroidism and hypoparathyroidism.
- localized bone pains
  These are features of hyperparathyroidism and osteoporosis.

EXAMINATION

Only those parts of the physical examination relating to endocrine disorders and only those techniques of examination peculiar to the endocrine system will be discussed in this section.

General inspection

There are specific 'clue' signs on general inspection which should direct the doctor's attention to the endocrine system. Many of these are extensions of normality rather than specific abnormalities. It often strikes the examiner that something is 'odd' about the patient. Some of these 'odd' appearances are listed in Table 23.

Body size and proportions

Accurate height and weight measurements should be made in all patients with suspected endocrine disease. These are helpful not only in diagnosis but in monitoring the patient's response to treatment. Many endocrine disorders interfere with growth and development and recording changes is essential.

Accurate measurement of height and weight is especially important in the endocrine assessment of the child and adolescent. These results should be recorded at each visit on the type of growth chart shown in Section 1, pp. 46–51.

Overall size may be *increased* in:

- gigantism
- eunuchoidism

may be *decreased* in:

- pituitary dwarfism
- hypothyroidism in childhood

In addition to overall size there may be alterations in body proportions. Normally the span is slightly shorter than the height and the ground-to-pubic distance is slightly shorter than that between pubis and crown. In eunuchoidism and gigantism these relationships may be reversed.

Increase in the size of certain parts of the body occurs in acromegaly (see CP-29–33, 36 and X-7, X-8):

- hands
- feet
- jaw
- facial features

Deformities may occur, especially in endocrine disorders which affect bone. Table 24 shows some examples.

Table 24. Endocrine disorders causing bone deformity

Deformity	Cause
Kyphosis	Osteoporosis, hyperparathyroidism
Bow legs	Rickets
Deformed skull, sabre tibia	Paget's disease

Table 23. Physical clues to endocrine disease

Body size	Body shape	Age	Sexuality	Skin	Mental state	Voice	Local features
Too tall	Long limbs	Too young looking	Masculine looking female	Too dark	Too jumpy and fidgety	Too high	Eyes too prominent
Too short	Big extremities		Effeminate male	Too pale	Too slow	Too low and coarse	Neck swollen
Too fat	Peculiar fat distribution		Sexually underdeveloped	Hairless	Inappropriate behaviour		Round or hump backed
Too thin	Too puffy		Sexually overdeveloped	Too hairy			

Fat distribution

The usual patient with simple obesity has a fairly even distribution of adipose tissue (CP-4, CP-21).

The significance of abnormal distribution of fat is seen in Table 25.

Table 25. The significance of abnormal distribution of body fat

Disorder	Distribution
Cushing's syndrome	Centripetal pattern (CP-5)
Anorexia nervosa	Loss of body fat
Diabetic lipoatrophy or hypertrophic lipodystrophy (CP-6.1)	Localized decrease or increase of fat deposits at insulin injection sites

Skin thickness callipers may be used to measure local adiposity. These may be particularly useful in monitoring obese children.

Local atrophy of fat is sometimes seen at sites of insulin injection in diabetics (CP-6).

Skin

The skin is cold, coarse and dry in hypothyroidism and warm and sweaty in hyperthyroidism.

In severe hypothyroidism or myxoedema the skin is puffy and has a 'doughy' consistency often described as 'non-pitting' oedema.

The skin is thin and is easily bruised in Cushing's syndrome.

Brown pigmentation which is most marked in skin creases and in the buccal mucosa occurs in Addison's disease and in ectopic ACTH syndromes (CP-7, 8).

Striae occur with any rapid expansion of subcutaneous or underlying tissues. Obesity and pregnancy are good examples (CP-9, 10).

The combination of increased adiposity and fragile thin skin make large purplish striae one of the features of Cushing's syndrome.

Skin infections, especially fungal, are common in diabetics and also occur in autoimmune hypoparathyroidism, and acne, recurring in adult life, occurs in Cushing's syndrome.

Various atrophic lesions are seen on the shins of some diabetics and a raised, ruddy oedema (pretibial myxoedema) (CP-11) occurs in a small number of patients with Graves's disease (CP-12).

Nails

Fungal infections of the nails are common in hypoparathyroidism (CP-13).

Many patients with hyperthyroidism demonstrate recession of the nails from the distal nail bed – onycholysis or Plummer's nails (CP-14).

Hair

The types of hair normally found and their distribution are outlined in Table 26.

Table 26. Normal hair types

Non-sexual hair	Sexual or terminal hair
Lanugo	*Ambisexual hair (both sexes)*
Vellus	Arms
Scalp	Legs
Eyebrows	Chest
Eyelashes	Face
	Lower pubic
	Axillae
	Male sexual hair
	Upper lip
	Chin
	Chest
	Upper pubic
	Fingers and toes
	Nose
	Ears

Sexual hair replaces vellus at puberty and is

- coarser
- longer
- pigmented.

In endocrine disease hair can change in various ways as summarized in Table 27.

Table 27. Hair changes in endocrine disease

Disease	Amount of hair	Texture
Hypopituitarism	Decreased (CP-15, 16)	
Hypothyroidism	Decreased (CP-17)	Coarse, dull and brittle
Addison's disease	Decreased	Coarse, dull and brittle
Hirsutism	Increased	
Virilization	Increased	
Hyperthyroidism		Soft and fine
Cushing's syndrome		Soft and fine

Changes in hair distribution are mainly related to androgen levels. The major abnormality seen is the development of male sexual hair in a female due to conditions in which androgen production is increased.

Face

The clinical facial appearance may be very revealing of an underlying endocrine disorder (Table 28).

Table 28. Facial appearance in endocrine disease

Facial appearance	Cause
Pale, wrinkled and little hair	Hypopituitarism, eunuchoidism
Coarse features and prognathism	Acromegaly (CP-29–33)
Pale puffy and pasty	Hypothyroidism (CP-18)
Sweaty, wild-eyed and staring	Hyperthyroidism (CP-58)
Fat, florid and moon shaped	Cushing's syndrome (CP-19)
Thin and pigmented	Addison's disease

Eyes

There may be changes in the external appearance of the eyes in hyperthyroidism – especially that due to Graves' disease. These are due to:

- spastic changes
  lid retraction
  lid lag
- infiltrative changes
  exophthalmos (proptosis)
  oedema and bulging of the lids
  chemosis (oedema of the conjuctivae)
  conjunctival congestion
  ophthalmoplegia

Exophthalmos

Exophthalmos may be distinguished from simple lid retraction as shown in LD-44 (p. 95).

Various instruments are available for measuring the amount of eyeball protrusion beyond the lateral orbital rim – exophthalmometers – but a reasonably accurate measurement can be made using a simple transparent plastic ruler with a millimetre scale placed against the lateral orbital rim.

Exophthalmos is present when:

- the distance from the orbital rim is greater than 20 mm
- the difference in protrusion of the two eyes is greater than 2 mm.

Eye movement

Eye movements should be tested in all cases of Graves' disease since infiltration of the extraocular muscles often leads to impairment, particularly of upward and outward gaze.

Visual field testing

Careful visual field testing by the confrontation method should always be carried out in cases of suspected pituitary lesions where bitemporal hemianopia may follow pressure on the optic chiasma. If a pituitary tumour is diagnosed perimetry must be performed to pick up minor changes in visual fields which are easily missed by the confrontation method.

Ophthalmoscopy

Inspection of the lens, vitreous and fundi is an essential part of physical examination and may reveal:

- cataracts due to diabetes mellitus, or hypoparathyroidism
- retinopathy due to diabetes mellitus
- optic atrophy due to pituitary tumours

Neck

The thyroid gland is one of the few endocrine glands (the other being the testis) capable of being palpated. Careful palpation of the neck can reveal clinical details of the thyroid structure that can be further assessed by scanning techniques.

The technique is important. The patient should be:

- comfortably seated
- in a good light
- given a glass of water to aid swallowing

Inspection

This should be carried out from in front of the patient whose neck should be extended. The patient should be asked to swallow a few times. Since the thyroid gland is attached to the trachea it moves upwards on swallowing. The size and shape of any abnormalities should be noted.

Palpation

For palpation, the examiner stands behind the patient whose neck is now placed in a slightly flexed position to relax the sternomastoid muscles. The tips of the index, middle and fourth finger of each hand are used. First the isthmus is located by the fingers of both hands as the patient swallows. The fingers are then moved laterally over the two lobes again with the patient swallowing.

Next each lobe is carefully palpated in turn with the head turned and slightly tilted towards the lobe being palpated. The fingers of the other hand slightly displace the cricothyroid cartilage towards the side being palpated to bring the lobe into prominence. Thereafter the remaining areas of the neck are palpated for other abnormalities and for lymphadenopathy.

These steps are outlined in LD-43.

Using this technique experienced examiners can palpate the thyroid gland in about 40% of normal people. When a palpable abnormality is present careful examination will usually be able to distinguish between:

- diffuse goitre where the whole gland is enlarged
- uninodular goitre due to a solitary nodule
- multinodular goitre due to multiple nodules

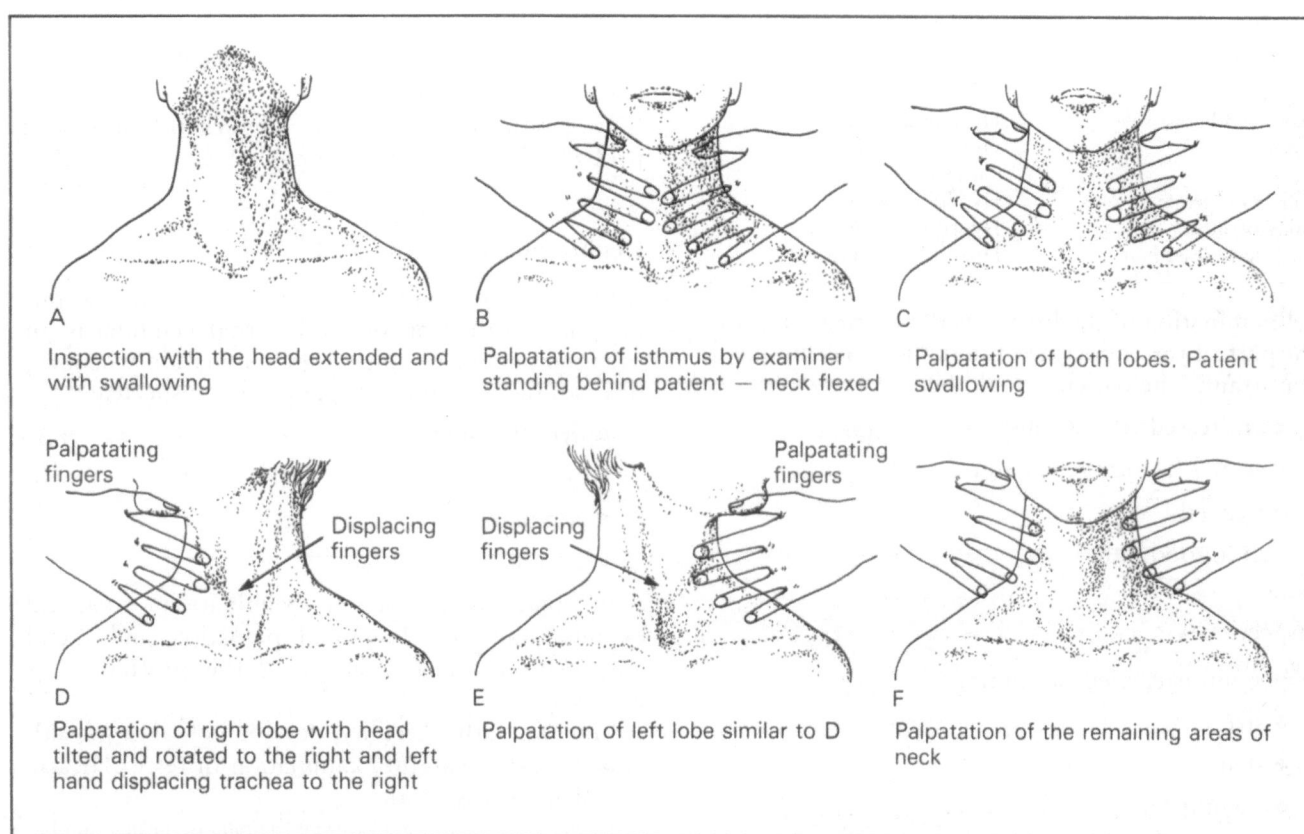

A — Inspection with the head extended and with swallowing

B — Palpatation of isthmus by examiner standing behind patient — neck flexed

C — Palpatation of both lobes. Patient swallowing

D — Palpatation of right lobe with head tilted and rotated to the right and left hand displacing trachea to the right
Palpatating fingers / Displacing fingers

E — Palpatation of left lobe similar to D
Displacing fingers

F — Palpatation of the remaining areas of neck
Palpatating fingers

LD-43. *Palpation of the thyroid gland.*

Occasionally a thyroglossal duct remnant or cyst will be palpated in the midline above the isthmus. This can be confirmed by finding that the nodule moves upwards on protrusion of the tongue.

Auscultation

Finally, auscultation should be carried out with the breath held in expiration over both lobes, the isthmus and any palpable abnormality.

Bruits over the thyroid gland must be distinguished from:

- venous hum which changes with position
- murmurs transmitted from the heart
- bruits in underlying carotid arteries

The breast

Although not an endocrine organ the breast is a major target tissue for several hormones.

The endocrine diseases that cause breast abnormalities are summarized in Table 29.

Table 29. Endocrine diseases of the breast

Disease	Causes
Breast enlargement (gynaecomastia)	Excess oestrogen
Decreased breast size	Hypopituitarism Ovarian failure Virilism
Failure to lactate	Hypopituitarism
Galactorrhoea	Hyperprolactinaemia
Increased breast hair	Hirsutism, virilism

Examination of the breasts is therefore an important part of the endocrine examination. Technique is important. The patient should be:

- undressed to the waist
- seated facing the examiner
- relaxed
- in a good light

Inspection

Inspection is carried out noting:

- size
- shape
- asymmetry
- nipple inversion
- dimpling
- alteration of contour
- alteration of skin colour

Palpation

Palpation is carried out quadrant by quadrant followed by the axillary tail. Palpation of larger breasts is probably best carried out with the patient lying supine with the hands behind the neck. This spreads out the breast tissue.

Lumps should be examined for:

- size
- shape
- consistency
- fixture to the skin
- fixture to the underlying fascia

The nipple, areola and subareolar structures should be carefully examined in cases of:

- amenorrhoea
- infertility
- breast discharge

Galactorrhoea should be sought by forcefully 'milking' the breast with the examining fingers.

Finally all four walls of the axilla should be carefully examined.

A careful record of any abnormal findings is probably best made by a simple outline drawing.

Genitourinary system

Although commonly neglected on routine examination, examination of the external genitalia is an essential part of the physical examination whenever diseases of the following glands are suspected:

- hypothalmus
- pituitary
- adrenals
- gonads

In some cases internal examination – vaginal examination with bimanual palpation and rectal examination for assessment of the prostate – is necessary.

In males with infertility, developmental or growth abnormalities, careful examination of testicular size and shape is essential.

Ambivalent or confusing *external genitalia* occur in a number of rare

- inborn errors of adrenal metabolism

• disorders of sexual differentiation
• gonadal tumours during fetal life

Measurement of *testicular size* is best recorded by comparison with one of the currently available orchidometry sets.

Other components of physical examination

In any complete physical examination certain other components may give information pertinent to the endocrine system. These are summarized in Table 30.

Table 30. General features of physical examination influenced by the endocrine system

Component	Variation	Significance
Pulse	Slow	Hypothyroidism
		Hypopituitarism
	Rapid	Hyperthyroidism
		Sometimes phaeochromocytoma
		Hypoglycaemia
Blood pressure	Low	Addison's disease
		Hypopituitarism
	High	Primary hyperaldosteronism
		Phaeochromocytoma
		Occasionally hypothyroidism
Respiration	Hyperventilation	Diabetic ketoacidosis
Neurological system	Isolated cranial or other palsies	Diabetes mellitus
	Peripheral neuropathy	Diabetes mellitus
	Myopathy (proximal)	Hyperthyroidism
		Osteomalacia
	Periodic paralysis	Hyperthyroidism
		Hyperaldosteronism

SECTION III

The Presentation of Endocrine Disease and the Diagnostic Possibilities

INTRODUCTION

The patient, through his or her appearance, behaviour and background, supplies diagnostic clues to the physician who interprets them according to his or her own experience and training. The doctor then forms a hypothesis which may be accepted or rejected as the result of further investigation. At some stage, a diagnostic label may be applied and appropriate management instituted. The response to treatment may itself alter the original hypothesis.

The student must be aware of this and must learn to tolerate a degree of diagnostic uncertainty if he or she is to avoid self-delusion and spurious diagnostic precision on the one hand and fruitless overinvestigation of his or her patient on the other.

This section considers the approach to patients presenting with complaints which are often caused by endocrine disorders. Thus, we begin with the symptoms and signs and follow through to the possible diagnosis – a presentation-orientated approach. This approximates the clinical situation faced by the practising doctor. However, it must be appreciated that diagnostic clues are rarely presented singly, and the interrelationships created may themselves be diagnostic. This 'pattern recognition' is a function of experience.

Recommended for further reading

Hodgkin, K. (1978). *Towards Earlier Diagnosis in Primary Care.* 4th Edn. (Edinburgh: Churchill Livingstone).

This is a valuable book for general practitioners. It gives an indication of the incidence of different diseases in modern Britain.

SHORT STATURE

Causes

These are:

- constitutional
    without skeletal and pubertal delay
    with skeletal and pubertal delay
- chronic systemic disorders
- abnormalities of chromosomes or genes
- intrauterine growth retardation
- bone and cartilage disorders
- psychological and social factors
- endocrine disorders
- iatrogenic

The patient

It is important to diagnose short stature at as early an age as possible if the best results of treatment are to be obtained.

Children with chronic organic disease usually do not grow at a normal rate and consecutive height measurements on such children may lead to a diagnosis which would not otherwise be suspected.

It is more usual for short stature to present in an otherwise apparently normal child. If the height falls in the shaded area between the 3rd centile line and the − 3 standard deviation line (see charts 3,4), it is adequate to monitor the height velocity over a year (see Charts 7,8) to ensure that it is normal. When the height falls below the − 3 standard deviation line (see Charts 3,4), or when the growth velocity over a year is abnormally slow, i.e. 25th centile or less (see Charts, 7,8), the child should be investigated fully.

Referring children for investigation because of short stature depends on awareness of the problem on the part of general practitioners and community medicine specialists working in infant or school clinics. Accurate height measurement is essential.

The best apparatus in the UK for measuring height is the Harpenden Stadiometer, which is accurate to the nearest millimetre. However, a cheaper alternative is to fix a scale on the wall, using a steel tape to mark correct heights above the floor, and to use a flat board on the child's head. When a child is being measured he should be looking straight in front and stretching up maximally without lifting his heels from the floor.

History

Enquire about:

- any abnormality during pregnancy
    Include details of the delivery
- birth weight
    The gestational age should be noted and whether the baby was light-for-dates.
- the neonatal period
- nutritional history during infancy and childhood
- any illnesses during childhood
- intelligence and the kind of school attended
- any adverse psychological factors during childhood

Family history

Enquire about:

- parents' heights

- heights of siblings
- any familial disorder causing short stature

Social history

Enquire about:

- father's occupation
- mother's occupation, if employed
- social class

Examination

Measure:

- height
- weight
- skinfold thickness over triceps and subscapular regions
- stages of puberty
- bone age
  This is described in Section I, p. 43.

Certain specific features which may be revealed on examination may help in the diagnosis of short stature.

Central nervous system (c.n.s.)

Routine examination should include ophthalmoscopy and an assessment of the visual fields by the confrontation test.

Cardiovascular system

Look particularly for evidence of congenital or rheumatic heart disease.

Respiratory system

Exclude chronic disease – asthma, cystic fibrosis or tuberculosis.

Alimentary system

Coeliac disease or clinical evidence of chronic liver disease should be sought.

Renal system

Examine for enlargement of the kidneys.

Haemopoietic system

Check for enlargement of the liver, spleen or lymph nodes.

Locomotor system

Examine for skeletal disproportion with shortening of the limbs or asymmetry of the trunk.

Skin

The presence of eczema may be of significance.

Metabolic

Check for the clinical manifestations of metabolic disease e.g. mucopolysaccharidoses.

Endocrine system

Check for the clinical manifestations of endocrine disease e.g. hypothyroidism, growth hormone deficiency.

Investigations

Check the chromosomes in girls. Radiological assessment of the skull and pituitary fossa, followed if necessary by e.e.g. and CAT scanning, will be required if neurological or pituitary disease is suspected.

Chest X-ray and an e.c.g. should be carried out if cardiac or respiratory disease is suspected. A sweat test will help to eliminate a diagnosis of cystic fibrosis.

Jejunal biopsy or liver function tests should be carried out if coeliac or liver disease is suspected.

Urinary infection should be excluded and routine urinanalysis undertaken. Check the plasma urea, electrolytes and creatinine; if renal disease is suspected, an intravenous pyelogram (IVP) as a preliminary investigation should be ordered.

The haemoglobin should be estimated and a blood film examined.

If a bone or cartilage disorder is suspected then, in the first instance, the skull, spine, hips and left arm and left leg should be X-rayed.

If the child has eczema, excessive use of steroid creams may lead to growth retardation. A Synacthen test should be carried out in such circumstances to assess adrenocortical reserve.

Check for any metabolic disorder suspected clinically.

Endocrine

Investigations are summarized in Table 31.

In the majority of patients after investigation it should be possible to make a definitive diagnosis of the cause of short stature.

Table 31. Endocrine investigation of short stature

	Hormone investigation	Disorder
Thyroid axis	Serum TSH thyroxine	Hypothyroidism
Gonadal axis	Serum LH FSH testosterone oestradiol prolactin	Delayed puberty (p. 81)
Adrenal axis	Plasma cortisol levels at 23:00 and 08:00 to demonstrate diurnal variation. Synacthen test in the morning to demonstrate adrenal reserve	Hyperfunction of the adrenal cortex Adrenocortical insufficiency (p. 123)
Growth hormone	Standardized exercise test Insulin hypoglycaemia as final definitive test	Growth hormone deficiency (p. 111)

TALL STATURE

Causes

These are:

- constitutional
  - without skeletal and pubertal advance
  - with skeletal and pubertal advance
- disorders of the central nervous system
- abnormalities of chromosomes or genes
- those associated with obesity
- endocrine disorders

The patient

Tall stature is present if the height falls above the 97th centile line on Charts 3,4.

History

Enquire about:

- when the tall stature was first noted
- epilepsy
- family and social history
  The details are given above, under Short stature.

Examination

The same general and systematic approach should be followed as outlined above, under Short stature. Look in addition for evidence of obesity or mental retardation.

From this information a diagnosis of constitutional tall stature will be made in the majority of patients.

Investigations

From the following groups of diseases the investigations which could be useful have been detailed in Table 32.

Table 32. Initial investigations to determine cause of tall stature

Suspected system/disease	Initial investigations
Genetic	Chromosome analysis
Neurological disease e.g. Sotos's syndrome	X-ray skull and pituitary fossa e.e.g. CAT scan to detect hydrocephalus
Marfan's syndrome	Check for: subluxation or dislocation of the lens, aortic aneurysm or regurgitation, chest deformity, long thin extremities and fingers
Inborn error of metabolism	Homocystinuria

Hormonal investigations that may be used are shown in Table 33. With the above investigations it is usually possible to determine the rare organic causes of tall stature.

Table 33. Endocrine investigation of tall stature

	Hormone investigation	*Disorder*
Thyroid axis	Serum thyroxine	Thyrotoxicosis
Gonadal axis	Serum LH FSH testosterone oestradiol prolactin	Precocious puberty (p. 81)
Adrenal axis	Serum dehydroepiandrosterone (DHA) androstenedione (testosterone) 17α-OH-progesterone to exclude 21-hydroxylase deficiency ACTH	Excess adrenal androgen
Growth hormone	Glucose tolerance test with growth hormone estimations to demonstrate an elevated basal GH level which remains unsuppressed by the glucose load	Excess growth hormone

DELAYED OR PRECOCIOUS PUBERTY

Causes

When investigating delayed or precocious puberty, the causes should be kept in mind. (See below, pp. 81 and 119).

The patient

The first signs of puberty in boys are growth of the penis and enlargement of the testes, from the pre-pubertal size of 2–3 ml to 4 ml. Puberty begins in the average boy at the age of 12 years with a normal variation of ±2 years. If no development takes place by the age of 14 years, the boy has delayed puberty. If development occurs before 10 years, the boy has precocious puberty (CP-39.3).

The first sign of puberty in girls is breast enlargement. This begins in the average girl at the age of 11 years with a normal variation of ±2 years. If no development takes place by the age of 13 years, the girl has delayed puberty. If development occurs before 9 years, the girl has precocious puberty. The menarche occurs at the age of 13 years with a normal variation of ±2 years. Early menstruation may be the only evidence of precocious puberty.

Adolescent boys and girls are very sensitive regarding delayed puberty and seek advice when they feel different from their friends.

Precocious puberty is more usually a matter of parental concern as tall stature is a feature of the condition and the children stand out as very different from their classmates.

History in delayed puberty

Enquire about:

- associated short stature
  Time of onset of growth retardation if known.

- evidence of chronic systemic disease

- associated obesity

- evidence of a c.n.s. disorder
  For example, visual disturbances, meningitis, encephalitis, brain injury, mental retardation, cerebral palsy, epilepsy.

- undescended testes in boys

- abdominal or pelvic disease in girls

History in precocious puberty

Enquire about:

- associated tall stature
  Time of onset of growth acceleration if known.

- associated obesity

- evidence of a c.n.s. disorder
  As described above for delayed puberty.

In boys

Enquire also about:

- Time of onset of enlargement of penis and testes and growth of pubic hair
- Gynaecomastia
- Speed of development through the stages of puberty

In girls

Enquire also about:

- Time of onset of breast enlargement and growth of pubic hair
- Time of menarche
- Regularity of periods
- Speed of development through the stages of puberty

Family history

Enquire about:

- parents' heights
- heights of siblings
- any familial tendency to delayed or precocious puberty

Social history

Enquire about:

- occupation of parents
- social class

Examination

General

This should include:

- height
- weight
- stages of puberty assessed by Tanner method (see p. 52)
- testicular examination for descent and for tumours
- bone age
- chromosome analysis

Systematic

This follows the same approach as described above for short or tall stature. It should include examination of the optic discs and visual fields. An e.e.g., X-ray of skull and pituitary fossa and a CAT scan are also useful.

Investigations

For hypothalamic–pituitary–gonadal function, the investigations are:

- in males
   LH
   FSH
   testosterone
- in females
   LH
   FSH
   oestradiol

An elevated prolactin level may indicate hypothalamic dysfunction.

For adrenal function the investigations are:

- dehydroepiandrosterone (DHEA)
- androstenedione
- 17-OH-progesterone.

Stimulation tests indicated:

- the LHRH test
   may be used in the assessment of delayed or precocious puberty
- the HCG test
   may be of value in diagnosing delayed puberty due to testicular disease

In girls with delayed puberty, a laparoscopy to determine the state of the ovaries should precede the introduction of oestrogen/progestogen therapy. Further, this technique may be of value in diagnosing a suspected ovarian tumour.

Delayed puberty may be associated with short stature, and precocious puberty and tall stature may coexist. When formulating an investigation plan it may be necessary to investigate the combined disorder.

MENSTRUAL IRREGULARITY

The average menstrual cycle lasts 28 days: 5 days bleeding with a dry interval of 23 days. Menstrual irregularity is of two kinds:

- menorrhagia
   Here the bleeding occurs at regular, approximately 28 day intervals, but the bleeding phase and/or the rate of blood loss is increased.
- metrorrhagia
   Here the bleeding becomes irregular in inci-

dence, rate and duration, varying from a few hours spotting to many weeks of occasionally heavy loss.

Other abnormalities, such as amenorrhoea, although technically menstrual irregularities, are really different from heavy or irregular periods and not considered here.

There is a good deal of variation in the normal cycle from woman to woman and month to month. Cycle lengths of 24–32 days are within the normal range and so are bleeding phases of 1–7 days.

There is a strongly subjective element in the determination of excessive menstrual loss. What appears as a heavy period to one woman may seem quite normal to another.

Causes

Occasionally mechanical factors, such as endometrial polyps or submucous fibroids, are evident. These enlarge or distort the uterine cavity.

Rarely, menstrual irregularity is the presenting sign of serious disease such as cervical or endometrial cancer. Most commonly, however, it signals an endocrine abnormality. Here alterations in the production of the ovarian hormones oestrogen and progesterone are the cause.

Endocrine

Endocrine menstrual irregularities consists of two basic types:

- ovulatory
- anovulatory

Ovulatory

Here ovulation still occurs more or less regularly. In general there is menorrhagia with the bleeding phases prolonged to 7–10 days but occurring at regular intervals.

The basal body temperature is biphasic, plasma progesterone rises above 20.0 mmol/l in the second half of cycle and the endometrium shows secretory changes.

Anovulatory

Here ovulation occurs only rarely, although cystic follicles are common. There is metrorrhagia, with no regular bleeding pattern and periods of amenorrhoea interspersed with bleeding episodes, often prolonged and severe.

The endocrine changes are:

- irregular peaks of plasma oestradiol sometimes exceeding 1.0 mmol/l, but occasionally there are persistent low levels
- plasma progesterone below 10 mmol/l
- a proliferative endometrium often showing cystic glandular hyperplasia. Sometimes it is atrophic.

The patient

Ovulatory menorrhagia commonly presents in parous women in their late thirties or early forties. The onset is insidious and they may become anaemic from iron loss. They become debilitated and have difficulty in coping with household management. The uterus is bulky and is sometimes distorted by fibroids.

Anovulatory metrorrhagia commonly occurs at the extremes of reproductive life, just after the menarche or before the menopause. There is often a strong psychological overlay. These patients are usually demanding, with a multiplicity of complaints. The condition may sometimes present, especially in young women, as severe bleeding episodes necessitating hospitalization and occasionally blood transfusion. There is seldom an overt pelvic lesion.

History

A careful menstrual history should elucidate:

- amount of blood loss
- pattern of blood loss
- type of blood loss

The age of menarche and details of any previous pregnancies and abortions should be obtained. Evidence of pituitary and thyroid disease should be sought.

Examination

In addition to a complete general examination a thorough pelvic examination should be made, to rule out local mechanical factors. The presence of hirsutism or virilism should be noted.

Investigations

Measurements of plasma progesterone or dilation and curettage are often required to determine the cause.

Specific endocrine diseases include:

- Stein–Leventhal syndrome
- hypothyroidism
- Cushing's syndrome

The appropriate function tests should be undertaken.

GYNAECOMASTIA

Gynaecomastia is the term given to unilateral or bilateral breast enlargement in males. This feature may cause considerable emotional and social difficulties especially in younger patients. It is due to either an increase in the glandular tissue – the ducts and lobules – or proliferation of the stromal supporting tissue. The former tends to be related particularly to the effects of oestrogens, and the latter more usually to the anterior pituitary hormone prolactin, and possibly to growth hormone which is structurally similar.

Breast enlargement may occur for other reasons:

- deposition of adipose tissue in the breast area may commonly be mistaken for gynaecomastia
- increase in breast size may rarely be the result of carcinoma of the male breast or neurofibromatosis

Causes

These are tabulated on p. 164.

In patients with high circulating oestrogen or prolactin levels an obvious aetiological relationship may exist. However, it is important to realize that gynaecomastia may be related to many conditions in which no abnormality of hormones can be detected.

The patient

Age

Gynaecomastia due to primary testicular failure is commonly diagnosed at an early age. Further visible breast enlargement occurs in about 50% of pubertal boys probably as a consequence of sex hormone production (CP-20). In later life it is more likely to be related to:

- drug side-effects (see CP-72)
- liver disease
- an association with a non-endocrine tumour such as bronchogenic carcinoma

History

In young patients a history of primary testicular disease, e.g. bilateral torsion or orchitis, is of value.

In all patients careful attention to past and present illnesses, alcohol ingestion and drug therapy is important. Gynaecomastia is common in patients taking oestrogenic drugs for prostatic carcinoma – it is frequently observed in cirrhosis and it is well recognized in other endocrine disorders such as thyrotoxicosis.

Examination

The degree of breast enlargement is important. Objective assessment must be made, to confirm true glandular enlargement. This may range from a small cone of tissue to a breast of normal female proportions.

Unilateral gynaecomastia is more often non-endocrine.

Genitalia

Examination of genitalia may reveal evidence of hypogonadism with small soft testes. Small testes associated with gynaecomastia and eunuchoid body habitus is suggestive of Klinefelter's syndrome. Conversely unilateral testicular enlargement may indicate the presence of a testicular seminoma or teratoma.

The presence of one small testis and one of normal size may suggest the presence of an interstitial cell tumour in the normal sized testis which is secreting oestrogen causing gonadotrophin suppression and atrophy of the contralateral testis.

General

In patients with gynaecomastia associated with hepatic cirrhosis there may be spider naevi and palmar erythema. Clinical features can suggest endocrine disorders such as acromegaly or hyperthyroidism. In patients receiving medications causing gynaecomastia there may be signs suggestive of the primary disease for which they are being treated, e.g. congestive cardiac failure, nephrotic syndrome.

Investigations

Plasma oestradiol levels are normal in physiological gynaecomastia of puberty. In the presence of oestrogen-secreting tumours of the testis or adrenal plasma oestradiol levels may be markedly raised.

In hepatic disease and gynaecomastia associated with other systemic disorders oestrogen levels may be minimally raised but, more frequently, are normal or occasionally reduced. Clearly there is no consistent pattern in these cases.

Patients with primary testicular failure such as Klinefelter's syndrome characteristically demonstrate high plasma gonadotrophin levels.

Prolactin levels may be helpful.

Further investigations which may be of value include a buccal smear and karyotype (if the patient is infertile), skull and chest X-ray, plasma testosterone estimation and liver and thyroid function tests.

Commonly, however, no abnormalities are detected.

GALACTORRHOEA

Persistent lactation may occur spontaneously or follow pregnancy.

Causes

Galactorrhoea is frequently the consequence of hyperprolactinaemia and is often associated in female patients with amenorrhoea. In males it is a rare feature since galactorrhoea occurs only in oestrogen-primed breasts exposed to prolactin excess.

The patient

Presentation

Females usually present with galactorrhoea often accompanied by:

- amenorrhoea
- infertility

Males usually present with:
- galactorrhoea
- impotence
- infertility

History

A detailed history with careful attention to current medications and past history is important. A history of previous pituitary surgery with possible stalk section will obviously be of value. Likewise previous thoracic surgery or herpes zoster may give rise to persistent lactation in the 'suckling reflex'.

Various drugs, particularly psychotrophic agents such as phenothiazines and tricyclic antidepressants, may give rise to galactorrhoea by affecting biochemical pathways in the hypothalamus. The release of prolactin inhibitory factor is prevented.

Oestrogen in oral contraceptives may result in an 'oversuppression syndrome' with amenorrhoea and galactorrhoea by triggering prolactin secretion.

Examination

There may be typical features to suggest a functioning tumour of the pituitary, e.g. acromegaly, Cushing's syndrome.

More commonly a functioning anterior pituitary tumour will produce only prolactin. Such lesions may exhibit no clinical features other than amenorrhoea and sometimes galactorrhoea. Galactorrhoea should be specifically sought for in all patients complaining of secondary amenorrhoea.

In the early phase such lesions are small (microadenomata) and rarely produce pressure or space-occupying complications. It is thought that chromophobe adenomas, traditionally regarded as non-functional, may frequently develop from a prolactin secreting microadenoma.

There may be clinical signs suggestive of hypothyroidism. TRH directly stimulates prolactin as well as TSH release, and high levels can result in galactorrhoea.

Signs suggestive of bronchogenic neoplasm may indicate ectopic prolactin secretion.

Investigations

All patients exhibiting sustained galactorrhoea should have serial plasma prolactin estimations carried out over a period of several days. In functional hyperprolactinaemic states a single estimation may not identify an established abnormality.

If the prolactin level is raised, screening should include:

- lateral skull X-ray
- pituitary fossa coaxial tomography
- thyroxine and serum TSH estimation
- chest X-ray

Other investigations can be undertaken according to specific clinical features.

OBESITY AND INCREASE IN WEIGHT

The problem of excessive weight is nearly always due to obesity (CP-21 and CP-4). Remember that weight exceeding the 'ideal' for a given sex, height and age may also result from an accumulation of oedema or less commonly from transverse enlargement of the bony skeleton as in acromegaly, or by an abnormal increase in muscle mass as produced by virilizing tumours in women.

What is obesity?

The problem is really one of defining what is 'normal' weight.

There are two basic approaches to this problem – defining 'normal weight' as either:

- average weight or
- ideal weight (best or desirable weight)

Average weight

This is the mean weight of people who are not obviously obese, grouped according to age and body build.

The problem with this method is that the values differ depending on where and when this is done. Average weight in Western societies has been steadily increasing.

Ideal weight

Ideal, best or optimum weights are taken from actuarial tables and basically define what life insurance companies have found to be related to longevity.

Ideal weights are the most useful indications of health. Obesity may be said to exist if body weight is more than 10% in excess of the ideal weight (Tables 34, 35).

Table 34. Ideal weights* in indoor clothing (metric)

Height (cm)	Weight (kg)		
	Small frame	Medium frame	Large frame
Men aged 25 and over			
158	50.5–54.5	53.5–58.5	57.0–64.0
160	52.0–56.0	55.0–61.0	58.5–65.0
163	53.5–57.0	56.0–62.0	60.0–67.0
165	55.0–58.5	57.5–63.0	61.0–69.0
168	56.0–60.0	58.5–65.0	62.5–71.0
170	58.0–62.0	56.0–66.5	64.5–73.0
173	60.0–64.0	62.5–69.0	66.5–75.0
175	62.0–66.0	64.5–71.0	68.5–77.0
178	63.5–68.0	66.0–72.5	70.0–79.0
180	65.0–70.0	68.0–75.0	72.0–81.0
183	67.0–71.5	70.0–77.0	74.0–83.5
185	69.0–73.5	71.5–79.0	76.0–85.5
188	71.0–76.0	73.5–81.5	78.5–88.0
190	72.5–77.5	76.0–79.0	80.5–90.5
193	74.0–79.0	78.0–86.0	82.5–92.5
Women aged 25 and over			
147	42.0–44.5	43.5–48.5	47.0–54.0
150	42.5–46.0	44.5–50.0	48.0–55.0
152	43.5–47.0	46.0–51.0	49.5–56.5
155	45.0–48.5	47.0–52.5	51.0–58.0
158	46.5–50.0	48.5–54.0	52.0–59.5
160	47.5–51.0	50.0–55.0	53.5–61.0
163	49.0–52.5	51.0–57.0	55.0–62.0
165	50.5–54.0	52.5–58.5	57.0–64.5
168	52.0–56.0	54.5–61.0	58.5–66.0
170	53.5–57.5	56.0–63.0	60.0–68.0
173	55.0–59.5	58.0–65.0	62.0–70.0
175	57.0–61.0	60.0–66.5	64.0–71.5
178	58.5–63.5	62.0–68.5	66.0–74.0
180	61.0–65.0	63.5–70.0	67.5–76.0
183	63.0–67.0	65.0–72.0	69.5–78.5

* After (1959) *Metropolitan Life Statistical Bulletin*, No. 40.

Some authors accept 15 or 20% in excess of the base level. It should be noted that the 10% allowance still allows the 178 cm tall (5 ft 10 in) medium-build male to carry up to 7.5 kg (16 lb) of excess fat before being labelled obese.

Skin fold thickness is another measure of obesity but is seldom used in adult clinical practice.

Causes

Obesity invariably results from an imbalance between the input and output of energy. This may occasionally be conditioned by endocrine disease which either stimulates the appetite or reduces the expenditure of energy.

Phases of normal growth are, of course, controlled by the endocrine glands, such as the increase in muscle which occurs in boys and in fat in girls at the time of puberty.

Table 35. Ideal weights* in indoor clothing (imperial)

Height (ft/in)	Weight (lb)		
	Small frame	Medium frame	Large frame
Men aged 25 and over			
5′ 2″	112–120	118–129	126–141
5′ 3″	115–123	121–133	129–144
5′ 4″	118–126	124–136	132–148
5′ 5″	121–129	127–139	135–152
5′ 6″	124–133	130–143	138–156
5′ 7″	128–137	134–147	142–161
5′ 8″	132–141	138–152	147–166
5′ 9″	136–145	142–156	151–169
5′ 10″	140–150	146–160	155–174
5′ 11″	144–154	150–165	159–179
6′ 0″	148–158	154–170	164–184
6′ 1″	152–162	156–175	168–189
6′ 2″	156–167	162–180	173–194
6′ 3″	160–171	167–185	178–199
6′ 4″	164–175	172–190	182–204
Women aged 25 and over			
4′ 10″	92– 98	96–107	104–119
4′ 11″	94–101	98–110	106–122
5′ 0″	96–104	101–113	109–125
5′ 1″	99–107	104–116	112–128
5′ 2″	102–110	107–119	115–131
5′ 3″	105–113	110–122	118–134
5′ 4″	108–116	113–126	121–138
5′ 5″	111–119	116–130	125–142
5′ 6″	114–123	120–135	129–146
5′ 7″	118–127	124–139	133–150
5′ 8″	122–131	128–143	137–154
5′ 9″	126–135	132–147	141–158
5′ 10″	130–140	136–151	145–163
5′ 11″	134–144	140–155	149–168
6′ 0″	139–148	144–159	153–173

* After (1959) *Metropolitan Life Statistical Bulletin*, No. 40.

Similarly, in pregnancy, progesterone in particular leads to the deposition of metabolically active fat mainly in the trunk and around the kidneys. This fat tends to disappear during breast feeding, but suppression of lactation allows it to remain and indeed increase with successive pregnancies.

Endocrine disease

Endocrine disease is rarely the cause of ordinary obesity but can cause increase in weight and abnormal fat disposition in some instances:

- Cushing's syndrome
- insulinoma
- hypothyroidism
- hypogonadism

Cushing's disease or syndrome

An example of increased appetite in endocrine disease is seen in Cushing's syndrome in which, under the influence of the adrenal cortical steroids, excess fat tends to be laid down on the trunk, face and around the shoulders and hips. The rest of the limbs are spared. At the same time the muscularity of the body is reduced.

Insulinoma

Another example of increased appetite due to an endocrine cause occurs when an excess secretion of insulin, such as from an insulinoma, leads to a craving for carbohydrate resulting in increased fat deposition.

Hypothyroidism

In contrast, a reduced energy expenditure occurs in hypothyroidism or myxoedema. Here there is a retention not only of water but also of associated glycoprotein.

Hypogonadism

This leads in the male to poor muscle development with its replacement by fat. If the hypogonadism is present during growth the body proportions would be eunuchoid with relatively long limbs and a short trunk.

The patient

When obesity is the presenting complaint there are a number of points in the history and physical examination which help in making a diagnosis and in the successful management of the patient.

History

In the history the following are important:

- the pattern of weight increases and losses in the past
- distribution of weight gained
- the pattern of food intake
- a family history of obesity
- social circumstances
- motivational factors
- symptoms suggestive of endocrine disease

Examination

In the physical examination pay special attention to:

- distribution of fat
- presence of oedema
- other complications – e.g. hypertension, diabetes mellitus

It is important to realize that 99% of obesity is exogenous in origin with no underlying cause. If a careful history and physical examination reveals no evidence of an endocrine or metabolic cause of obesity no further investigation is necessary.

The average practising doctor sees many obese patients who are not complaining of (and indeed may not be aware of) their obesity. In view of the documented consequences of obesity (see Section IV), it is important to point out their obesity to them and advise them on management.

HYPERTENSION

Elevated blood pressure is a common presenting feature in some of the less common endocrine disorders. It must be distinguished from purely systolic hypertension found in high output cardiac states, e.g. hyperthyroidism.

Endocrine causes

Hypertension of endocrine origin is found most frequently in association with selective hyperfunction of the adrenal gland:

- Cushing's syndrome
- Conn's syndrome (primary hyperaldosteronism)
- phaeochromocytoma

The degree of elevation of blood pressure is usually mild to moderate. Accelerated hypertension with 'malignant' phenomena is extremely rare.

The ultimate diagnosis may be obvious after a careful history and physical examination. In some cases, the typical clinical features may be absent and the diagnosis is made on the basis of laboratory investigation.

The patient

Cushing's syndrome

Cushing's syndrome or adrenocortical hyperfunction

associated with excessive secretion of cortisol (and adrenal androgens) usually produces a distinctive clinical picture of:

- weight gain
- lassitude
- muscle weakness
- emotional disturbance

Physical examination reveals:

- truncal obesity
- thinning of the skin
- abdominal striae
- muscle wasting

Conn's syndrome

In primary hyperaldosteronism the commonest presenting feature is hypertension. Associated symptoms include:

- polyuria
- polydipsia
- muscle weakness

Apart from finding a raised blood pressure, there are usually few other physical signs on examination.

Congenital adrenal hyperplasia

This is a rare condition resulting from a defect in adrenocortical steroid synthesis due to an enzyme deficiency. 21-hydroxylase deficiency is the commonest defect (90% cases) but only the rarer deficiency of C-11-hydroxylase leads to hypertension. (Refer to Section I.)

Phaeochromocytoma

Hypertension is by far the commonest presentation of phaeochromocytoma. Although the classical description is of labile blood pressure, this is less common than stable hypertension. Blood pressure elevation results from excessive secretion of catecholamines. Occasionally the hypermetabolic effects or hyperglycaemic effects of adrenaline are prominent features and the patient may have:

- tachycardia
- sweating
- weight loss
- glycosuria.

Investigations

Prior to detailed investigation of Cushing's syndrome support for the diagnosis can be obtained by simple screening tests:

- loss of the diurnal rhythm of cortisol
- the dexamethasone suppression test

In primary hyperaldosteronism a useful outpatient screening test is an estimation of the plasma electrolytes on 3 consecutive days. In many cases, a characteristic electrolyte pattern is observed:

- low K^+
- elevated Na^+
- elevated HCO_3

In congenital adrenal hyperplasia the enzyme defect is deduced by measurement of hormone precursor levels in urine (see p. 129).

Screening for phaeochromocytoma is achieved by estimation of catecholamines in a 24 hour urine collection. Measurements of adrenaline or noradrenaline are the most useful but in practical terms their metabolite vanillylmandelic acid is most commonly measured. It is important that the patient be on a low catecholamine containing diet prior to and during the urine collections.

MUSCLE WEAKNESS

Generalized or local muscle weakness which may be constant or episodic is not an uncommon finding in patients with untreated endocrine disease. Rarely it may be the presenting feature and the demonstration of glandular dysfunction may require careful clinical and laboratory assessment.

Clinical acumen, physical examination and simple investigation techniques will be sufficient in most instances to distinguish muscle weakness due to an endocrine disturbance from other myopathic syndromes.

Unfortunately, endocrine muscle weakness is not always reversible even when adequate control of the primary endocrine disturbance has been established, e.g. endocrine ophthalmoplegia and Cushing's syndrome.

Causes

The commoner causes of generalized muscle weakness are summarized in Table 36.

Table 36. Causes of generalized muscle weakness

	Endocrine	Non-endocrine
Episodic	Hyperthyroidism Hyperaldosteronism	Familial periodic paralysis Myaesthenia gravis
Constant	Hyperthyroidism Hypothyroidism Diabetes mellitus Addison's disease Cushing's syndrome	Muscular dystrophies Inflammatory myopathies Toxic myopathies Many others

Hyperthyroidism

Three types of muscle weakness occur in hyperthyroid Graves' disease:

- generalized
- localized
- periodic

Generalized muscle weakness

Up to 80% of patients with hyperthyroidism have muscle weakness by objective measurement. However, as a presenting feature of thyrotoxicosis muscle weakness is uncommon.

Generalized loss of muscle bulk is found, but the limb girdle muscles may be affected more than the rest of the body. Thus patients may be unable to rise unaided from the squatting position.

Localized muscle weakness

Involvement of the extraocular muscles may lead to exophthalmos and ophthalmoplegia. In the early stages ophthalmoplegia may affect only certain eye movements, usually elevation and abduction, but it may progress to complete paralysis.

It is important to realize that the condition is not always associated with hyperthyroidism but may appear in association with any level of thyroid function, and does not necessarily improve with correction of the abnormal thyroid status.

Periodic muscle weakness

Hypokalaemic periodic paralysis has been reported in association with hyperthyroidism. It is particularly common in people of Asian extraction. It ceases to be a problem when the patient is made euthyroid.

Hypothyroidism

Myopathic changes in limb girdle muscles may be

found in hypothyroid subjects. Patients commonly present with muscle cramps and stiffness.

The most commonly demonstrable muscle abnormality, however, is sluggishness of the tendon reflexes which is best demonstrated by the ankle jerk. This is due to an alteration in the contractile mechanism.

These features are reversible with thyroxine treatment.

Hyperaldosteronism (Conn's syndrome)

The predominant clinical finding is usually hypertension but, in a few cases, profound muscle weakness is the presenting complaint. The weakness may be episodic and of variable degree. It is a consequence of the intra- and extracellular hypokalaemia produced by excessive secretion of aldosterone.

Addison's disease

Muscle weakness is a common presenting feature within the vague symptom complex of patients with chronic adrenocortical insufficiency.

There is no evidence of a true myopathy and symptoms are relieved by correction of the salt and water imbalance.

Electromyogram (e.m.g.) and biopsy are normal. It is considered that the weakness is the result of electrolyte disturbance associated with combined aldosterone and cortisol deficiency.

Cushing's syndrome

Excessive secretion of cortisol is associated with true myopathic changes especially involving the limb girdle muscles. Structural changes can be seen in the muscle.

The hypercatabolic effect of excessive amounts of cortisol is the main cause of muscle wasting and weakness but the weakness may be accentuated by associated hypokalaemia.

Muscle wasting and weakness are found in patients on long-term corticosteroid therapy. This side-effect is most marked when synthetic analogues carry a fluorine atom in the 9a position of the molecule, e.g.:

- triamcinolone
- dexamethasone
- betamethasone.

Diabetes mellitus

Muscle weakness with wasting may be a presenting feature of undiagnosed diabetes mellitus, especially in the young insulin dependent patient. Loss of muscle bulk is a consequence of the use of protein as an alternative source of energy. The muscle involvement is, however, usually overshadowed by the more prominent symptoms of polyuria and polydipsia.

Isolated muscle weakness or paralysis, often reversible, is quite common in diabetic subjects and is usually secondary to a neurological lesion – peripheral neuropathy.

Primary muscle disease in diabetes is uncommon and, when it occurs, it is usually in elderly patients with apparently mild disorders of carbohydrate tolerance. Weakness, atrophy and fasciculation of the limb girdle muscles are found. This *diabetic amyotrophy* is a true myopathy showing normal nerve conduction and abnormalities on the e.m.g. It usually recovers with careful control of the diabetes.

Glycogen storage disease

Muscle weakness associated with cramping on exercise is a feature of two of the very rare glycogen storage diseases:

- Type V (McArdle's syndrome)
- Type VII

Hypercalcaemia

Muscle weakness in association with elevated levels of ionized calcium is due to decreased neuromuscular excitability.

Hypogonadism

Poor muscular development and limited exercise tolerance is a characteristic feature of male hypogonadism from whatever cause. It is rarely, if ever, the presenting complaint (CP-22).

Testosterone promotes protein anabolism. Diminished muscular development is a direct consequence of testosterone deficiency.

Investigations

Muscle weakness is occasionally the presenting feature of a number of endocrine conditions.

Following a careful history and physical examination the endocrine disorder is usually obvious and can be confirmed by the appropriate tests (Table 37).

Table 37. Tests for endocrine disorders

Disorder	Tests
Thyroid disease	Serum T4 or T3, serum TSH
Cushing's syndrome	Diurnal rhythm of cortisol
	Urinary free cortisol
	Dexamethasone suppression
Addison's disease	ACTH stimulation
Hypogonadism	Serum testosterone

If hypertension accompanies the muscle weakness, Conn's syndrome should be suspected and this may be confirmed by the typical serum pattern of hypokalaemia, hypernatraemia and alkalosis.

Sometimes the diagnosis is not clinically obvious and specific tests may have to be ordered, e.g. serum calcium, oral glucose tolerance test.

Clinical evidence of neuropathy may be the first manifestation of diabetes and sensory changes are usually the first to appear.

Check glucose tolerance in patients with undefined peripheral neuropathy.

LASSITUDE

By definition, lassitude means excessive tiredness or exhaustion. Conventional usage has diluted the specificity of its meaning, and care must be taken to define accurately what is meant if the symptom is offered by a patient. It must be distinguished, by a careful history, from muscle weakness.

Lassitude is a common presenting complaint of a wide variety of physical and psychological disorders. As a result, its discriminant value as a diagnostic guide is so poor that ultimate accurate diagnosis will depend on the detection of other clinical and laboratory abnormalities.

Although seldom the major presenting symptom in endocrine disease, it is often present, even early in the course of the disease, when specifically asked for.

Endocrine causes

Lassitude is a particularly prominent symptom in:

- hypothyroidism
- Addison's disease
- hyperthyroidism

but is usually also present to some degree in:

- hypopituitarism

- hyperparathyroidism
- Cushing's disease
- acromegaly

LOSS OF WEIGHT

Loss of weight is a common presenting symptom. It is well recognized by patients as a feature of disease and indeed may be the first sign of serious illness.

The doctor should always take the complaint seriously and attempt to elucidate the cause for the weight loss. A loss of weight of more than 5% should be regarded as significant. In a 10 stone (140 lb) patient this is equivalent to a loss of about 7 lb or in a 70 kg patient 3.5 kg.

The magnitude of the weight loss reflects the duration and severity of an illness. The loss may be ignored where it occurs in response to dieting in an obese patient or where it is due to loss of fluid in an oedematous patient.

Causes

These are outlined in Table 38.

Table 38. Endocrine causes of weight loss

Weight loss with	System	Disorder
Increased appetite	Endocrine	Hyperthyroidism
		Diabetes mellitus
	Other	Malabsorption
Normal or poor appetite	Endocrine	Hypopituitarism
		Addison's disease
		Hyperparathyroidism (occasionally)
	Gastrointestinal	Malabsorption
		Malignancy
	Chronic infections	Tuberculosis
		Brucellosis
	Malignancies	
	Psychiatric illness	

A large number of conditions may be associated with loss of weight. In general, a patient's weight falls whenever there is negative calorie balance and where there are insufficient calories to meet the metabolic needs. This results:

- when dietary intake is insufficient
- when the calories assimilated by the body are inadequate, e.g. steatorrhoea
- when the calorie expenditure is excessive, e.g. in hyperthyroidism.

The patient

Only patients with endocrine causes of weight loss are considered here.

The young diabetic patient may present with weight loss, polyuria and polydipsia. Diabetes should be considered when a young patient presents with rapid loss of weight. Always test the urine for sugar.

In hyperthyroidism weight loss is a fairly common presenting feature usually accompanied by increased appetite. It is usually associated with other clinical features of the disease (CP-23).

Weight loss is commonly found in patients with adrenocortical insufficiency (Addison's disease). The absence of significant weight loss in these patients indicates a need to reconsider the diagnosis of Addison's disease. Associated symptoms such as tiredness, lassitude and vague gastrointestinal complaints are usually found.

Weight loss due to hypopituitarism is nowadays seldom striking though previously it was described as a constant feature of pituitary cachexia. Presumably the diagnosis is currently made at an earlier stage of the condition before severe weight loss develops.

In other endocrine conditions weight loss is seldom a prominent presenting feature though occasionally it occurs due to anorexia produced by the disease effects, e.g. the hypercalcaemia produced by hyperparathyroidism.

Investigations

The investigations are those of the appropriate suspected endocrine disorder.

Anorexia nervosa

Uncommon in males, this condition usually occurs in adolescent females or young women. Although in many ways it would appear to be a primary hypo-thalamic endocrine disorder, its basis is almost certainly psychiatric. The endocrine features are secondary to changes at higher cerebral levels.

The predominant feature is profound weight loss accompanied by refusal to eat. Amenorrhoea is the rule and finding a low basal metabolic rate, hyper-cholesterolaemia and delayed ankle jerks indicates at least a secondary effect on the endocrine system.

The most consistent endocrine abnormalities are decreased levels of LH and FSH. There is often loss of the sleeping pulsatile secretion of these hormones which is the first hormonal indication of the onset of puberty. These patients thus revert hormonally to the stage of prepuberty. In addition the levels of growth hormone and serum T_3 may be low. These hormonal changes return to normal with refeeding.

The condition is currently relatively common probably due to slimness being fashionable. The cause of the aversion to food is not known, though psychologically an exaggerated fear of becoming what current fashion dictates as obese plus a rejection and fear of sexuality appear to be common features.

PAIN IN THE NECK

Pain in the neck is common and has many causes. Only rarely does neck pain have an endocrine cause but disorders of the thyroid gland are probably the commonest cause of pain and tenderness in the anterior neck.

Endocrine causes

Endocrine associated pain is relatively uncommon, but may be encountered in conditions affecting the thyroid gland

- subacute thyroiditis
- haemorrhage into a thyroid cyst or nodule
- acute bacterial thyroiditis

Pain may also be due to infection involving an abnormality such as a thyroglossal cyst or fistula.

Subacute thyroiditis

This is an acute viral infection of the thyroid gland often associated with or following upon an upper respiratory tract infection. Mumps virus has been incriminated in some cases. This is described on p. 137.

Haemorrhage into a thyroid cyst

This may be mistaken for subacute thyroiditis. The onset of pain is usually more sudden with the appearance of swelling limited to one lobe of the thyroid. Pain is rarely severe as in subacute thyroiditis and there is no systemic upset. The pain normally remits within a few days and the swelling more slowly. Recurring attacks are unusual.

Acute bacterial thyroiditis

This is very rare.

Infection of a thyroglossal cyst or fistula

Thyroglossal cysts or fistulae may become infected. Diagnosis is usually easy. Drainage will be required. Once settled, definitive surgery should be undertaken.

Non-endocrine causes

Infections and musculoskeletal disorders account for most of the other causes of pain. These are summarized in Table 39.

Investigations

The investigations are those appropriate to the suspected disorder.

SWELLING IN THE NECK

Swelling in the neck is a common clinical presentation and is occasionally the sign of a serious condition.

Causes

Neck swelling or lumps may be, for practical purposes, considered under those arising in:

- the thyroid gland – goitre
- thyroidal development remnants
- other structures

Goitre

A goitre is a swelling of the thyroid gland. Clinically goitres present as:

- diffuse enlargement – diffuse goitre
- solitary lump – single nodule
- multiple lumps – multinodular goitre

These may be distinguished one from the other by careful palpation and a thyroid scan.

It is not normally difficult to distinguish thyroid swellings from those lying outside the gland. They:

- lie in the characteristic area
- are closely attached to the thyroid gland
- move upwards with swallowing

Table 39. Non-endocrine causes of neck pain

	Local	Systemic
Infections	Acute 　　cellulitis 　　lymphadenitis 　　Ludwig's angina 　　parapharyngeal abscess 　　osteomyelitis Chronic 　　tuberculosis 　　actinomycosis	Acute 　　bacterial meningitis 　　viral meningitis
Musculoskeletal disorders	Whiplash injuries Cervical spondylosis Metastatic disease Cervical ribs Scalenus anterior syndrome Costoclavicular syndrome	

Diffuse goitre

The patient with diffuse goitre may be:

- hyperthyroid
  Graves' disease
  thyroiditis, subacute thyroiditis, silent or painless

- euthyroid
  colloid or simple goitre
  Hashimoto's thyroiditis

- hypothyroid
  Inborn errors of thyroid hormone metabolism.
  Hashimoto's thyroiditis

Thyroid function testing may be necessary to define these more accurately.

Single nodule

Solitary nodules often present problems in diagnosis and management. The patient is usually euthyroid, though occasionally large autonomous adenomas of the thyroid gland produce enough hormone to cause hyperthyroidism.

Ultrasound can confirm the clinical impression of whether the nodule is:

- solid
  these are sometimes malignant

- cystic
  these are only occasionally malignant

Thyroid radioisotope scanning may occasionally reveal that what appears to be a solitary nodule is in fact one in a multinodular gland. This technique will also reveal whether the nodule is a:

- 'cold' nodule
  These do not concentrate as much isotope as the surrounding normal tissue.

- 'warm' nodule
  These are indistinguishable from the normal surrounding tissue.

- 'hot' nodule
  These concentrate the iodine more avidly than the surrounding tissue which in fact may be so 'suppressed' that it takes up no isotope at all.

Malignancy is rare in hot nodules, uncommon in warm nodules but commoner in nodules which are cold on isotope scanning.

Needle biopsy may help in differentiating malignant from non-malignant nodules but histological interpretation is often difficult and the diagnosis is only definitively made by excisional biopsy.

Multinodular goitre

These are common in older people. They are seldom malignant and the patient is:

- usually euthyroid
- occasionally hyperthyroid
- rarely hypothyroid

Thyroid isotope scanning usually confirms the multinodularity and thyroid function tests together with the clinical findings will establish the patient's thyroid status.

Thyroidal developmental remnants

These include thyroglossal duct remnants and cysts. Their development is described in LD-19 (p. 23).

Normally the thyroglossal duct involutes, though its distal part may remain as the pyramidal lobe of the thyroid. Failure of involution may give rise to swellings which may lie anywhere along the thyroglossal duct from the root of the tongue – a lingual thyroid – to the isthmus.

The cystic swelling is normally midline except in the region of the thyroid cartilage where it comes to lie on one or other side. Diagnosis is rarely difficult and it is characteristic of a thyroglossal cyst that it rises up on swallowing or on protruding the tongue.

Thyroglossal cysts may become infected. If they discharge spontaneously to the surface or are incised and drained then a thyroglossal fistula results. It is subject to recurring attacks of infection.

Treatment of a thyroglossal cyst involves removal, not only of the cyst, but of the entire thyroglossal duct from isthmus of thyroid to its origin often at the root of the tongue.

Non-thyroidal swellings

Congenital causes of these include:

- giant lymph nodes
- cystic hygroma
- solitary lymph cyst
- haemangioma
- branchial cysts and fistulae
- sternomastoid tumour

Acquired causes of these include:

- acute infections
- cervical adenitis
  including bacteria tuberculosis, viruses and toxoplasmosis

● submandibular and parotid disorders

● lymphadenopathy – primary and secondary

● carotid body tumour – chemodectoma

These conditions are not usually difficult to distinguish from goitres because of their:

● situation

● consistency

● mobility

Knowing their specific characteristics may help in suspecting the preoperative diagnosis but often excisional biopsy is required to make a specific diagnosis.

Giant lymph nodes are usually several times normal size and have been present since childhood. They are soft and spongy, usually single and of no clinical consequence.

A *cystic hygroma* may present at birth or during infancy, and arises from lymphoid tissue sequestered during organogenesis. It appears in the lower third of the neck and extends towards the ear.

A *solitary lymph cyst* is a variant of cystic hygroma and arises during adult life.

A *branchial cyst* is commonly first seen in adult life but arises from an abnormality of development in the 2nd branchial cleft. It presents as a solitary, painless, smooth cystic swelling in the upper third of the neck under the sternomastoid muscle. Branchial fistulae represent persistent 2nd branchial clefts. They are rare.

A *sternomastoid tumour* is a fusiform swelling usually occurring in the lower third of the muscle. It eventually leads to shortening of the muscle and torticollis.

Investigations

The investigations are those appropriate to the underlying pathology. Often biopsy is required to make a definite diagnosis.

EXOPHTHALMOS

Exophthalmos (or proptosis) is an abnormal protrusion of the eyeball from its socket. It is associated with a large number of systemic and local conditions and may be unilateral or bilateral. It must be distinguished from the staring expression produced by lid retraction.

Clinically, exophthalmos can be recognized by the presence of sclera between the lower border of the iris and the upper margin of the lower lid (LD-44).

The protrusion of the eye can be measured as the distance from the lateral orbital rim to the anterior corneal surface. Exophthalmos is present if the distance is greater than 20 mm (CP-24).

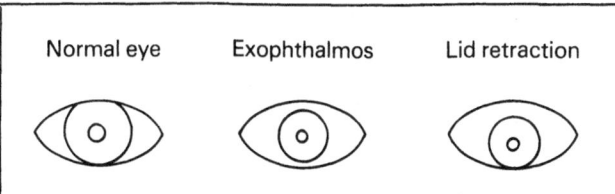

Normal eye   Exophthalmos   Lid retraction

LD-44. *Exophthalmos.*

Asymmetry of 2 mm or more between the eyes is pathological.

The patient or a relative may have noticed that one or both eyes has become more prominent and the patient may be described as having a 'staring' expression (CP-25).

Unilateral exophthalmos may be wrongly diagnosed, where the palpebral fissure is asymmetrical as in ptosis, facial nerve paralysis or lid retraction.

Measurement of exophthalmos can be carried out more easily using a Luedde exophthalmometer which is basically a thick ruler with a millimetre scale on each side. These markings are superimposed on each other and lined up with the anterior surface of the cornea (CP-24).

Causes

Thyroid disease (in the form of Graves' disease) is the commonest cause of bilateral or unilateral exophthalmos in adults. It is commonly asymmetrical. The patient may be hyperthyroid or have a past history of hyperthyroid Graves' disease but occasionally is euthyroid or rarely hypothyroid.

Exophthalmos does not occur in hyperthyroidism due to a multinodular goitre or autonomous adenoma. The finding of the associated eye changes listed in Table 40, below, suggests that Graves' disease is the cause of the exophthalmos.

Exophthalmos may also be found in other endocrine diseases, notably:

● Cushing's syndrome

● acromegaly

Where an endocrine cause is found, further ophthalmological investigation is not required.

The non-endocrine causes of exophthalmos include:

● lymphoma

● ophthalmic vein thrombosis

● cavernous sinus thrombosis

● retrobulbar haemorrhage

● metastatic carcinoma

● haemangioma and lymphangioma

In a significant proportion of patients no cause is found and the exophthalmos may be labelled idiopathic.

In most idiopathic cases the disease resolves spontaneously without surgical exploration and there is no opportunity for a pathological diagnosis.

Less common causes include:

- primary neural tumour
- meningioma
- rhabdomyosarcoma
- lachrymal gland tumour

The patient

History

The history is often not too helpful unless thyroid symptoms are present.

Examination

A detailed clinical examination of the orbit and the patient's vision should be made.

Thyroid function should be assessed both clinically and using the appropriate function tests. Other features of thyropathic eye disease may be present – lid retraction and limitation of upward and outward movement of the eye (CP-26).

Eye signs of Graves' disease are shown in Table 40.

Table 40. Eye signs in Graves's disease

Spastic	Infiltrative
Lid retraction	Exophthalmos
	Oedema and bulging of eyelids
Lid lag	Chemosis (CP-27, CP-28)
	Conjunctival injection
	Ophthalmoplegia (CP-26)

Orbital palpation around the globe will sometimes reveal the presence of a mass. A bruit of ocular pulsation suggests an arteriovenous shunt as the cause for the exophthalmos.

Investigations

Sometimes thyroid function tests are normal in patients with endocrine exophthalmos. However, the following abnormalities may be present:

- failure of the thyroid uptake of radioactive iodine to be suppressed following administration of tri-iodothyronine.

- failure of the patient's pituitary to respond to administration of thyrotropin releasing hormone (TRH).
- presence of antithyroid autoantibodies in the serum

If the clinical diagnosis is not obvious and endocrine exophthalmos has been ruled out, some help may be obtained from further investigations. These may include:

- blood count
- X-ray of orbit
- ultrasonography
- isotope scanning
- arteriography and venography
- computerized axial tomography (CAT scan)

Where there is no demonstrable endocrine disease, the patient presents a more difficult problem and should be referred for an ophthalmological opinion, especially if the exophthalmos is unilateral.

Recommended for further reading

Smith, M.E. (1967). The differential diagnosis of unilateral exophthalmos. *Int. Ophthalmol. Clin.*, 7, 911.
Grove, A.S. (1975). Evaluation of exophthalmos. *N. Engl. J. Med.*, 292, 1005.

PSYCHIATRIC DISTURBANCE

Generally patients fear or become depressed about possible serious organic illness. More commonly in the endocrine clinic patients may be anxious about abnormalities of appearance or body habitus:

- hirsutism
- gigantism
- dwarfism
- gynaecomastia

However, specific psychiatric states may represent part of the recognizable symptom complex of endocrine disorders. Elderly patients, in particular, may present with psychiatric symptoms and few features of the primary endocrine disease.

The patient

The psychological features are frequently non-specific but the following are commonly encountered:

- anxiety states
- depressive illness
- psychosis and behavioural abnormalities
- delirium and confusion

Anxiety states

Typically anxiety is associated with sympathomimetic features:

- tachycardia
- sweating
- tremulousness
- apprehension

Most commonly these features are integral features of:

- thyrotoxicosis
  Severe cases verge on hypomania. In clinical practice hyperthyroid patients may receive tranquilizing drugs for emotional lability for some time before thyroid disease is suspected.

- phaeochromocytoma
  This characteristically occurs with intermittent 'panic' attacks or acute anxiety states due to secretion of sympathomimetic amines by the tumour.

- Cushing's disease
  This is often associated with marked emotional lability.

Depressive illness

This occurs in:

- hypothyroidism
  The gradual onset of features of apathy, tiredness, intellectual impairment and withdrawal from social contact may closely mimic depression. It is mandatory to look for hypothyroidism in patients presenting with a first attack of depression in middle age.

- chronic adrenal insufficiency
  Due to primary adrenal failure (Addison's disease) or secondary to panhypopituitarism. This is commonly associated with vague psychiatric symptoms suggestive of depression.

- hypercalcaemia in hyperparathyroidism
  This is associated not only with generalized muscle fatigue but also features of depressive illness.

Psychosis and behavioural disturbance

Florid psychotic behaviour may be associated with hyper- or hypothyroidism:

- hypomania in thyrotoxicosis
- 'myxoedema madness' in hypothyroidism

Psychosis and severe behavioural abnormalities are probably more often associated with the other less common endocrine abnormalities:

- hypercalcaemic crisis in hyperparathyroidism
  May be associated with catatonic stupor or a severe confusional state.

- in Cushing's Syndrome
  There is characteristic emotional lability which can progress to frank psychotic behaviour.

- in spontaneous hypoglycaemia
  Bizarre behavioural abnormalities may occur. These symptoms may vary from confusion with aggressive behaviour to serious psychopathic states. In non-diabetic patients such features may be the first indication of hypoglycaemia due to rare insulin secreting tumours. For causes of hypoglycaemia, see Section IV, p. 173).

Delirium and confusion

Such vague symptoms may occur to varying degrees in many hormonal abnormalities. The following are probably of most clinical significance:

- acute and chronic adrenal insufficiency
- hypothyroidism
- hypercalcaemia
- diabetic ketoacidosis
- hypoglycaemia
- Cushing's disease

Investigations

The occurrence of psychiatric symptoms in patients with endocrine disease is usually associated with other clinical features to suggest the primary pathology. However, hormonal screening of large populations of psychiatric patients has revealed surprisingly large numbers with unsuspected endocrine disorders. Awareness of this fact and a reasonable degree of suspicion by the clinician would considerably reduce that number.

UNDIAGNOSED COMA

Causes

Endocrine causes of coma need to be differentiated from the many other sorts of coma which a patient with known endocrine disease may acquire. The differential diagnosis must often depend largely on the physical signs and biochemical tests.

Non-endocrine causes

The non-endocrine causes of coma can be divided for convenience into the following categories:

- metabolic
  uraemia
  hepatic failure
  water intoxication
- neurological
  postepileptic
  head injury
  cerebrovascular occlusion
  brain tumour
- drugs
  alcohol
  hypnotics
  analgesics
  carbon monoxide
- infections
  meningitis
  encephalitis

Coma in diabetes mellitus

Diabetics are particularly prone to several types of coma. These are in decreasing order of frequency due to:

- hypoglycaemia

- ketoacidosis
- hyperosmolarity
- lactic acidosis

The last three usually coexist with varying emphasis on one or more factors. The main points in differential diagnosis are given in Table 41.

In using the table for differential diagnosis it should be noted that various other causes of coma may confuse the issue. Thus there is often hyperventilation with the coma of uraemia and salicylate intoxication. Similarly brain injury may produce a rise in blood sugar, but this is transient. Starvation in any coma and especially after alcohol may produce moderate but never heavy ketonuria.

In clinical practice it should not be difficult to distinguish between the four types of coma so long as they are each borne in mind. The only factor that they have in common is that they occur in association with diabetes.

Coma in other endocrine disorders

Coma in association with other endocrine disease is uncommon. It may occur in severe hypothyroidism (myxoedema coma). The diagnosis is usually obvious on clinical grounds alone and is confirmed by the finding of hypothermia. If it is suspected, blood should be taken and sent for thyroid function tests but treatment should be commenced immediately without awaiting the results.

Severe adrenocortical insufficiency or hypopituitarism may lead to hypoglycaemic coma if two or more consecutive meals are omitted. Such patients lack the normal ability to break down protein to sugar. When suspected, blood should be taken for cortisol and ACTH levels and immediate treatment with glucose and intravenous corticosteroids started. A Synacthen test carried out subsequently will confirm the diagnosis.

Table 41. Differential diagnosis of coma in a diabetic

	Hypoglycaemia	Ketoacidosis	Hyperosmolarity	Lactic acidosis
Speed of onset	Minutes or hours	Hours or days	Hours or days	Hours or days
Dehydration	○	+ +	+ + +	○ or +
Air hunger	○	+ +	○	+ +
BP	N or ↑	↓	↓	N or ↓
Blood sugar	↓	↑	↑↑	N or ↑
Ketonuria	○	+ + +	N or tr	○
Plasma pH	N	↓	N	↓

Coma may occur in the later stages of water intoxication with the syndrome of inappropriate ADH secretion. The diagnosis is confirmed by finding a normal or high urine osmolality in the face of a low serum sodium or hypotonic plasma.

HYPERCALCAEMIA

One result of automation in clinical biochemistry is that increasing numbers of patients are now being identified with hypercalcaemia. Many of these patients have symptoms which would not ordinarily have suggested a disorder of calcium metabolism and some have no symptoms at all. It is important that a diagnosis should be made in these cases as hypercalcaemia may be the result of a serious but remediable underlying disease. It may cause renal damage and a wide variety of symptoms.

Causes

The commonest causes of hypercalcaemia are:

- hyperparathyroidism
- malignant disease

Thiazide therapy commonly causes a mild degree of hypercalcaemia. Less common causes of hypercalcaemia in adults include:

- vitamin D poisoning
- sarcoidosis
- hyperthyroidism
- the milk alkali syndrome
- adrenocortical hypofunction (Addison's disease)

In infants, hypercalcaemia can be due to hyperparathyroidism but this is very rare. More common is 'idiopathic hypercalcaemia of infancy', a disorder of unknown aetiology.

Malignant disease may cause hypercalcaemia in two different ways. Hypercalcaemia is associated with multiple secondary deposits in bone. The commonest disorders to do this are breast carcinoma and myeloma, but many other primary carcinomas can spread to bone. Hypercalcaemia is also caused when the primary tumour secretes a substance which behaves like parathyroid hormone. The most frequent causes of this syndrome (sometimes known as 'pseudohyperparathyroidism' or the ectopic PTH syndrome) are carcinomas of the bronchus or kidney, but most other malignancies can, on occasion, behave in the same way.

The milk alkali syndrome occurs in patients with a peptic ulcer who ingest excessive amounts of calcium-containing antacid or milk and any antacid.

It is possible that some milder cases of idiopathic hypercalcaemia of infancy seen in the 1950s represented an oversensitivity to vitamin D at a time when infant foods were heavily fortified.

The patient

History

Hypercalcaemia of any cause may result in depression and irritability. More rarely neurological symptoms such as ataxia occur. Polyuria and thirst are common symptoms and reflect the effect of the hypercalcaemia on the distal renal tubule. Patients with hypercalcaemia may present with acute pancreatitis.

In most cases of hypercalcaemia NOT due to hyperparathyroidism the underlying cause can be elucidated by a careful history and physical examination. From the history of dietary and medicinal intake eliminate:

- thiazide therapy
- vitamin D poisoning
- milk alkali syndrome

Bone pain may occur with malignancies and can occasionally occur with hyperparathyroidism.

Examination

Physical examination may reveal evidence of malignancy or systemic evidence of sarcoidosis, hyperthyroidism or hypoadrenalism.

Investigations

If the cause is not obvious investigations should include:

- repeat serum calcium and albumin levels
- serum phosphate and alkaline phosphatase
- erythrocyte sedimentation rate (ESR)
- serum protein electrophoresis
- skeletal survey or bone scan

MUSCLE AND BONE PAIN

The patient presenting with multiple aches and pains

often does not know whether the pain comes from bones, joints or muscles. Disorders of any of these should be considered in such a patient.

Causes

Causes of muscle pain

These include:

- viral infections such as influenza
- epidemic myalgia (Bornholm disease) resulting from the Coxsackie B virus

Another important cause of muscle pain is polymyalgia rheumatica, a disorder of unknown aetiology which responds readily to steroids.

Causes of joint pain

The commonest joint disorders causing widespread pain are:

- rheumatoid arthritis
- osteoarthropathy
- rheumatic fever (in children) – now becoming rare

Causes of bone pain

The most common cause of bone pain is malignant disease, notably metastases, for which the most common primary tumours are carcinomas of the breast and prostate.

Myeloma is also a common cause of bone pain and primary bone tumours should be considered when the pain comes from a single area.

Benign bone disorders causing bone pain include:

- vitamin D deficiency (osteomalacia in adults and rickets in children)

  The pain of osteomalacia is often in the arms and legs, often worse at night and is sometimes accompanied by other clinical features of osteomalacia such as muscle weakness and bone tenderness. Pain in the knees is a common presentation of rickets in adolescents.

- osteoporosis with crush fractures of vertebrae

  Here the pain is usually in the spine and probably follows the crush fractures. Each painful episode lasts only 4–6 weeks.

- Paget's disease of the bone (rarely)

  This is common among the middle-aged and elderly (3%) but only a small minority have pain. Severe localized bone pain in a patient with Paget's disease might be the first sign of sarcoma.

- hyperparathyroidism (very rarely)

Investigations

The approach to investigating these diseases is described under the relevant disorder.

INFERTILITY

No pregnancies after a year of married life with regular intercourse (2–3 times per week) and no contraception are sufficient grounds for a diagnosis of infertility.

Causes

Some of the common causes of infertility are outlined in Table 42.

Table 42. Causes of infertility

Causes of male infertility	Causes of female infertility
Failure of penetration Impotence	
Obstruction to passage of spermatozoa: absence of vas deferens prostatovesiculitis	Obstruction to passage of ova: fallopian uterine
Testicular failure: congenital – e.g. Klinefelter's cryptorchidism orchitis	Ovarian failure: congenital – e.g. Turner's syndrome surgical idiopathic – menopause praecox
Failure of testes stimulation: panhypopituitarism isolated gonadotrophin deficiency	Failure of ovarian stimulation: panhypopituitarism gonadotrophin deficiency idiopathic

In 40% of infertile couples the fault lies with the male partner. This has important practical relevance since male infertility is usually easy to demonstrate with a seminal analysis.

Causes in the male

In the male, testicular failure, either primary or secondary, is a more common cause than obstruction. Where the pituitary is at fault, the patient usually also has subnormal development of secondary sex characteristics.

Causes in the female

In the female, the commonest cause is failure of the signals which cause maturation of the follicle, release of the ovum or hormone synthesis by the corpus luteum. Such a defect presents as anovulation or as a failure of nidation.

Obstructive lesions also play a larger part in the female than the male. The ovum, unlike the sperm, has no inherent motility, and must rely on the wave motion of the cilia in the Fallopian tube for its passage. This is a vulnerable process and ciliary function may be disrupted without gross anatomical obstruction. Commonly, though, the fimbria are glued together or the body of the tube blocked by adhesions following from inflammation. As with the vas, the deficiency may be congenital:

- long thin tubes
- imperforate tubes
- absent segments

The incidence of any particular cause of female infertility depends so much on the population being surveyed, that detailed figures have no validity. It would be widely agreed that anovulation is common, followed closely by tubal obstruction, with all other causes such as testicular feminization, absent uterus, cervical incompatibility etc. taken together not amounting to as much as the first two.

The patient

Infertility presents as a problem of two people and should always be approached as such. Infertile patients seldom present themselves as a couple. Although in the first instance access may be to the female partner only, the male role should be evaluated from the very beginning. History taking should cover such elements as frequency of and success of penetration and ejaculation.

History

A failure of consummation is more frequent than might be supposed and an enquiry into sexual habit should always be made.

A detailed menstrual history will often point to the cause of infertility:

- primary amenorrhoea
  This suggests developmental failure of uterus or ovaries.

- secondary amenorrhoea
  This suggests anovulation.

- oligomenorrhoea
  This may not necessarily be associated with anovulatory cycles but if a woman is ovulating only three or four times a year her fertility is *ipso facto* reduced.

Regular cycles, particularly if accompanied by dysmenorrhoea, are seldom anovulatory but women with metrorrhagia usually do not ovulate.

Ovulation can often be induced in a variety of anovulatory states. This cause for female infertility carries a relatively favourable prognosis. Paradoxically, couples who present no abnormality in their history may be very difficult to deal with. Oligomenorrhoea in a young woman with a fertile husband probably has the best prognosis.

Examination

Physical examination in the male is seldom helpful. Although hypogonadal states and congenital testicular defects such as Klinefelter's syndrome are easily recognized, they are rare, and usually the infertile male is physically normal. It is generally sufficient to note:

- the degree of development of secondary sex characters such as pubic hair

- that the testes are descended and are of normal size

- that there is no hernia or varicocoele

Infertility due to a hypogonadal state is more common in the female and the physical examination should include:

- examination of breast development

- the presence of galactorrhoea

- growth and distribution of pubic hair

A full vaginal examination is necessary: both digital and per speculum. Points to note are:

- the degree of resistance to examination and the introital state
  Women with vaginismus or an introitus which

does not admit two fingers are unlikely to permit complete penetration.

- the presence of any vaginitis or vaginal discharge
- the state of the cervix
- the uterine size
- position and the presence of any adnexal tumours

Although physical examination of the female partner of an infertile couple is often helpful, e.g. in displaying the hirsutism and ovarian enlargement of the polycystic ovarian syndrome, it is very difficult to assess the importance of many common gynaecological abnormalities such as a cervical erosion or a uterine retroversion.

Investigations

At what stage investigations should be undertaken depends on the age of the couple and the strength of their desire for a child. The younger the woman the longer you can wait.

Laboratory investigations on both male and female partners should proceed simultaneously.

The wife should be instructed how to measure and chart her basal body temperature. At first visit blood samples are taken from the wife for estimation of:

- oestradiol
- progesterone
- follicle stimulating hormone
- prolactin

The husband should have a seminal analysis. If there are any features of hypopituitarism, that condition should be investigated.

The next stage of investigation is admission to hospital for:

- tubal insufflation
- cervical dilatation and curettage
- laparoscopy, if necessary

This should be delayed until the results of seminal analysis are available as there is no point in testing for tubal patency in the wife if her husband is infertile.

Oestradiol levels may be helpful in the diagnosis of hypogonadism. A single progesterone estimation, if done during the luteal phase, will tell whether or not the patient is ovulating.

Some infertile patients have high FSH levels indicative of a perimenopausal state and in any case the

FSH levels are useful in deciding the likely response of attempts to induce the ovulation with menopausal gonadotrophin.

Hyperprolactinaemia is a treatable cause of anovulation and must therefore be sought for at an early stage.

Laparoscopy is done as a routine in some infertility clinics but is probably best reserved for those patients where there is reasonable doubt about tubal patency or some suspicion of visible intrapelvic pathology.

Male infertility is difficult to define in terms of sperm density and motility and the criteria vary from centre to centre.

HIRSUTISM

The amount of hair and its distribution varies enormously in normal women and between one race and another.

'Excessive' facial and body hair is called hirsutism. The associated psychological disturbance is often a major problem in the management of affected patients.

The types of normal hair are outlined in Table 26 and Section II.

Mediterranean races have more body hair than Scandinavian women. Women with black hair may appear to have more hair than they would consider desirable.

Minimal hair on the chin and around the nipples and extension of the pubic hair to the umbilicus may be normal findings.

Causes

These are outlined in Table 43.

Table 43. Causes of hirsutism

Cause	Disorder
Idiopathic	
Menopausal	
Adrenal	Congenital adrenal hyperplasia
	Cushing's syndrome
	Carcinoma
Ovary	Polycystic ovary syndrome (Stein–Leventhal)
	Masculinizing tumours
Iatrogenic	Androgens
	Anabolic steroids
	Glucocorticoids
	Phenytoin

The patient

Age

The patient's age affects diagnostic probability. Patients with idiopathic hirsutism and the polycystic ovary syndrome (Stein–Levethal syndrome) often notice increase in body hair during their teens.

An increase in facial hair is common with ageing and at the menopause. Congenital adrenal hyperplasia may only become manifest after puberty.

Family

Increased hair may be found in the parents and/or siblings of patients with idiopathic hirsutism.

History

Usually women presenting with hirsutism are in good general health. They come to their doctor because they are embarrassed about their appearance and need reassurance.

The extent and rapidity of hair growth should be defined. Whether or not the patient shaves and the number of times she shaves per day is important.

The menstrual history is very important. Patients with idiopathic hirsutism usually have regular, normal periods whereas in the polycystic ovary syndrome irregular periods or secondary amenorrhoea are the rule. Androgen-producing ovarian and adrenal tumours are usually associated with secondary amenorrhoea.

Details of change in sexual drive may be valuable additional information.

A drug history is important. Anabolic steroids and certain testosterone-containing preparations used in the treatment of menopausal symptoms cause hirsutism.

Other drugs may produce hirsutism by an androgen-like effect:

- progesterone
- streptomycin
- phenytoin

Examination

The extent and distribution of hair should be confirmed. Evidence of virilization should be sought:

- enlargement of the clitoris
- increase in muscle bulk
- receding temporal hair
- breast atrophy

If significant virilization is apparent there is more likelihood of finding a treatable abnormality. It occurs in:

- congenital adrenal hyperplasia – when it is often marked
- ovarian and adrenal androgen-producing tumours
- Cushing's syndrome

The patient may be obese. Obesity is a feature of Cushing's syndrome and is often associated with the polycystic ovary syndrome.

Abdominal or pelvic examination may reveal bilaterally enlarged ovaries in the polycystic ovary syndrome or unilateral enlargement in association with an ovarian tumour.

If hirsutism is mild, the periods regular, and abdominal and pelvic examination normal, the patient probably has idiopathic hirsutism.

Investigations

In patients with excessive hair growth only, standard tests frequently fail to identify any treatable cause. Abnormalities when identifiable are usually manifest as changes in androgen metabolism. Normally only small amounts of androgen are secreted by the ovary. However, in certain disease states, either a potent androgen such as testosterone is secreted or a less potent hormone with less virilizing effect, which is then converted to a more potent agent elsewhere in the body, is produced, e.g. androstenedione. This is an intermediate product of oestrogen synthesis and may be converted to testosterone in liver or other androgen sensitive tissue such as skin. The adrenal gland may also produce these hormones in excess quantities.

Measure:

- serum testosterone
- androstenedione
  A significantly elevated level is suggestive of ovarian rather than adrenal disease.
- dehydroandrostenedione (DHA)
  This is a weak androgen but high levels may be an indicator of disordered steroid production.

If an abnormality is identified, it may be of value to proceed to dynamic tests to distinguish the source – adrenal or ovarian. However, dynamic function tests in this situation are extremely difficult to interpret.

In the absence of an autonomously functioning tumour, androgens produced by adrenal glands will be:

- increased by ACTH
- decreased by betamethasone

Androgens produced by the ovaries will be:

- increased by HCG
- decreased by oestrogen

POLYDIPSIA AND POLYURIA

Thirst and polydipsia result from cellular dehydration. They may arise from the excessive loss of fluids from the body from any route and are only combined with polyuria if the route of fluid loss is through the kidneys.

Thus polydipsia without polyuria occurs with increased sweating (the term is relative depending on climate) and with increased gastrointestinal loss. Acute internal or external haemorrhage may also stimulate thirst.

Polyuria exists when the daily urine output exceeds 2 litres. One early symptom is nocturia. Polyuria invariably leads to polydipsia in an attempt to maintain normal fluid balance.

Causes

The only form of polyuria not associated with ill health is that due to psychogenic polydipsia. Apart from this, the main mechanisms of excessive fluid intake include:

- increased solute concentration in the glomerular filtrate causing an osmotic diuresis
- failure of the kidney's concentrating mechanism

- increased glomerular filtration rate due to increased circulatory rate

The causes of polyuria are summarized in Table 44.

The commonest endocrine causes of polydipsia and polyuria, in order of frequency, are:

- diabetes mellitus
- hyperthyroidism
- hyperparathyroidism
- diabetes insipidus

The patient and investigations

It is important to establish that:

- the patient has true polydipsia and does not take frequent sips because of a dry mouth
- the patient has true polyuria and not just frequency

A careful history and physical examination may reveal an obvious cause. If not, serum osmolality and urine osmolality (or specific gravity) should be measured. In many situations where there is failure of the kidneys to concentrate urine in the tubules, the osmolality of the urine becomes fixed at 380–420 sm/l (sp.gr. 1.012–1.014).

The concentration of glucose, urea, calcium and potassium in blood should be measured.

If the history and examination reveal nothing and the tests are all normal, then a water deprivation test will distinguish between psychogenic polydipsia and diabetes insipidus. The administration of vasopressin (ADH) will distinguish neurogenic diabetes insipidus from nephrogenic diabetes insipidus.

GASTROINTESTINAL SYMPTOMS OF ENDOCRINE DISEASE

Patients with endocrine disorders seldom present with

Table 44. Causes of polyuria

Mechanism	Common endocrine causes	Common non-endocrine causes
Osmotic diuresis	Diabetes mellitus	Chronic renal failure Excess salt in diet
Failure to concentrate	Diabetes insipidus Hyperparathyroidism (hypercalcaemia) Hyperaldosteronism (hypokalaemia)	Chronic renal failure Nephrogenic diabetes insipidus
Increased GFR	Hyperthyroidism	Anaemia

gastrointestinal symptoms as their major complaint. Such symptoms, however, are common as part of the overall picture of some endocrine disorders. The major alimentary symptoms of endocrine disease are:

- disturbances of appetite
- nausea and vomiting
- diarrhoea
- acute abdominal pain

Disturbances of appetite

Anorexia

This is a common symptom of many acute and serious chronic diseases. It may occur and is occasionally the presenting feature in:

- hypopituitarism
- hypoadrenalism (Addison's disease)
- hyperparathyroidism

Occasionally difficulty in swallowing – dysphagia – despite a good appetite is a presenting symptom with a large goitre.

Hyperphagia or polyphagia

This means excessive appetite and is a symptom which should always require the exclusion of an endocrine disease. It occurs in:

- hyperthyroidism
  This is commonly and often accompanied by weight loss.
- diabetes mellitus
  The onset is acute, often with weight loss.
- acromegaly and gigantism
- Cushing's syndrome – occasionally

Acute hunger is a common feature of hypoglycaemia from any cause.

Nausea and vomiting

These are only occasional symptoms of endocrine disease. They may occur in hyperthyroidism.

If hypercalcaemia is present, nausea and vomiting may also be a feature of the acute phase of diabetic ketoacidosis or hypoadrenalism (Addison crisis). They are, however, more commonly due to the intercurrent illness which often precipitates these events.

Diarrhoea

Usually defined as an increased frequency of bowel movements, diarrhoea may be a feature of:

- hyperthyroidism
- hypoadrenalism

and occasionally of:

- diabetes insipidus
- hyperparathyroidism
  constipation is more common
- diabetes mellitus

The diarrhoea which occurs in diabetes mellitus is due to autonomic neuropathy, is characteristically nocturnal and usually only occurs when the disease has been longstanding.

Acute abdominal pain

It must be remembered that occasionally diabetic ketoacidosis can present as an acute abdomen with:

- pain
- vomiting
- tenderness
- guarding and rigidity

The issue is further clouded by finding a high white blood cell count which frequently accompanies the condition.

A careful history, physical examination and a blood sugar will distinguish these cases from those with intra-abdominal pathology.

Hypercalcaemia due to any cause, including hyperparathyroidism, can present with a bewildering array of vague gastrointestinal symptoms:

- anorexia
- nausea
- vomiting
- abdominal pain
- constipation
- diarrhoea

It is, in fact, associated with a high incidence of true peptic ulcer disease.

Hypercalcaemia should be kept in mind on assessing a patient with vague gastrointestinal symptoms where the cause is obscure.

TEST-YOURSELF QUESTIONS
Sections I, II and III

1. Would a general practitioner in the United Kingdom expect to see each year
a) One new patient with a pituitary disorder?
b) Seven new cases of thyroid disease?
c) Ten new patients with diabetes mellitus?
d) One new case of thyroid cancer?
e) Two patients with obesity due to an identifiable endocrine abnormality?

2. List five causes of inadequate hormone action.

3. Match the substance on the left with the effects on the right.
a) Dopamine
b) Thyrotrophin releasing hormone
c) LHRH

1) Prolactin release
2) Prolactin inhibition
3) FSH release
4) Reduction of insulin release
5) Increase in insulin release

4. Which of the following statements are true concerning anterior pituitary hormones?
a) ACTH shares large parts of its structure with β-lipoprotein
b) TSH, LH and FSH have the same structure in their β chain
c) Growth hormone reduces lipid synthesis
d) Prolactin has a well defined physiological function in the human male

5. Which of the following statements are true concerning the posterior pituitary hormones?
a) Both contain nine amino acids
b) The major controlling effect on ADH release is blood volume
c) Oxytocin is important in suckling
d) ADH and vasopressin share a number of actions in common

6. List five major metabolic actions of glucocorticoids.

7. Which of the following actions are
i) α-adrenergic responses?

ii) β-adrenergic responses?
a) Constriction of veins
b) Inhibition of insulin release
c) Increase in heart rate
d) Dilation of bronchi

8. Which of the following statements are true regarding thyroxine binding blobulin (TBG)?
a) Pregnancy increases TBG
b) Diphenylhydantoin increases TGB
c) Androgens decrease TBG
d) Oral contraceptives decrease TBG

9. Name one *major* tissue action of each of the following:
a) Parathormone
b) Calcitonin
c) 1,25-Dihydroxycholecalciferol (DHCC)

10. Regarding gonadal differentiation and function, which of the following are true?
a) The presence of the 'y' chromosome causes the medulla of the primitive gonad to develop
b) Androgens from the primitive testis cause regression of the paramesonephric ducts
c) The Leydig cells produce androgens
d) The mid cycle drop in LH and FSH causes the Graafian follicle to rupture

11. List three hormones involved in breast development for lactation.

12. Match the hormones on the left with the actions on the right.
a) Insulin
b) Glucagon
c) Growth hormone
d) Adrenaline
e) Glucocorticoids

1) Increases lipolysis
2) Increases glycogenesis
3) Increases glycogenolysis
4) Increases gluconeo-genesis

13. Using the tables on pages 46–51, indicate whether the following children should be investigated and give your reason.
a) A boy aged 10 whose height was 100 cm
b) A girl aged 7 weighing 40 kg

c) A 14-year-old boy who has had a growth spurt of 11 cm over the past year

d) A girl aged 15 with no menarche, stage 2 + breast and pubic hair development

14. Which of the following should be considered organ-specific auto-immune disorders?

a) Graves' disease
b) Subacute thyroiditis
c) Rheumatoid arthritis
d) Polyarteritis nodosa

15. Which of the following types of hormone assays is the most sensitive?

a) Cytochemical assays
b) Whole animal bioassays
c) Enzyme-linked immunoassays
d) Radioimmunoassay

16. For which of the following purposes is thyroid isotope scanning useful?

a) Assessment of thyroid size
b) Detection of metastases from thyroid cancer
c) Detection of ectopic thyroid tissue
d) Assessment of functioning of thyroid nodule

17. Which of the following may be presenting symptoms of an endocrine disorder?

a) Heat intolerance
b) Bone pain
c) Increased pigmentation
d) Galactorrhoea
e) Carpal tunnel syndrome

18. Match the following facial features with the conditions listed.

a) Pale, wrinkled and sparse hair
b) Prognathism
c) Thin and pigmented
d) Fat, florid and moon-shaped

1) Diabetes insipidus
2) Addison's disease
3) Cushing's syndrome
4) Acromegaly
5) Hypopituitarism
6) Hyperthyroidism

19. A patient complains of bitemporal headaches and visual disturbance. Which of the following symptoms do you think it is important to enquire about?

a) Nervousness
b) Weight loss
c) Change in glove or shoe size
d) Decrease in libido

20. What other aspects of the history in the patient referred to in Question 19 would you expect to yield useful information?

a) Past history of growth
b) Dental history
c) Family history
d) Social and personal history

21. In this patient, which of the following might yield important information on physical examination?

a) Examination of the external genitalia
b) Visual field examination
c) Examination of the hands and feet
d) Chest examination

22. Which of the following drugs may be associated with endocrine disease?

a) Hydrochlorothiazide
b) Chlorpromazine
c) Iodine
d) Lithium

23. List three endocrine conditions which affect the eyes and indicate the eye problems they cause.

24. Match the disorders on the left with the endocrine causes on the right.

a) Gynaecomastia
b) Decreased breast size
c) Failure to lactate
d) Galactorrhoea

1) Hyperprolactinaemia
2) Excess oestrogen secretion
3) Hypopituitarism
4) Virilism

25. Which of the following conditions may be associated with hypertension?

a) Hypopituitarism
b) Primary hyperaldosteronism
c) Phaeochromocytoma
d) Addison's disease

26. When should a child with short stature be investigated?

27. Which of the following menstrual patterns are abnormal?

a) Interval varying from 25 to 32 days and lasting 5 days
b) Interval of 28 days, periods lasting 10 days
c) Interval varying from 24 to 30 days and lasting two to 12 days
d) Alteration from usual pattern of 20 days lasting 5 days to 28 days lasting 8 days

Which of the above constitute

i) menorrhagia
ii) metrorrhagia?

28. Match the following two conditions with the appropriate statements.

a) Galactorrhoea
b) Gynaecomastia

1) Oestrogen priming is necessary
2) Caused by oestrogen excess
3) May be drug induced
4) Prolactin is a common cause
5) Infertility is common

29. Which of the following are true regarding obesity?

a) About 25% of cases are due to endocrine disease
b) In a 173 cm tall man a weight of 74 kg would be considered obese
c) The distribution of fat is important
d) Full investigation is warranted

30. Which of the following statements are true about precocious puberty?

a) In the girl it applies to development occurring before 11 years
b) The only clinical evidence may be early menstruation
c) Tall stature may be a feature
d) The stages of puberty can be assessed by Tanner method

31. Which of the following statements are true regarding endocrine hypertension?

a) It accounts for 5% of hypertension encountered in general practice
b) Polyuria and polydipsia would cause Cushing's disease to be suspected
c) The commonest adrenal enzyme deficiency causing hypertension is 21-hydroxylase deficiency
d) Episodic hypertension is the most common way a phaeochromocytoma presents

32. Name 4 endocrine disorders which may cause muscle weakness.

33. Which two endocrine disorders may be associated with loss of weight and increased appetite?

34. What are the two commonest causes of pain in the thyroid?

35. Which of the following statements are true regarding thyroid swellings?

a) Hashimoto's thyroiditis is a common cause of painful swelling
b) Ultrasound may be of value diagnostically
c) Malignancy is unlikely in nodules which are 'hot' in scanning

d) Multinodular goitres are frequently associated with hypothyroidism

36. A patient presents with exophthalmometry readings of 21 mm left eye and 23 mm right eye. Which of the following statements are true?

a) Exophthalmos is present
b) Thyroid function tests will show hyperthyroidism
c) Elevation and abduction of the eyes may be impaired
d) The patient probably has a multinodular goitre

37. Complete the following table using + or − to indicate likely presence or absence.

	Hypoglycaemia	Ketoacidosis	Hyperosmolar coma
Dehydration			
Air hunger			
Blood pressure			
Ketonuria			
Plasma pH			

38. Hypercalcaemia is characteristically found in which of the following conditions?

a) Hypothyroidism
b) Vitamin D poisoning
c) Sarcoidosis
d) Hypoparathyroidism
e) Malignancy

39. Which of the following statements are true of infertility?

a) In the male a detailed physical examination is usually helpful
b) Infertility due to a hypogondal state is more common in the male
c) In over one-third of couples the problem lies in the male partner
d) Male infertility is usually easy to demonstrate with seminal analysis

40. In which of the following conditions may hirsuitism be a feature?

a) Cushing's syndrome
b) Thyrotoxicosis
c) Congenital adrenal hyperplasia
d) Polycystic ovary syndrome
e) Hyperprolactinaemia

41. List 4 endocrine causes of polyuria.

SECTION IV

Description of
Specific Diseases

PITUITARY RELATED DISORDERS

ACROMEGALY

Acro (distant parts) megaly (enlargement) is a clinical disorder due to excessive growth hormone secretion from the anterior pituitary gland.

Causes

Although there is evidence, in some cases, to implicate a hypothalamic disorder, an autonomous pituitary tumour (usually an eosinophilic adenoma) is the most common cause. In a few cases, this adenoma may be associated with adenomata or hyperfunction of the parathyroid or pancreas. This is described under multiple endocrine adenomatosis (p. 188).

Clinical features

The clinical features are produced by:
- hormonal effects
- local tumour effects

Hormonal effects

The classical features are produced by excessive growth hormone secretion. They are summarized in LD-45. Skeletal overgrowth is invariable. In adults,

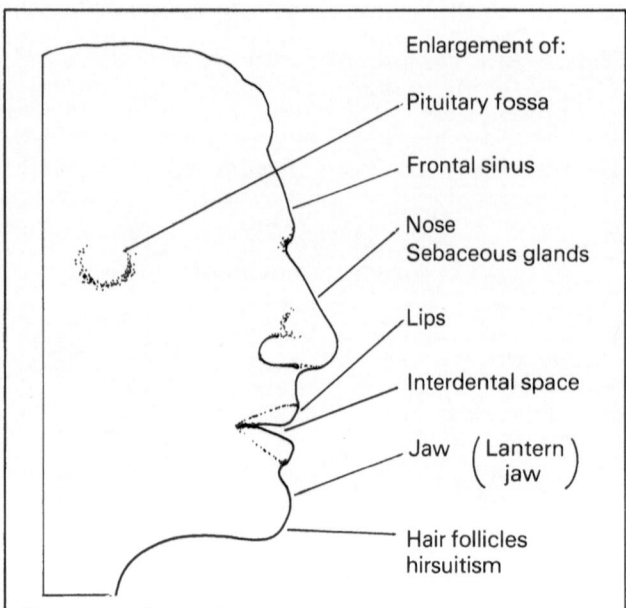

LD-45. Summary of the features of acromegaly.

since the epiphyseal plates have fused, the visible enlargement involves the:
- head and jaw
- hands
- feet

Prognathism develops and widening of teeth spacing occurs. Soft tissue hypertrophy results in:
- broad hands (spade-like)
- thick greasy skin
- large feet
- coarsening of facial features

The soft tissue enlargement in certain confined spaces may encroach upon nerves, producing tingling of the fingers or feet (carpal and metatarsal tunnel syndromes). Hypertrophy of the heart, thyroid gland and joints may also occur.

Apart from being involved by a specific (acromegalic) cardiomyopathy, the vascular system may be affected by hypertension. Heart failure, cardiovascular and cerebrovascular deaths frequently occur in untreated acromegaly.

Growth hormone is an insulin antagonist and impaired glucose tolerance is seen in at least 25% of acromegalics.

Local tumour effects

Headaches are common but are frequently non-specific. The optic chiasma (or nerves) may be compressed and a visual field defect – bitemporal hemianopia – may be found, particularly by perimetry. Encroachment on the third, fourth or sixth cranial nerves is uncommon but results in diplopia and strabismus.

Pressure by the tumour may also result in (usually partial) hypopituitarism. With decreased gonadotrophins, reduced libido commonly occurs early. Involvement of ACTH and TSH production is uncommon. Prolactin levels are often elevated.

In many patients, the changes are so gradual that they are not noticed as being abnormal and the diagnosis is made incidentally by an astute physician (CP-29–33 and X-7, X-8).

Clinical features may be more obvious when compared with old photographs, shoe and hat sizes.

Investigations

Acromegaly is basically a clinical diagnosis. The diagnosis is confirmed by the demonstration of elevated growth hormone levels which do not suppress during an oral glucose tolerance test.

X-ray abnormalities include:

- erosion or expansion of the sella turcica (X-7)
- tufting of the terminal phalanges
- thickening of the heel pad (X-8)

Serum phosphate levels may be elevated during the active phase.

Management

Surgery

Hypophysectomy, either transfrontal or transethmoidal, is the treatment of choice where there is evidence of extension of tumour outside the pituitary fossa. It is the only feasible therapy when the optic chiasma or nerves are compressed. In a few cases microsurgery via the transsphenoidal approach can remove microadenoma, leaving the remaining pituitary gland intact.

Radiotherapy

Pituitary implantation of radioactive yttrium or gold is a specialized invasive technique with fewer complications than surgery. It is, however, generally less successful than surgical removal. Elevated growth hormone levels frequently persist (CP-34–36).

Heavy particle therapy is somewhat more successful but is available in only a few centres.

Drug therapy

Bromocriptine (an ergot derivative) may successfully lower growth hormone levels in uncomplicated acromegaly although its long-term value remains to be established.

Replacement therapy (cortisol, thyoxine, sex hormones and vasopressin) will usually be required by patients following hypophysectomy. Cerebrospinal fluid (CSF) leakage and infection are rare complications of surgery.

Response to treatment

Untreated acromegalic patients have a markedly reduced life expectancy but, with successful therapy, there is a striking resolution (frequently 1–2 years following irradiation) of soft tissue changes including the facial appearance. Hypertension and diabetes may improve and there is a reduction in mortality.

Patients with a sustained remission of the disease usually have a marked fall, frequently to within the normal range, in growth hormone levels.

Frequent follow-up assessments are required to:

- assess response to therapy
- assess progress of the disease
- detect late development of hypopituitarism
- check adequacy of replacement therapy

GIGANTISM

Growth hormone excess occurring before fusion of the epiphyseal plates of long bones and vertebrae results in gigantism. This rare condition is characterized by tall stature with normal skeletal proportions.

Features of acromegaly such as:

- growth of distal parts
- coarse features
- radiological evidence of cortical thickening and tufting of phalanges

may also be present and become more prominent following epiphyseal fusion. Unlike constitutionally overgrown children, the bone age is normal. The condition is usually clinically distinguishable from other causes of tall stature (p. 80) and the more common cerebral gigantism (p. 113) by the latter condition demonstrating:

- dysmorphic features
- excessive growth, particularly in the first year
- variable degrees of mental retardation

Diagnosis and management are similar to those of acromegaly. In the absence of extension of the pituitary tumour, L-dopa therapy may be preferable as irradiation given in childhood is associated with a high incidence of hypopituitarism.

SHORT STATURE AND GROWTH HORMONE

Inadequate production of growth hormone from the anterior pituitary gland causes short stature when it occurs during the growing period. Fasting growth hormone levels in the plasma are normally low and stimulation tests are necessary to diagnose the condition.

Growth hormone deficiency has been defined as 'severe' in the United Kingdom when provocative tests result in growth hormone levels below 7 mu/l and 'partial' when they result in growth hormone levels between 7 mu/l and 15 mu/l.

Causes

Primary deficiency of the secretion of growth hormone from the anterior pituitary gland, e.g. in chromophobe adenoma, is relatively rare. Usually the pathology lies in the hypothalamus where it is postulated that there is impaired production of growth hormone releasing hormone or possibly excess release of growth hormone inhibitory hormone. Hydrocephalus or tumours affecting the hypothalamus, e.g. craniopharyngioma, provide a structural basis for growth hormone deficiency but usually no demonstrable cause can be found and the diagnosis is 'idiopathic' growth hormone deficiency.

Irradiation of the brain affecting the hypothalamus, e.g. local irradiation of a craniopharyngioma or tumour of the optic chiasma or more general irradiation used prophylactically in the treatment of acute lymphoblastic leukaemia, may lead to growth hormone deficiency, and such patients should be screened for an inadequate growth hormone response to provocative tests.

Occasionally, growth hormone deficiency is familial, suggesting a genetic cause. There is recent evidence in favour of birth injury, including damage at breech delivery, as another possible aetiological factor.

Clinical features

The usual presentation of growth hormone deficiency in childhood and adolescence is short stature (CP-37.1). The child looks immature and is proportionately small, and associated obesity of greater or lesser degree is common. The bone age is usually delayed but relatively less so than in hypothyroidism.

Apart from the short stature, the child is very well clinically and, for this reason, unless attention is paid to the child's growth, the condition may be missed. Children with growth hormone deficiency are not small at birth and the diagnosis in severe congenital growth hormone deficiency is made because of hypoglycaemia. If children are measured annually, it should be possible to pick up patients with the common, less severe form of growth hormone deficiency by the age of 5 or 6 years, considerably earlier than the usual age of diagnosis at the present time. Treatment with growth hormone results in a better final height the earlier it is begun (CP-37.2, 37.3).

Investigations

The clinical investigation of short stature has been outlined in Section III (p. 78). If growth hormone deficiency is suspected, an exercise test should be done in the first instance (see Glossary).

The exercise must be severe enough to make the child tired, hot and sweaty, the blood sample for the growth hormone level being taken exactly 25 minutes after the commencement of the exercise. A growth hormone level of 15 mu/l or more excludes growth hormone deficiency and, in practice, this test separates off the majority of the children who might have growth hormone deficiency.

The definitive test at present in use in the United Kingdom is the growth hormone response to insulin hypoglycaemia (see Glossary).

In partial deficiency, a second definitive stimulatory test should be performed to corroborate the diagnosis, e.g. subcutaneous glucagon, oral clonidine, oral L-dopa or intravenous arginine. 'Partial' deficiency is diagnosed if the growth hormone response is between 7 mu/l and 15 mu/l in both tests.

Growth hormone deficiency is usually an isolated phenomenon but it may be associated with abnormalities of release of other anterior pituitary hormones.

Further assessment of:

- adrenal function
- gonadal function
- thyroid function

should be undertaken.

The prolactin level may be raised in association with structural and functional hypothalamic disease.

In hypothyroidism, it is necessary to reassess the patient for growth hormone deficiency when euthyroid, as hypothyroidism causes blunting of the growth hormone response to hypoglycaemia. Normal gonadal function cannot be assumed with certainty until puberty occurs.

Management

Treatment is by the use of human growth hormone which, in the United Kingdom, is supplied on the recommendation of a special committee to which applications must be sent from recognized 'growth' centres. At the present time, growth hormone is usually given in doses of 4 units intramuscularly three times weekly.

The response to treatment is reviewed annually by comparing height velocity with the normal centiles of height velocity for age and sex and to the height velocity during the pretreatment year. In addition, the bone age is reviewed as well as the puberty status. A substantial increase in height velocity confirms the diagnosis and the first year of treatment acts as a therapeutic test.

Treatment is continued throughout the growing period which may extend beyond the normal age of 16 years for girls and 18 years for boys because patients with growth hormone deficiency usually have a delayed bone age and they continue growing until the epiphyses fuse. Development of antibodies to the growth hormone preparation may rarely cause failure of therapy at any time during treatment.

TALL STATURE AND GROWTH HORMONE

Excess growth hormone secretion from an adenoma of the anterior pituitary gland is a rare cause of tall stature in older children and adolescents and the results of hormone tests are the same as in adult acromegaly.

Cerebral gigantism (Sotos' syndrome)

Tall stature affecting infants and young children in whom excessive length and weight are present at birth is known as cerebral gigantism (Sotos' syndrome). Normally, the very rapid growth seen in fetal life decelerates after birth, especially during the first year. Delay in this process is seen in cerebral gigantism although the growth rate usually diminishes to normal after several years. The condition is not a single entity from the aetiological, clinical or hormonal point of view. In some, a genetic abnormality has been demonstrated. The possible role of growth hormone awaits further clarification.

The majority of reports in the literature quote negative results in tests for growth hormone excess. However, during a glucose tolerance test, some patients have shown failure of suppression of growth hormone or even a paradoxical rise of growth hormone early in the test when the blood glucose level was rising, suggesting an abnormality of growth hormone homeostasis. Some may have glucose intolerance. The basal prolactin levels may be raised. It is likely that there is underlying hypothalamic dysfunction.

The children are of excessive length and weight at birth and typically grow very quickly after birth especially in the first year. They have long, narrow heads, large hands and feet, and clumsy movements and are mentally retarded. Some have hydrocephalus and some are epileptic. Gradually the growth rate becomes more normal although precocious puberty may occur with an early pubertal growth spurt.

At present there is no treatment available to influence the excessive postnatal growth. Therapy may be indicated for hydrocephalus or epilepsy and the majority require special schooling.

HYPERPROLACTINAEMIA

The recognition of clinical syndromes associated with high prolactin secretions has been one of the significant developments in endocrinology in recent years.

Patients may present with symptoms directly related to high levels of prolactin or with features of other specific disorders in which hyperprolactinaemia is an associated phenomenon.

Causes

These are:

- Physiological – pregnancy
- Functional or idiopathic–pituitary adenomas
- Psychoneuroendocrine – including effects of psychotropic drugs such as chlorpromazine
- Secondary causes – e.g. hypothyroidism, acromegaly, trauma resulting in pituitary stalk damage and surgery

The control of prolactin secretion is summarized in LD-7.

Clinical features

Directly related to high prolactin levels

In females:
- amenorrhoea
- infertility

are the commonest presenting features. Galactorrhoea is present in about 30% of such patients.

In males:
- loss of libido
- impotence
- infertility

may be the major symptoms.

Galactorrhoea is rare in affected males since this tends to occur only in oestrogen primed breasts.

The hypogonadal features related to hyperprolactinaemia follow the dual effect of prolactin in hypothalamus and ovary. High prolactin levels result in loss of pulsed secretion of LH from the anterior pituitary because of an effect on hypothalamic releasing factors. In addition, prolactin acts on the ovary reducing the sensitivity to gonadotrophins.

Related to the primary lesion

In patients with pituitary tumour symptoms related to space occupying effects may predominate:

- headache
- visual disturbance

Hypothyroidism or acromegaly with secondary hyperprolactinaemia will present with features typical of the primary condition.

Investigations

Which patients should be screened for hyperprolactinaemia?

The combination of amenorrhoea (or impotence) and galactorrhoea is a mandatory indication. However, since the majority of affected patients do not have galactorrhoea, all females with amenorrhoea and all males with impotence and infertility should be screened. It is important to emphasize that a single estimation of serum prolactin is not adequate. Up to three interval specimens should be assayed to confirm normality. Further a single minimally raised value probably does not represent pathology.

The level of prolactin when raised may be useful diagnostically – pituitary prolactinomas give rise to markedly raised prolactin levels sometimes as high as 20 000–30 000 mu/l.

Further investigation may depend on associated conditions. A good screening routine should include:

- CAT scanning of the pituitary fossa
- thyroid function tests

Management

It is important to identify hyperprolactinaemic patients because of the readiness with which the associated hypogonadism will respond to the dopamine-like drug bromocriptine. This therapy is reserved for patients with moderately raised prolactin levels and with no recognizable abnormality of the pituitary fossa.

When structural change is evident on skull tomographs or CAT scanning of the pituitary fossa suggesting the presence of a pituitary macro- or microadenoma, then transphenoidal microsurgery for selective adenoma removal is the treatment of choice. The alternative is combined irradiation of the pituitary fossa and administration of bromocriptine.

Successful treatment of hypothyroidism or acromegaly usually results in reduction of raised prolactin levels and disappearance of clinical features.

PITUITARY TUMOURS

Although small chromophobe adenomas of the anterior pituitary are frequently seen in routine post-mortem examination, especially in the elderly, clinically important pituitary tumours producing either endocrine effects or local pressure effects are uncommon.

The features of hormone-secreting tumours:

- hyperprolactinaemia
- acromegaly
- Cushing's syndrome

are described elsewhere but these tumours may present, as do some non-hormone-secreting tumours with:

- headache
- visual defects
- varying degrees of hypopituitarism

Tumours arising adjacent to and encroaching upon the pituitary may produce similar effects (Table 45).

Chromophobe tumours are the commonest and are usually very slow growing. They may expand the sella turcica and eventually extend above the sella. It is now

Table 45. Pituitary tumours

Primary adenoma	Hormone-secreting:
	  chromophobe – prolactin
	  eosinophilic – growth hormone
	  basophilic – ACTH
	Non-hormone-secreting:
	  chromophobe adenoma

Secondary tumours	Space occupying lesions arising in or adjacent to the pituitary:
	  craniopharyngioma
	  granulomatous lesions
	  glioma

recognized that a considerable proportion of these tumours are not 'inactive' as previously thought but secrete prolactin.

Craniopharyngiomas arise from Rathke's pouch, are cystic and are commonly calcified on X-ray. They usually cause hypopituitarism or pressure symptoms.

Clinical features

Hormonal

Clinical features attributable to excess production of:

- prolactin
- growth hormone
- ACTH

Clinical features attributable to inadequate production of:

- growth hormone
- ACTH
- TSH
- gonadotrophin

Pressure effects include:

- headaches
  These are variable in site and may be frontal.
- visual disturbance
  Loss of vision may result from pressure on the optic chiasma producing bitemporal hemianopia. Papilloedema due to tumour compression of the third ventricles producing hydrocephalus is uncommon.
- others
  Progressive cranial nerve (III, IV and V) palsies and disturbance of hypothalamic control of thirst, appetite or sleep are rare and result from large locally invasive tumours.

Investigations

Visual fields

Detailed perimetry is necessary to identify pressure damage to the optic chiasma or nerves.

Endocrine assessment

Assessment of hyper- or hypofunction is essential.

Radiology

A plain skull X-ray with tomography of pituitary fossa may show asymmetry or a double floor (X-7). Air encephalography can demonstrate suprasellar extension. Computer-assisted tomography (CAT scanning) is now (X-7a) replacing the more invasive technique of encephalography.

Management

The treatment of hormone-producing tumours has been dealt with in the preceding pages.

Small chromophobe adenomas are slow-growing and may be removed surgically (transphenoidal approach) with little trauma to the gland.

Large tumours with an upward extension are managed by surgery (transfrontal approach) which may be only palliative for some invasive tumours and craniopharyngiomas.

Subsequent radiotherapy may help to prevent recurrence of an adenoma.

Gliomas and metastases from carcinoma elsewhere are usually treated by radiotherapy. This may further damage the pituitary or hypothalamus and hormone deficiency states may then become apparent.

HYPOPITUITARISM

The clinical picture of total pituitary hormone deficiency with marked wasting is rare. More frequently, a variable degree of deficiency of one or more hormones is seen. With an expanding lesion such as a pituitary tumour, gonadal function is usually impaired first, due to decrease in gonadotrophin secretion or the effects of hyperprolactinaemia. The other hormones are affected later and in varying degrees. Vasopressin deficiency rarely occurs.

Causes

The common causes of hypopituitarism are:

- tumours
- trauma including surgery and radiotherapy
- infarction – postpartum
- granulomatous lesions
- isolated deficiencies – GH, LH, FSH, TSH, ACTH

Tumours, together with the effects of their management, may be responsible for a variety of deficiency states. In these patients, the hypopituitarism may fluctuate, particularly following treatment, and require regular assessment.

During pregnancy, the highly vascular hypertrophied anterior pituitary may undergo infarction due to spasm and thrombosis of pituitary vessels. Recent evidence suggests that in some cases disseminated intravascular coagulation (DIC) may play a role. This usually occurs with a massive uterine haemorrhage, often occurring in the postpartum period. There is then:

- failure of lactation
- breast atrophy
- amenorrhoea
- loss of pubic and axillary hair
- the skin becomes pale, wrinkled and aged
- Hypoadrenalism occurs and is usually followed by the slow development of hypothyroidism. Hypopituitarism from this cause is called Sheehan's syndrome. (For original description see Sheehan, H.L. (1939). *Q. J. Med.* 8, 277.)

Granulomatous lesions affecting the pituitary:

- sarcoidosis
- tuberculosis
- histiocytosis

are usually part of a generalized condition.

Isolated deficiency states affecting growth hormone or gonadotrophins are seen in childhood or early adulthood. These may be associated with other developmental abnormalities. An example of this is Kallman's syndrome which consists of gonadotrophin deficiency and loss of smell.

Clinical features

These are dependent on the patient's age and the number and severity of the hormone deficiencies (CP-15, 16, 22, 38, 39).

Gonadotrophin deficiency

If deficiency develops in childhood or adolescence there may be:

- failure to enter puberty
- undescended testes
- obesity
- eunuchoidism

Loss of libido may be associated with:

- infertility
- amenorrhoea
- oligospermia
- progressive loss of secondary sex characteristics

Growth hormone deficiency

If this occurs in children there is failure of longitudinal growth. In adults, the only manifestation may be a tendency to hypoglycaemia.

TSH deficiency

This leads to the development of hypothyroidism or cretinism if it occurs in the newborn. The thyroid is impalpable. The degree of hypothyroidism is seldom great in adults but may be quite severe in childhood.

ACTH deficiency

The features here are similar to those of primary hypoadrenalism except that there is a decrease in pigmentation rather than an increase and there is little or no disturbance in electrolytes due to the presence of an intact aldosterone-secreting mechanism.

Investigations

These are directed towards discovering the underlying cause as well as to determining the extent and severity of the hormonal deficiencies.

Basal estimations are required of:

- luteinizing hormone (LH)
- follicle stimulating hormone (FSH)
- thyrotrophin stimulating hormone (TSH)
- ACTH

- growth hormone (GH)
- prolactin
- thyroxine
- cortisol
- oestradiol
- testosterone

These will usually have to be supplemented by provocative tests which measure the pituitary response to control from higher centres, e.g. the effects of insulin-produced hypoglycaemia on growth hormone and the effects of the various hypothalamic releasing hormones such as TRH, LHRH on the pituitary.

Failure of response to all or some of these tests indicates the extent of the pituitary damage. A partial response or a normal response on repeated stimulation suggests that the primary lesion lies in the hypothalamus. Such patients may have associated hyperprolactinaemia or disturbance of fluid or food intake.

The differential diagnosis between hypothalamic disorders and hypopituitarism may be difficult. Primary hypofunction of the thyroid, adrenals or gonads is confirmed by the finding of elevated levels of pituitary stimulating hormones in the presence of low levels of the target gland hormones.

Although anorexia nervosa may be associated with disturbance of hypothalamic function, the findings are usually specific and a dietary history further supports the diagnosis.

Management

When there has to be surgery for a tumour it should be preceded by adequate hormone replacement therapy. Thyroxine and cortisol are given because they are more easily administered than TSH and ACTH. Mineralocorticoids are not required but steroid dosage must be increased during stress and surgery.

There is no need to replace growth hormone once the epiphyses have fused but, if the deficiency occurs during childhood or adolescence, growth hormone injections must be given.

Where fertility is required, FSH, LH or human chorionic gonadotrophin (HCG) and, in certain cases, LHRH, can be given. If fertility is not required, testosterone given to males will increase libido and potency. A cyclical oestrogen/progesterone combination should be prescribed for premenopausal women.

DELAYED PUBERTY

Delayed puberty in comparison with 95% of the normal population is diagnosed in boys when there is no evidence of testicular or penile enlargement by the age of 14 years, and in girls when there is no evidence of breast development by the age of 13 years. Short stature, i.e. a height below the 3rd centile, may accompany delayed puberty because of the associated delayed growth spurt.

Causes

These are summarized in Table 46.

Table 46. The aetiology of delayed puberty

Group	Disorder
Constitutional	Without associated short stature
	With associated short stature – 'growth and puberty delay'
Idiopathic	Without associated hormonal deficiencies
	With associated hormonal deficiencies, e.g. growth hormone deficiency
Chronic systemic disease	
Associated with obesity	
Hypothalamic	Tumour, e.g. craniopharyngioma
	Post meningitis or encephalitis
	Post brain injury or radiation
	Mental retardation
	Cerebral palsy
	Abnormal e.e.g. ± clinical epilepsy
	Hydrocephalus
Anterior pituitary	Chromophobe adenoma
Gonadal	Undescended testes
	Gonadal agenesis
	Gonadal dysgenesis: e.g. Klinefelter's syndrome Turner's syndrome
Adrenal	Congenital adrenal hypoplasia

Constitutional delayed puberty is common and a family history of the condition in either parent lends support to the diagnosis.

In some patients no cause is found and in some the cause may be isolated LHRH deficiency. If associated growth hormone deficiency is found, a hypothalamic disorder is present.

Chronic systemic disease may be associated with delayed puberty and short stature, e.g. cystic fibrosis, renal failure or Crohn's disease (CP-41.1, 41.2).

Delayed puberty associated with obesity is unusual in that obesity is more often accompanied by earlier puberty than in the normal population. The Prader–Willi and Laurence–Moon–Biedl syndromes are examples of conditions with obesity and delayed puberty.

Underproduction of LHRH from the hypothalamus – hypogonadotrophic hypogonadism – may be due to a variety of pathological processes (see Table 46).

Primary anterior pituitary disease, e.g. chromophobe adenoma, with impaired LH and FSH secretion, is a less common cause of delayed puberty.

Clinical features

Boys

In boys, gonadal failure, partial or complete, may be associated with undescended testes.

Undescended testes

It is likely that there is an element of primary hypothalamic deficiency in this condition, the low fetal production of LH and testosterone leading to failure of testicular descent. In extreme circumstances, there may be complete gonadal agenesis in which case no testicular tissue can be found on surgical exploration.

The testes are present in the scrotum at birth in about 95% of full-term boys and, by the age of a year, they are present in the scrotum in 99% of boys. Failure of testical descent may be unilateral or bilateral and there may be an associated inguinal hernia. Some testes are ectopic, that is, in an abnormal position, but the majority are found along the normal pathway of descent, e.g. in the abdomen or inguinal canal or at the neck of the scrotum.

It is important to distinguish between true failure of descent and the condition known as 'retractile testes' by careful and, if necessary, repeated clinical examination.

There is evidence that testicular tissue degenerates if surgery is delayed and it is now usual to operate between the ages of 2 and 3 years. If the testis is palpable high in the neck of the scrotum, a trial of HCG therapy is worthwhile, as descent may occur and HCG therapy may be indicated prior to operation to enlarge an underdeveloped scrotum so that it can accommodate the testes. If, at operation, a testis cannot be brought down, it should be removed, in view of the increased risk of malignancy, and a prosthesis put in.

Impaired testicular function, both from the point of view of pubertal development and later fertility, is commoner in boys with bilateral undescended testicles. Ideally, however, all boys who have been operated on for undescended testicles should be seen at the age of 14 years when puberty ought to have begun. If it is delayed, appropriate investigations should be made and replacement therapy instituted, if indicated, without further delay.

Klinefelter's syndrome

With a chromosome constitution 47,XXY in its commonest form, it is associated with testicular dysgenesis affecting both the production of testosterone and fertility. If there is inadequate production of testosterone, high resting levels of LH and FSH even before the age of 12 years are found ('hypergonadotrophic hypogonadism').

Adrenal hypoplasia

This rare condition in boys is always associated with failure of pubertal development, suggesting that adrenal and gonadal function are closely linked at puberty.

Girls

Turner's syndrome

In girls, gonadal failure accompanies Turner's syndrome which, in the 'complete' form, is associated with the chromosome constitution, 45,XO. The ovaries fail to develop so that no oestrogen or progesterone is produced and the patient is sterile. High resting levels of LH and FSH are present even before the age of 11 years.

Investigations

The investigation of patients with delayed puberty is considered in Section III, and, if short stature is associated, full investigation of this condition is also necessary, as given in Section III.

A hypothalamic cause for the delayed puberty is likely when the resting levels of LH and FSH are low and the LHRH test show a sluggish anterior pituitary response of LH and FSH. In such patients, the prolactin level may be raised, indicating hypothalamic damage to the area producing prolactin inhibitory factor.

In primary anterior pituitary failure there is no response to LHRH.

If the resting levels of LH and FSH are high, and there is an above-normal response of LH and FSH to LHRH in association with low levels of testosterone or oestrogen, a primary gonadal cause is likely. This may be confirmed in boys by the short or prolonged HCG test in which there is a failure of testosterone response. In girls, with Turner's syndrome, it is usual to perform a laparoscopy in order to confirm ovarian dysgenesis prior to the introduction of oestrogen/progestogen therapy.

Management

If the condition is thought to be constitutional or idiopathic, and the LHRH and HCG tests give normal prepubertal results, the patient should be kept under review as spontaneous development of puberty may occur.

If, however, puberty does not develop, replacement therapy will be required. The age at which this should be begun varies with the individual clinical circumstances, but it is not usually later than 17 years for girls and 18 years for boys, and if failure of puberty is certain from the diagnosis, it should be begun at the normal time for puberty. Should a chronic systemic disorder, obesity, hypothalamic, or anterior pituitary disease be present, the primary cause should be treated if possible.

Boys

In boys, where the cause lies in the hypothalamus or anterior pituitary, an HCG test usually will show a good testicular response. HCG may be given therapeutically or replacement therapy with testosterone or one of its derivatives.

Primary gonadal dysfunction is a relatively common cause of delayed puberty in boys and the degree of testicular failure may be ascertained by the HCG test. If there is a poor testicular response, HCG therapy would be of no value and replacement therapy with testosterone or one of its derivatives is necessary.

Mesterolone (Pro-Viron) is an orally-active but relatively weak androgen. It may be necessary to give testosterone intramuscularly (Sustanon '100' or '250'), a more potent preparation to produce satisfactory growth of the penis and scrotum.

In the treatment of infertility, human menopausal gonadotrophin (HMG, 'Pergonal') may occasionally be of value as it may stimulate growth of the tubules and spermatogenesis.

Girls

In girls, delayed puberty associated with ovarian failure, e.g. gonadal dysgenesis proven by laparoscopy, may be treated by replacement therapy with a small dose of oestrogen along with a progestogen (e.g. Loestrin) given for 21 days and then withheld for 7 days in a cyclical manner.

PRECOCIOUS PUBERTY

Precocious puberty is diagnosed in boys when there is evidence of testicular or penile enlargement before the age of 10 years and, in girls, when there is evidence of breast development before the age of 9 years.

Tall stature, i.e. a height above the 97th centile, may accompany precocious puberty because of the associated early growth spurt. True precocious puberty, if untreated, progresses through the 'Tanner stages' to a fertile state – the emission of live sperms or regular ovular menstrual cycles. If the condition continues unchecked, the growth spurt is accompanied by a disproportionate advance in bone age, the epiphyses fuse early and, although the patients are relatively tall as children, they are of short stature in adult life. If, therefore, it is decided that no treatment is required for a girl with precocious puberty, reassurance may be given to the parents that she will not be excessively tall as an adult.

Causes

These are summarized in Table 47.

Constitutional precocious puberty is common and a family history of the condition in either parent lends support to the diagnosis. In some patients, no cause is found.

In idiopathic obesity early puberty is common but not usually early enough to be termed precocious puberty as defined above.

Overproduction of LHRH from the hypothalamus may be due to a variety of pathological processes (see Table 47, CP-39.3).

Excess production of LH and FSH from the

Table 47. The aetiology of precocious puberty

Group	Disorder
Constitutional	Without associated tall stature
	With associated tall stature
Idiopathic	Associated with obesity
Hypothalamic	Tumour, e.g. hamartoma
	Tumour of adjacent brain structures (CP-39.3)
	Postmeningitis or encephalitis
	Post brain injury
	Mental retardation
	Cerebral palsy
	Abnormal e.e.g. ± clinical epilepsy
	Hydrocephalus
	Polyostotic fibrous dysplasia
Anterior pituitary	Hypothalamic overstimulation producing hypertrophy
Gonadal	Testicular tumour
	Ovarian tumour
Increased sex organ receptor sensitivity	Possibly breast tissue in precocious thelarche
Adrenal	Precocious adrenarche
	Adrenal hyperplasia

anterior pituitary gland is secondary to overstimulation from the hypothalamus.

Production of testosterone or oestrogen from a gonadal tumour, before the normal age of puberty, produces virilization or feminization, that is, the development of the secondary sex characters. However, the condition is not true precocious puberty as the patients do not become fertile.

In girls, partial pubertal development may occur at an early age, e.g. enlargement of the breasts (precocious thelarche) or precocious menstruation presenting as isolated clinical abnormalities. Hormonal imbalance may be detectable but often this is not so, and increased end organ cell receptor sensitivity to oestrogen has been postulated. Precocious thelarche is usually a benign self-limiting condition but precocious menstruation is more likely to go on to precocious puberty in the fullest sense and is more often associated with demonstrable pathology.

Precocious adrenarche, i.e. early development of pubic hair as an isolated finding, before the age of 8 years in boys or 7 years in girls, is due to increased secretion of adrenal androgens. It has been postulated that adrenal androgen secretion, as well as being influenced by ACTH, is controlled by a separate anterior pituitary hormone, 'adrenal androgen stimulating hormone' which, like ACTH, is probably controlled by a hypothalamic factor.

In adrenal hyperplasia, if hydrocortisone and fludrocortisone therapy do not suppress the excess production of adrenal androgens adequately, precocious adrenarche will occur and this condition may lead on to true precocious puberty. The relationship between the dramatic increase in adrenal androgen production and the commencement of puberty is a subject of current research, as the rise in adrenal androgens precedes a rise in LH, FSH or the gonadal hormones by a few years.

Clinical features (CP-39.3)

These have been described where relevant with the appropriate disorder.

Investigation

The investigation of patients with precocious puberty is considered in Section III and, if tall stature is associated, investigation of this condition is also necessary, as considered in Section III.

A hypothalamic cause for precocious puberty is likely when resting levels of LH and FSH are high and the LHRH test gives an exaggerated rise of LH and FSH. Sometimes there is a rise of prolactin in such patients indicating damage to the hypothalamic centre producing prolactin inhibitory factor.

A gonadal cause is likely when high levels of testosterone or oestrogen are found in association with normal levels of LH and FSH and a normal response for age to LHRH.

Although precocious puberty is commonly constitutional or idiopathic in origin, it is essential to investigate the possibility of a hypothalamic or, more rarely, a gonadal aetiology, in all patients in whom there is a deviation from normal of more than 1 or 2 years. This is particularly true of boys with precocious puberty of this order; a brain tumour is the cause in about 50% of these patients. A CAT scan should be done in all children about whom there is any doubt. Various abnormalities of the e.e.g. are common in precocious puberty and the e.e.g. may be the only evidence for a neurological cause.

Management

If precocious puberty is thought to be constitutional or idiopathic in origin, the patient is kept under review so that the speed of advance through the various stages of puberty may be recorded. Often in these 'physiological variants of normal' the pubertal advance is slow and, if so, lends support to the diagnosis of a benign condition. No treatment is indicated in such patients.

If a hypothalamic disorder is diagnosed, the primary cause should be treated as appropriate. Brain tumours of adjacent structures require operation and/or radiotherapy.

LHRH production has been demonstrated recently by immunofluorescent techniques in hamartomas of the hypothalamus. In spite of this, however, a conservative approach to the therapy of hamartomas is desirable as the effect of the LHRH excess can be counteracted by drug therapy on a long-term basis.

The rare testicular or ovarian tumours should be dealt with surgically.

When precocious puberty is present more than 3 years in advance of the normal age, e.g. at 7 years in boys or 6 years in girls, drug therapy to delay the pubertal advance should usually be given. Such children are usually very tall in comparison with their classmates and the sexual development is an embarrassment leading to psychological difficulties. It is also important to delay the onset of fertility.

Boys

In boys, cyproterone acetate (Androcur) is the drug of choice. It acts partly by inhibiting LH and FSH production and partly by directly antagonizing testosterone. The advance of puberty is arrested and there is suggestive evidence that, with adequate dosage, the rapid advance in bone age associated with puberty is diminished with consequent improvement in the final height. The adult height attained is important as, even in the presence of a hypothalamic tumour, e.g. a hamartoma, the outlook for survival is good. Gynaecomastia may occur as a complication of treatment. The drug may be given on a long-term basis but, as it has a glucocorticoid action and suppresses the hypothalamic–pituitary–adrenal axis, it should be stopped gradually to give time for the axis to recover once the epiphyses have fused.

Girls

In girls, cyproterone acetate may also be used. The advance of puberty is arrested and menstrual periods cease. The drug may be less effective than in boys in delaying bone age advance. Medroxyprogesterone acetate (Provera) is another drug used in the treatment of girls with precocious puberty, with results similar to those obtained with cyproterone acetate.

DIABETES INSIPIDUS AND RELATED DISORDERS

Two peptide hormones, oxytocin and vasopressin (antidiuretic hormone – ADH), are secreted by the posterior pituitary. They are initially synthesized in the hypothalamus and migrate as neurosecretory granules along axonal nerve fibres to secretory terminals in the pituitary stalk, median eminence and the posterior pituitary (see Section I, p. 5).

Vasopressin deficiency produces diabetes insipidus and represents extensive damage usually involving the hypothalamus. No syndrome attributable to oxytocin excess or deficiency has yet been described.

Causes

The causes include:

- tumour
  - craniopharyngiomas
  - pituitary tumours
  - metastatic carcinoma
- inflammation
  - meningitis
  - tuberculosis
  - syphilis
- granuloma
- trauma
  - skull fracture
  - head injury
  - surgery
- vascular lesions
- idiopathic
- nephrogenic

Clinical features and investigations

The marked symptoms of:
- polyuria (5–20 l/day)
- thirst
- polydipsia

may lead to:
- severe dehydration
- exhaustion
- coma

The onset and severity vary considerably and the condition may be transient following head trauma or neurosurgery.

Diabetes insipidus must be distinguished from other conditions causing polyuria and polydipsia:

- diabetes mellitus
  Glycosuria and hyperglycaemia are present.

- chronic renal failure
  Nocturia is prominent and there is biochemical evidence of renal insufficiency

- hysterical polydipsia
  Withholding fluids for 24 h with regular weighing (to detect surreptitious drinking) will lead to a concentrated urine (> 600 mosmol/kg). In addition, compulsive water drinking tends to cause a fall in plasma osmolality where, as in true diabetes insipidus, plasma osmolality tends to be raised (> 300 mosmol/kg).

- electrolyte disturbances
  Hypercalcaemia, hypokalaemia, diuretic or lithium therapy are remedial causes of secondary nephrogenic diabetes insipidus.

The correct diagnosis is usually made following a water deprivation test (terminated if weight falls by more than 3 kg), demonstrating continued polyuria, raised plasma osmolality but low urinary osmolality. Vasopressin injection causes a rise in the urinary osmolality. Failure to respond to vasopressin suggests nephrogenic diabetes insipidus.

Having established a firm diagnosis, one must consider the underlying pathology although frequently none is found. There may be clinical or radiological evidence of tumour including metastases from lung or breast.

Management

Although some patients with mild diabetes insipidus may be successfully managed by a moderate increase in fluid intake, the majority require substitution therapy.

Pitressin is the drug of choice in short-term situations (e.g. head injury). Pitressin, or pitressin in oil, which has a longer action, have been superseded by desmopressin (DDAVP). This has almost pure vasopressin activity with no vasoconstrictor activity. The duration of effect varies from 12 to 18 hours and it is given once a day intranasally or twice daily in sufficient dosage to maintain a urinary output of 1–2 litres with a normal plasma osmolality.

Chlorpropamide may be useful where some endogenous secretion still persists. It is unclear how this acts but it may do so by either stimulating release of vasopressin or by sensitizing the renal tubules to its effect.

Nephrogenic diabetes insipidus

This is a familial and congenital condition in which the distal tubules are refractory to the water reabsorptive action of vasopressin.

Paradoxically, this condition frequently responds to thiazide diuretics but these agents may cause potassium depletion. This should be prevented by giving potassium supplements.

Syndrome of inappropriate antidiuretic hormone secretion (SIADH)

Excess vasopressin may be secreted from certain tumours, particularly bronchial, or its release may be stimulated by pulmonary infection or by drugs.

Failure to excrete water leads to a dilutional hyponatraemia, confusion and neurological disturbances.

In this condition, the urinary osmolality is high in the face of a low plasma osmolality. Serum sodium is decreased often to as low as 115 mEq/l.

Fluid restriction or hypertonic saline may be required to restore homeostasis. If the primary cause is not remediable, demethylchlortetracycline, which enhances the peripheral action of vasopressin, may be used.

ADRENAL

This section on the adrenal gland covers a number of disorders:

- adrenocortical insufficiency – acute and chronic
- adrenocortical hyperfunction
- inborn errors of metabolism
- hyperaldosteronism
- phaeochromocytoma

Inadequate secretion of adrenocortical hormones may be secondary to pituitary insufficiency. The following subsection, however, is devoted to primary adrenocortical failure. This may be:

- acute
- chronic

ACUTE ADRENOCORTICAL INSUFFICIENCY

Causes

Acute adrenocortical failure may occur as a result of bilateral adrenal haemorrhage or may complicate chronic failure as a terminal event (Addisonian crisis).

A similar clinical picture may occur in patients on long-term synthetic corticosteroid drug therapy if the drugs are stopped suddenly, or if the patient fails to receive additional steroid cover during a 'stress' situation. These patients must be advised not to alter steroid drug dosage except on medical advice. They should carry a card or other evidence recording the nature of their treatment.

Bilateral adrenal haemorrhage is caused by:

- Gram-negative septicaemia (Waterhouse–Friderichsen Syndrome)
- adrenal vein thrombosis secondary to retroperitoneal haemorrhage

Clinical features

The clinical features of acute adrenocortical failure are the direct result of sudden reduction in the secretion of the two major hormones, cortisol and aldosterone.

The dominant clinical picture is one of:

- profound hypotension
- vomiting
- diarrhoea
- abdominal pain
- pyrexia

Signs of dehydration are prominent but the patient usually remains conscious and alert.

Clues to the specific diagnosis may be apparent from clinical examination – septicaemic rash or pigmentation.

Investigations

Several laboratory parameters are of value in suggesting the diagnosis:

- low Na^+
- elevated K^+
- low HCO_3^-
- elevated urea
- low glucose
- eosinophilia, lymphocytosis

However, only the demonstration of low plasma cortisol plus the finding of an elevated endogenous ACTH level, will confirm the diagnosis.

Management

Acute adrenocortical failure is a critical medical emergency. Treatment must not be delayed until laboratory results become available.

Immediately the clinical diagnosis is suspected, blood should be withdrawn for the later measurement of serum electrolytes and cortisol and the patient should receive 200 mg hydrocortisone intravenously. At the same time, an intravenous infusion should be begun with 0.9% sodium chloride, giving 1 litre in the first hour. This may be supplemented with 50 g glucose and a further 100 mg hydrocortisone.

The volume of fluid and the nature and amounts of additives (such as potassium and hydrocortisone) given after the first litre are titrated against the patient's clinical condition and serial biochemical estimations.

Over the course of 24 hours, the intravenous drug regime may be altered to an intramuscular route using reducing doses of hydrocortisone and fludrocortisone. When the patient is capable of taking oral medication, a drug regime can be established for long-term maintenance treatment (see following subsection on chronic adrenocortical insufficiency – Addison's disease).

Additional specific therapy will be required if there is clinical or laboratory evidence of other pathology.

CHRONIC ADRENOCORTICAL INSUFFICIENCY

The clinical condition is known as Addison's disease.

Thomas Addison (1795–1860). The paper read by Addison in 1849 to the South London Medical Society described a combination of pernicious anaemia and the disease of the adrenals. He was the first to show that the adrenal glands were necessary for life (CP-40).

Causes

Several pathological conditions can lead to chronic progressive adrenocortical insufficiency.

These include:

- autoimmune adrenalitis
- tuberculosis
- sarcoidosis
- metastatic carcinoma
- amyloidosis

Autoimmune adrenalitis is currently the commonest cause in developed countries.

The clinical manifestations and biochemical abnormalities are largely a consequence of progressive reduction in cortisol and aldosterone secretion. The clinical condition may develop over a period of several months or even many years and may remain undiagnosed until it results in an acute Addisonian crisis.

Clinical features

History

During the early stages of the disease the symptoms are often non-specific:

- lassitude
- weakness
- irritability
- depression

As the condition progresses, they become more striking. Anorexia and weight loss are extremely common and are often associated with gastrointestinal manifestations such as:

- nausea
- vomiting
- diarrhoea
- abdominal pain

Muscle weakness increases and amenorrhoea is common.

Examination

During the early phase of the disease, physical examination may reveal little apparent abnormality. Later, two dominant signs are detected. These are:

- hyperpigmentation
- hypotension

Other clinical findings include:

- muscle weakness
- alteration in mood
- increased taste sensitivity
- reduction in body hair

These are less obvious than the two major signs described above.

Hyperpigmentation

This is the more outstanding of the major signs.

Pigmentation is usually first evident on exposed areas of skin such as the face and in areas of friction such as:

- axillae
- abdominal wall
- palmar skin creases (CP-41, CP-8)

and in

- scar tissue

Eventually, it may become generalized, involving mucous membranes such as the gingival margins (CP-7). Scars inflicted after the onset of the disease become pigmented and are a useful pointer to the diagnosis and to dating the onset.

The low circulating level of cortisol in plasma, operating through the negative feedback control system, generates an increase in output of ACTH and β-lipotrophin from the anterior pituitary. Although these compounds have only a small stimulating effect on pigmentation, levels become high enough to cause the hyperpigmentation.

Hypotension

At first, this may only be detected following postural changes but, latterly, it becomes apparent even in the recumbent position. In the fully developed case of chronic Addison's disease, a systolic blood pressure greater than 100 mmHg is unusual.

The hypotension is multifactorial in origin. There is reduced intravascular volume due to mineralocorticoid deficiency and impaired myocardial contractility. In addition, the low cortisol level reduces the vasomotor tone produced by endogenous catecholamines.

Signs of other pathology

In addition to the findings related to adrenocortical insufficiency, there may be evidence of the causal pathology such as:

- tuberculosis
- sarcoidosis
- a chronic disease predisposing to amyloid infiltration

The finding of another organ-specific autoimmune condition, e.g. myxoedema, will heighten the likelihood of an autoimmune pathology for adrenal failure. The detection of adrenal autoantibodies in serum will confirm the diagnosis of autoimmune adrenalitis. The absence of autoantibodies does not, however, refute the diagnosis.

Investigations

Laboratory

Several laboratory phenomena are indirect markers of adrenocortical failure. These include the classical serum electrolyte pattern:

- low Na^+
- high K^+

Other abnormalities include:

- high urea
- hypoglycaemia
- hypercalcaemia
- inability to excrete a water load

In many situations, however, little abnormality in these tests may be found until late in the disease.

Direct evidence for the existence of Addison's disease derives from failure of repeatedly administered ACTH to produce a significant rise in the level of cortisol in the plasma or urine.

In adrenocortical failure secondary to pituitary disease or long-term corticosteroid drug therapy, serial ACTH injections will eventually restore glandular activity resulting in an increase in the plasma cortisol level. In primary adrenocortical failure (Addison's disease), no increase in cortisol output occurs even after repeated ACTH administration.

Additional investigations include:

- plasma ACTH measurements
- adrenal autoantibodies
- tuberculin test
- abdominal radiology
  adrenal calcification occurs in chronic tuberculous cases
- screening for occult neoplasia

Management

Lifelong replacement therapy is instituted using oral cortisol (hydrocortisone) usually in a dose of 20 mg in the morning and 10 mg in the early evening to mimic the natural diurnal variation in cortisol levels.

Most patients also require a small dose of a potent mineralocorticoid such as fludrocortisone in a dose of 0.05–0.2 mg daily. Adjustment in the dosage of both drugs is based on assessment of the patient's well-being and the abnormalities in the electrolyte pattern.

Patients with Addison's disease are at risk in stress situations. Careful explanation of the need for additional steroid dosage during even minor illnesses is essential. Accidental trauma or surgery must be covered by a doubling or trebling of the daily steroid dose. Failure to provide such cover will greatly increase morbidity and mortality in these patients.

Once the acute stress is over, the dose should be tapered back to the maintenance dose over a period of days.

The patient must be instructed to carry at all times notification of the nature of his illness and details of treatment.

Specific treatment for the active underlying disease should be instituted when its nature is defined.

ADRENOCORTICAL HYPERFUNCTION (CUSHING'S SYNDROME)

Causes

Hypersecretion of cortisol may be due to one of several causes:

- Cushing's disease
  A disturbance of the hypothalamic control of the negative feedback system results in excessive secretion of ACTH from the anterior

pituitary. Pathological features include bilateral adrenocortical hyperplasia and a basophil adenoma of the pituitary.

- ectopic ACTH syndrome
  Elaboration of polypeptides with ACTH-like activity can occur with malignant tumours such as the oat-cell carcinoma of the bronchus. Bilateral adrenal hyperplasia results.

- an autonomous functioning nodule of the adrenal cortex
  The remaining cortex and the contralateral gland undergo secondary atrophy due to suppression of ACTH secretion from the pituitary; the negative feedback system remains intact

- primary carcinoma of the adrenal
  This is very rare.

The pathology is represented in LD-46.

Harvey Williams Cushing (1869–1939). The original case described in the *Bulletin of the Johns Hopkins Hospital* (1932), 50, 137, had a basophil adenoma of the pituitary.

The most common cause of the development of the clinical features of Cushing's syndrome is the administration of the synthetic glucocorticoids.

Clinical features

Cushing's syndrome is the clinical disorder caused by excess adrenocortical hormones from any cause. Cushing's disease is that type due to hypothalamic or pituitary disease.

The clinical features of both are summarized in (LD-47). It may evolve over several years (CP-42).

Fat metabolism

Obesity, classically centripetal (CP-4), but sometimes generalized, is almost invariable in patients with Cushing's syndrome. Moon facies (CP-19, CP-43, 44) and the 'buffalo hump' are characteristic findings.

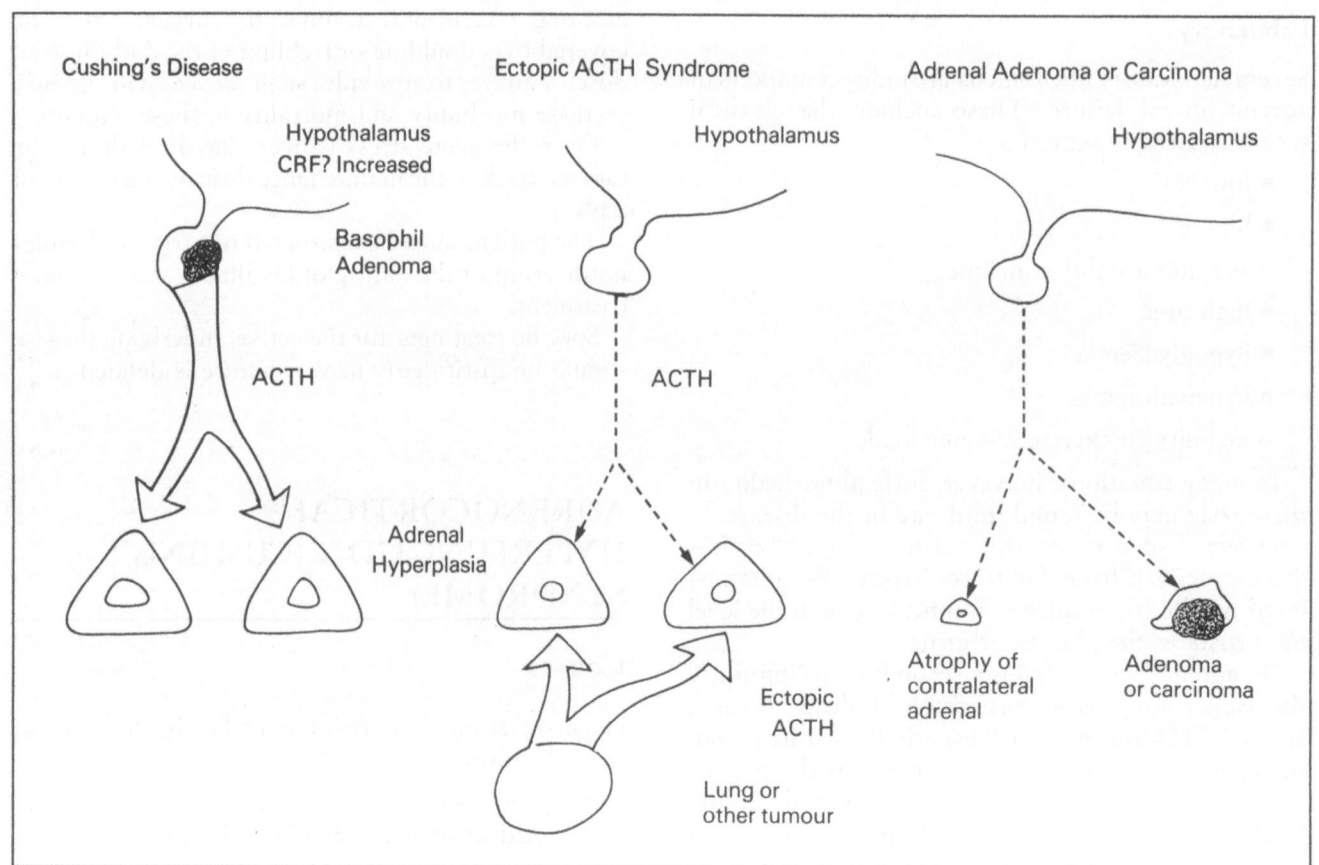

LD-46. *Pathophysiology of adrenocortical hyperactivity.*

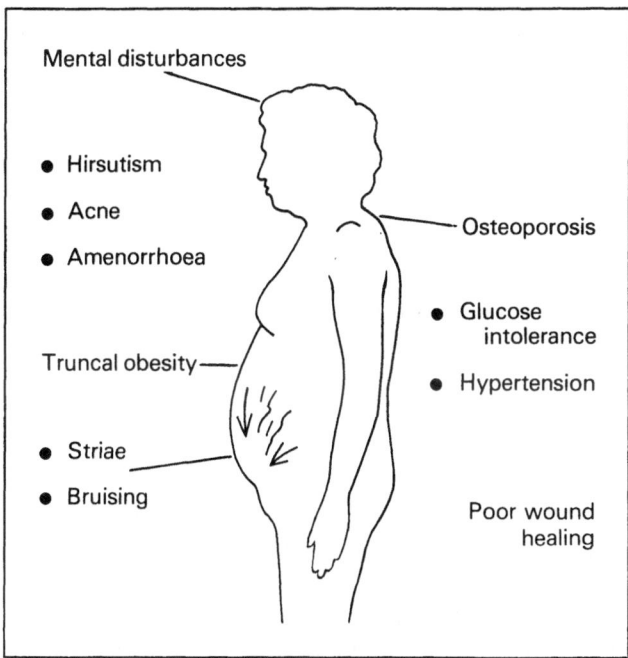

LD-47. *Clinical features of adrenocortical hyperactivity.*

Labels in figure:
Mental disturbances

- Hirsutism
- Acne
- Amenorrhoea

Truncal obesity

- Striae
- Bruising

Osteoporosis

- Glucose intolerance
- Hypertension

Poor wound healing

Protein metabolism

Increased protein catabolism is the result of increased cortisol production. There are several characteristic clinical manifestations:

- skin atrophy
- abdominal striae (CP-9, 10)
- osteoporosis with vertebral collapse (X-9)
- loss of capillary integrity leading to bruising
- impairment of wound healing
- muscle wasting

Androgen effects

Under normal circumstances, the adrenal androgens have little biological effect. When secreted in excess, as in some cases of Cushing's syndrome, in particular those due to an adrenal adenoma, varying degrees of virilization may be evident especially in female patients. The most striking abnormality is hirsutism. Acne may occur.

Psychological changes

In addition to the physical abnormalities, a wide variety of psychological disturbances may be produced. Psychotic and paranoid states may be very severe.

Investigations

Sodium and water retention is a result of the intrinsic mineralocorticoid activity of excessive circulating cortisol. Hypertension is the principal clinical manifestation.

The kaliuresis associated with sodium retention may accentuate the muscle weakness produced by muscle wasting.

Because of the rapid development of the ectopic ACTH syndrome, the biochemical abnormalities in that condition often predominate before the full-blown Cushingoid appearance occurs. Thus, hypokalaemic alkalosis may be the first and/or only indication of an ectopic ACTH syndrome.

In addition to the demonstration of excessive corticosteroid production, several related phenomena can be detected. These include:

- diabetic glucose tolerance curve
- lymphopenia and eosinopenia
- electrocardiographic evidence of hypertension or hypokalaemia

Laboratory

Laboratory investigation of patients with Cushing's syndrome has three main aims:

- demonstration of excessive corticosteroid production
- elucidation of the cause
- localization of the pathological tissue

Excess cortisol production is demonstrated by finding elevated levels of cortisol in plasma with loss of the normal diurnal variation. Measurement of urinary 17-hydroxycorticoid levels has now been largely replaced by measurement of plasma or urinary levels of cortisol.

That the high levels of cortisol are at least partly autonomous is demonstrated by failure of suppression by 2 mg of dexamethasone – the low-dose dexamethasone test.

The cause of the high cortisol levels may be elucidated by:

- measuring ACTH level:
  High in Cushing's disease and ectopic ACTH cases
  Low in cases of adrenal adenoma and carcinoma.
- failure of suppression by 8 mg dose of dexamethasone

There is little or no suppression by a high dose of dexamethasone in adrenal adenoma and carcinoma.

Localization of pathological tissue includes:

- skull X-ray and CAT scan
  Pituitary tumours

- chest X-ray
  Carcinoma is often found in the ectopic ACTH syndrome.

- intravenous pyelography

- adrenal angiography

- inferior vena caval catheterization with adrenal vein sampling

- adrenal scanning using either ultrasound or cholesterol labelled with a radioisotope e.g. selenium
  This localizes adrenal tumours.

Management

Surgery

Most cases of Cushing's disease are due to a microadenoma which can be removed by microsurgery using the transphenoidal route.

Sometimes, however, no well-localized pituitary tumour can be demonstrated. In these cases, management is often by:

- bilateral adrenalectomy with or without pituitary radiotherapy

- subtotal adrenalectomy
  Larger tumours and those causing radiological destruction of the sella require open surgery by the transfrontal route.

About 10% of cases treated by bilateral adrenalectomy alone develop a rapidly-growing pituitary tumour (often with devastating effects on vision) and marked pigmentation – Nelson's syndrome (CP-45).

Adrenal adenomas and carcinomas should be treated by unilateral adrenalectomy.

Radiotherapy

Some patients can be successfully treated with radiation (either conventional or proton-beam) to the pituitary gland alone.

Drug therapy

In selected cases – usually adrenal carcinomas – 'medical adrenalectomy' can be achieved by:

- aminoglutethamide
  This blocks normal hormone formation.

- o,p'-DDD
  This selectively destroys adrenal cells.

- trilostane
  This is a new competitive enzyme inhibitor believed to have significant advantages in leaving other physiological hormone production intact.

ADRENAL CORTEX – INBORN ERRORS OF METABOLISM

Several inborn errors of metabolism affect the pathways of adrenal steroid synthesis described in Section I (p. 15) (LD-11).

- cortisol synthesis
  This is impaired by reduction of enzyme activity affecting 21-hydroxylase, 11β-hydroxylase, 17α-hydroxylase or Δ^5, 3β-hydroxysteroid dehydrogenase.

- aldosterone synthesis
  This, as well as being reduced in a proportion of patients with 21-hydroxylase deficiency, is affected also by lack of 18-hydroxylase or 18-dehydrogenase. It is also impaired in 3β-hydroxysteroid dehydrogenase deficiency.

- adrenal androgen synthesis
  This is impaired in deficiency of 17α-hydroxylase or 3β-hydroxydehydrogenase.

- general impairment
  All of the corticosteroids and androgens are diminished if there is deficiency of the 20,22 desmolase enzyme necessary for the conversion of cholesterol to Δ^5-pregnenolone.

The clinical features of these enzyme defects may be logically worked out from knowledge of the various pathways of metabolism and of the biological effects of the steroids involved. 21-hydroxylase deficiency is the only condition which is common; the next most common is 11β-hydroxylase deficiency. The others are extremely rare and will not be considered further.

21-hydroxylase deficiency

This is an autosomal recessive condition. The block occurs after 17α-OH-progesterone, so that excess of this steroid is secreted (LD-11). An abnormally high level of 17α-OH-progesterone may be detectable during the first week of life but, in milder forms of the condition, it will be necessary to stimulate the adrenal cortex with ACTH in order to demonstrate the enzyme defect.

Adrenal hyperplasia

This, when it is due to 21-hydroxylase deficiency, is associated with a low secretion of cortisol. In about half of the patients, there is also a low secretion of aldosterone. Because of failure of the cortisol feedback mechanism on the hypothalamic production of corticotrophin releasing factor, pituitary ACTH secretion is increased. The androgen pathway, which is unaffected by the enzyme deficiency, is excessively stimulated by ACTH. Increased adrenal androgens are formed in the hypertrophied adrenal cortex and some peripheral conversion of these to the very potent androgen, testosterone, occurs.

Clinical features

The clinical manifestations are those of low cortisol production with an inadequate response to stress which may be life-threatening. Low aldosterone production with salt loss may occur.

Estimation of the plasma electrolytes reveals a low plasma sodium and a high plasma potassium and there may be an associated fall in blood pressure.

The excess androgen production leads to virilization in both males and females. Affected boys look normal at birth and may die undiagnosed of a salt-losing state. If they survive without treatment, the penis begins to enlarge and pubic hair develops about the age of 3 years. There is no associated increase in testicular size as this is dependent upon raised levels of FSH and LH. Virilization of girls may begin during fetal life but, in milder forms of the enzyme deficiency, no virilization may be seen until childhood or even until puberty. When virilization during fetal life has occurred, the sex at birth may be in doubt (see below, p. 160) (CP-46,1, 46.2).

In the majority of girls the clinical features are those of ambiguous genitalia, so that the abnormality is easily recognized. However, it should be remembered that in the most extreme virilization of a female fetus, the urethra traverses the enlarged clitoris and the labia fuse, so that the infant looks male with an underdeveloped scrotum and undescended testicles.

In the differential diagnosis of 21-hydroxylase deficiency, 11β-hydroxylase deficiency must be con-

sidered. In this condition, the block occurs just after 11-deoxy-cortisol and therefore this steroid is raised in the blood and its metabolites are raised in the urine. On the aldosterone pathway the block is just after 11-deoxy-corticosterone so that this steroid is also secreted in excess (LD-11). As both of these precursor compounds are salt-retaining, there is usually associated hypertension. Impaired cortisol production leads to excess ACTH with overstimulation of the androgen pathway and consequent virilization, as in 21-hydroxylase deficiency.

Investigations

These include:

- chromosome analysis to determine the genetic sex
- measurement of the plasma sodium, potassium and 17α-OH-progesterone
- estimation of ACTH, cortisol, aldosterone, renin, DHA, androstenedione and testosterone

From these tests it should be possible to make the diagnosis of 21-hydroxylase deficiency because of a raised 17α-OH-progesterone level and, if there is any doubt, ACTH stimulation is used to clarify the diagnosis. Electrolyte studies show whether salt loss is present. A normal cortisol level may be maintained in association with a high ACTH and, similarly, normal aldosterone levels may be maintained by high renin secretion. The androgens, DHA, androstenedione and testosterone are usually raised.

Management

In the treatment of adrenal hyperplasia due to 21-hydroxylase deficiency, replacement hydrocortisone is given four times daily and if salt loss is present, fludrocortisone is given twice daily. The dosage employed is dependent on the severity of the enzyme defect and the degree of salt loss.

It is very important to ask the parents to report immediately if the child becomes ill, or is vomiting, as systemic hydrocortisone therapy is then required in at least four times the maintenance dose. Failure to give adequate hydrocortisone therapy during, for example, infections, may lead to death. From the clinical point of view, the child should be growing satisfactorily, but the bone age should not advance more rapidly than normal. Excessive hydrocortisone therapy will stunt growth and delay the bone age, whereas inadequate hydrocortisone therapy will lead to excessive growth and a disproportionate advance in bone age and hence to early closure of the epiphyses and short stature in adult life.

Biochemical investigations are a useful guide to therapy but normality may not be achieved. The most useful steroid determination is that of 17α-OH-progesterone. Occasional checks of the cortisol and ACTH levels before the morning dose of hydrocortisone and measurement of the adrenal androgens and testosterone may be useful. The adequacy of fludrocortisone replacement therapy may be monitored by routine measurements of the blood pressure and plasma electrolytes and occasional estimations of plasma aldosterone and renin.

Corrective surgery is required to remove the clitoris and open the vagina in girls who have been virilized during fetal life and ideally this surgery should be carried out before the age of 1 year. Further plastic surgery to the vagina may be required just before puberty.

With good management and patient co-operation, the prognosis of 21-hydroxylase deficiency is good with regard to growth and life expectancy and there are reports of normal fertility in men and successful pregnancy in women.

HYPERALDOSTERONISM

Excessive production of aldosterone by the zona glomerulosa of the adrenal cortex results in sodium retention and potassium loss. Primary hyperaldosteronism is the commonest endocrine cause of hypertension.

Causes

Hyperaldosteronism may be:

- primary
    hyperplasia
    adenoma (Conn's syndrome)
- secondary to
    congestive cardiac failure
    cirrhosis
    nephrotic syndrome
    hypertension

In primary hyperaldosteronism the lesion in the adrenal gland acts autonomously. In secondary hyperaldosteronism the rise in aldosterone is secondary to an increased production of angiotensin. The high circulating aldosterone levels act on the distal convoluted tubules to promote the retention of sodium and excretion of potassium.

John W. Conn, born New York, 1907. See Conn, J.W. (1955). Primary aldosteronism: a new clinical entity. *Trans. Assoc. Am. Physicians*, 68, 215.

Clinical features

The clinical features of secondary hyperaldosteronism relate to the underlying causative conditions.

The clinical features in cases of autonomous production of hyperaldosterone, i.e. primary hyperaldosteronism, are due to the effects of the two major biochemical consequences:

- sodium retention – hypertension
- potassium loss – hypokalaemia

Hypertension

This is often the only presenting feature. It commonly occurs in the younger age group. The mechanism of its production is unclear.

In spite of the retention of sodium, oedema does not occur. This appears to be due to the fact that there is an 'escape' from ever-increasing sodium retention before there is sufficient to produce oedema.

Hypokalaemia

This usually, but not invariably, accompanies the hypertension. Depending on the severity, it may give rise to:

- muscle weakness
- nocturia
- polyuria
- polydipsia
- paraesthesia
- paralysis

The suggested mechanisms for the sodium retention induced hypertension include:

- arteriolar constriction from:
    sodium and water content of the vascular wall potentiation of vascular sensitivity to catecholamines
- increased blood volume

Investigations

Laboratory

The classical electrolyte picture of primary hyperaldosteronism:

- high or high normal serum sodium
- low serum potassium
- high serum bicarbonate

is not seen in all cases. Even with this pattern, the diagnosis must be confirmed by finding a:

- low renin level
- high urinary aldosterone level
- high serum aldosterone level

There should be relatively little change in these levels:

- in the erect or supine position
- on a high sodium diet
- on a low sodium diet

The degree of autonomy in bilateral hyperplasia may be somewhat less than when the lesion is an adenoma or carcinoma.

Radiology and other tests

After it has been established that primary hyper-aldosteronism exists, ancillary tests are used to localize any tumour which may exist:

- intravenous pyelography
- computerized axial tomography (CAT scan)
- adrenal angiography
- adrenal scanning with radiolabelled cholesterol
- catheterization of the inferior vena cava with sampling of adrenal veins for aldosterone

Management

Surgery

Where an adenoma or carcinoma is suspected, surgical removal of the affected adrenal is the treatment of choice.

Both adrenals should be explored at the time of operation and temporary cover with fludrocortisone may well be necessary since the contralateral adrenal is often hypoplastic due to longstanding suppression. Glucocorticoid cover during and immediately following surgery is also required.

Removal of an adenoma most often leads to an eventual return to normal of blood pressure but this may take many months.

Surgical treatment of bilateral hyperplasia is often less satisfactory. The usual course is to remove all of one adrenal and most of the other.

In preparation for surgery, and in those patients in whom surgery is not possible, or has been unsuccessful, an aldosterone antagonist must be used, e.g. spironolactone.

PHAEOCHROMOCYTOMA

Cause

This is a rare tumour of chromaffin tissue. The clinical manifestations are the result of excessive secretion of adrenaline and/or noradrenaline. Most of the tumours are located in the adrenal medulla.

Phaeochromocytomas occur singly in about 90% cases; they are usually histologically benign though about 10% are malignant.

Clinical features

The clinical manifestations generally reflect α- or β-receptor stimulation, depending on which catecholamine is dominant. Some of these are shown in Table 48.

Noradrenaline has predominantly α-receptor agonist activity.

Adrenaline is an α- and β-receptor agonist.

Table 48. Clinical features of alpha and beta stimulation

Alpha effects	Beta effects
Hypertension	Facial flushing
Headache	Sweating
Angina	Palpitations
Facial pallor	Fever
	Hypotension (rare)

Excessive secretion of both noradrenaline and adrenaline is the commonest finding with noradrenaline effects predominating. Pure adrenaline secreting tumours are rare.

Presentation

The commonest clinical presentation is hypertension. Though it is classically described as paroxysmal, in many cases there is a sustained elevation of blood pressure which may show superimposed paroxysmal hypertensive crises. Of these, crises occur only in about 25% of cases. In addition to the clinical manifestations outlined in Table 48, other symptoms include nervousness and tremor. The clinical picture may resemble hyperthyroidism or an anxiety state.

Usually there are no specific clinical findings but occasionally ectodermal abnormalities are found including:

- multiple neurofibromatosis
- café-au-lait spots
- pigmented naevi

In a few cases there is a family history of phaeochromocytoma, inherited as an autosomal dominant.

Investigations

The diagnosis is based on the demonstration of excessive secretion of catecholamines. Usually this is by measurement of adrenaline or noradrenaline or their major common metabolite, vanillylmandelic acid (VMA), in urine, or by measurement of adrenaline and noradrenaline in plasma.

The blocking and provocative tests are rarely used now because they are dangerous.

- blocking:
  phentolamine (Rogitine) test leads to hypotension
- provocative:
  histamine, glucagon, tyramine tests lead to hypertension

When the diagnosis is established, localization of the tumour is essential prior to surgical exploration. This is achieved by:

- intravenous pyelography
- angiography
- inferior vena cava catheterization and sampling of adrenal vein blood for catecholamines

Management

Surgical removal is the treatment of choice. It is vital that the patient be protected against the effects of release of catecholamines during surgery (or during the localization procedures mentioned above) by treatment for 3 days prior to surgery with a β-blocker (e.g. propranolol) and an α-blocker (e.g. phenoxybenzamine).

ASSOCIATED CHROMAFFIN TUMOURS

Phaeochromocytomas are only one type of tumour arising from sympathetic nervous tissue. Although all are rare, one of these tumours – neuroblastoma – is one of the most common malignant tumours of childhood.

Neuroblastomas usually occur before the age of 5 years and are highly malignant. Although most produce catecholamine excess (detectable in urine), in many cases this does not cause symptoms.

The other tumours of this group are:

- neuroblastoma of the fetus and first year of life
- the rare ganglioneuromas of young adult life

THYROID

The thyroid disorders which will be discussed are:

- hypothyroidism
- non-toxic goitre
- subacute thyroiditis
- hyperthyroidism
- tumours of the thyroid gland

HYPOTHYROIDISM

This is a clinical condition associated with decreased function of the thyroid gland and a decrease in the circulating level of thyroid hormones.

In hypothyroidism a mucoid substance is deposited in the skin and elsewhere in the body. This is called 'myxoedema'. Strictly speaking, the term myxoedema should be used only where such changes in the skin or elsewhere are obvious but often the terms myxoedema and hypothyroidism are interchanged.

Prevalence

The disease is by no means rare and a general practitioner in the UK could expect to have two patients per 2500 in his practice. Women are affected four times as frequently as men. Although it can appear at any age, it commonly presents between 30 and 50 years.

Recent studies suggest that 'subclinical hypothyroidism', i.e. normal serum levels of thyroid hormones with elevated serum TSH levels, may occur in up to 5% of the population.

Causes

These are summarized in Table 49.

Table 49. Causes of hypothyroidism

Mechanism	Cause
Failure of the thyroid gland itself (primary hypothyroidism)	Can follow treatment for hyperthyroidism
	Primary atrophic hypothyroidism
	Autoimmune thyroiditis
	Goitrogens
	Enzyme defects of the thyroid gland (dyshormonogenesis)
	Endemic cretinism
	Dysgenesis of the thyroid gland
Failure of TSH production (secondary hypothyroidism)	Disease of the pituitary gland
Failure of hypothalamic drive (tertiary hypothyroidism)	Disease of the hypothalamus

Primary hypothyroidism is much commoner than secondary or tertiary hypothyroidism.

Primary atrophic hypothyroidism and autoimmune thyroiditis (Hashimoto's thyroiditis) may belong to a continuous spectrum of disease.

Clinical features

The clinical features result from two factors:

- the widespread action in the body of thyroid hormone

  This increases cellular metabolism. In hypothyroidism, where there is lack of hormone, cellular metabolism slows down.
- localized effects of accumulation of mucoproteins

  The precise mechanism for this is not known.

With these points in mind, many features of hypothyroidism can be predicted.

The onset may be very slow and may easily go unnoticed.

The most important symptoms diagnostically are:

- mental and physical slowness
- tiredness
- cold intolerance
- dryness of the skin and hair

Other symptoms include:

- gain in weight
- constipation
- disordered menstrual function
- hoarseness
- deafness
- vague generalized pains
- paraesthesiae in the fingers
- poor memory

Because of the hypothyroidism, the doctor may require patience when taking the history. The patient herself may not have complained and may have been persuaded to see a doctor by relatives or friends.

Patients with severe hypothyroidism may develop 'myxoedema'. Changes in hypothyroidism may develop so gradually that they are not apparent to the patient, her relatives or her doctor.

Temperature intolerance

As a result of the decrease in cellular metabolism, the complaint is often of feeling cold and preferring to sit near a fire or radiator. In warm weather the patient may wear excessive amounts of clothing.

Weight gain

Despite a decreased appetite, there is frequently a gain in weight and the patient may have noted difficulty fastening her clothes. However, the weight gain in hypothyroidism is usually mild; of the patients who present with obesity few turn out to be hypothyroid. Weight gain is not invariably found and indeed about 10% of patients with hypothyroidism lose weight.

Constipation

Constipation may be so marked that faecal impaction occurs and paralytic ileus may result.

Paraesthesia

The paraesthesia in the hands with the tingling and numbness of the fingers is due to compression of the median nerve by the mucoprotein in the carpal tunnel.

Facial appearance

The facial appearance in hypothyroidism is fairly typical (CP-18, 47, 48) and the diagnosis is often first suspected from the appearance of the patient. There is puffiness of the face, particularly in the periorbital region. The skin may appear to hang in sacks under the eyes, and is often pale and waxy-looking. A malar flush is present in about half of the patients.

Hair

There may be thinning of the hair (CP-17). It is often coarse and dry. Loss of hair may be the presenting feature and indeed the signs of hypothyroidism may first be seen by the patient's hairdresser.

Skin

The skin is dry and rough and this is often noticed at the elbows (CP-49). The skin may feel cold to the touch. Erythema ab igne follows sitting close to the fire (CP-50).

Pulse

The pulse rate is slow. In hypothyroidism the cardiac output is decreased often to one half of the normal value.

Achilles reflex

The relaxation phase of the Achilles reflex is prolonged. This prolongation can be measured and may be used as a test of thyroid function. In some patients, however, the change is so gross as to be easily detected clinically.

The following is an extract from the first description of hypothyroidism by Sir William Gull in 1873. (The full account can be read in Major, R.H. (ed.) (1938). *Classic Descriptions of Disease.* 2nd Edn. (Springfield, Ill.: C.C. Thomas; (1939) London: Baillière.)

'Miss B became insensible and more and more languid with general increase of bulk. This change went on from year to year, her face altering from oval to round . . . the cheeks tinted a delicate rose-purple, the cellular tissues under the eyes being loose and folded . . . the voice guttural . . . the mind which had previously been active and inquisitive assumed a gentle placid indifference corresponding to the muscular languor.'

Investigations

Even if you are confident clinically about the diagnosis, it is necessary to confirm this by laboratory tests.

Estimation of the serum thyroxine concentration is the most convenient and valuable single test. Estimation of the serum TSH is also of value and the finding of a raised serum TSH (in primary hypothyroidism) confirms the diagnosis. A low serum thyroxine with a normal or low serum TSH suggests a diagnosis of secondary hypothyroidism.

Other tests (though now seldom required) include:

- a low uptake of radioiodine by the thyroid
- a TSH stimulation test
- an abnormal TRH test

Once the diagnosis is made, the patient is committed to lifelong treatment. Moreover, unless there is laboratory evidence, another doctor may be tempted at a later date to stop the therapy.

Hypothyroidism in the child

Hypothyroidism present at birth due to absence or dysgenesis of the thyroid gland, or an inborn error of metabolism, is sometimes known as 'cretinism', but 'congenital hypothyroidism' is a preferable and less obnoxious term (CP-51.1, 51.2). Many countries now have screening programmes measuring serum TSH and sometimes thyroxine levels on heel-prick blood samples taken on the 4th or 5th days of life. These have shown an incidence of congenital hypothyroidism of the order of 1:3000 live births. Early detection and treatment prevent the severe or moderate mental defect and cerebellar damage likely to occur if the condition remains undiagnosed for several months (CP-51.3, 51.4).

During infancy and childhood, early features of hypothyroidism include:

- delayed clearance of neonatal jaundice
- feeding problems
- goitre (for example, if an inborn error of metabolism is present)
- failure of growth associated with severe delay in bone age and sometimes epiphyseal dysgenesis which commonly affects the hips
- constipation
- dry hair and skin

Later signs are (CP-51.5–51.11):

- coarse facial appearance with broad nose, thick lips, protruding tongue
- abdominal distension and umbilical hernia
- myxoedema pads over neck and shoulders
- sluggish behaviour

In older children, as in adults, hypothyroidism may be due to autoimmune thyroiditis and a goitre may be present. A few children have mild hypothyroidism due to insufficient TSH production.

In acquired forms of hypothyroidism, where thyroid function was normal early in life, and in the milder forms of hypothyroidism, where there is some residual thyroid function, there is no mental defect although a slowing down of the patient's activities may be a presenting feature which is reversible with treatment.

Management

Once the diagnosis has been confirmed, the cause for the hypothyroidism should be established as the treatment will depend on this.

An approach to the patient is presented in LD-48.

First, exclude previous destructive therapy to the thyroid.

Second, examine the patient to determine whether a goitre is present. Of the three possible causes, Hashimoto's or autoimmune thyroiditis is by far the commonest cause of hypothyroidism and a goitre.

In the absence of a goitre, consider whether the hypothyroidism is the result of primary thyroid failure or is secondary to pituitary or hypothalamic disease. This secondary hypothyroidism may be suspected clinically and biochemically by the finding of a low plasma thyroxine and a low plasma TSH level.

Clinically, the following features are suggestive of secondary hypothyroidism:

- a history suggesting a pituitary lesion
  A history of headaches may suggest a pituitary tumour. A history of blood loss post-partum may suggest Sheehan's syndrome.
- evidence of deficiency of other trophic hormones may be present
  e.g. amenorrhoea or loss of axillary or pubic hair
- finer and less coarse skin than in primary thyroidism.

In a patient with hypothyroidism secondary to pituitary disease, other hormone deficiencies may exist. If thyroid replacement therapy alone is given, the result may be serious and adrenal crisis and death may follow. In drug-induced hypothyroidism, all that is usually required is discontinuation of the drug.

Treatment

Once the diagnosis of primary hypothyroidism has been made, treatment is in most cases simple.

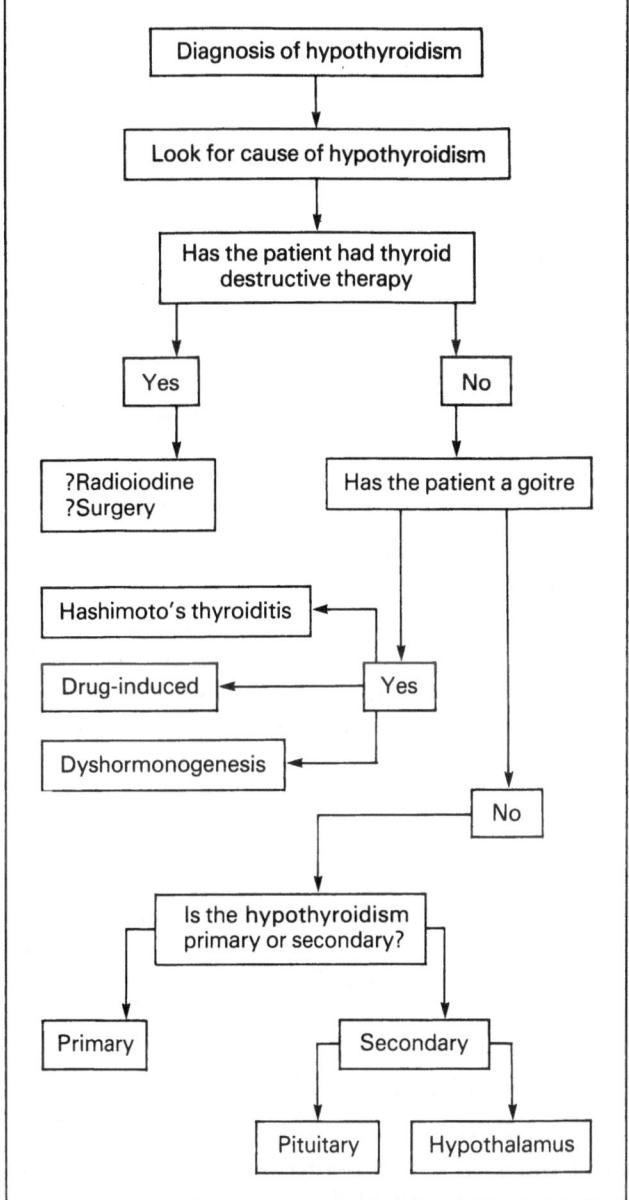

LD-48. *An approach to the patient with hypothyroidism.*

Thyroxine sodium (T4, tetraiodothyronine)

This is the preparation of choice. It can be taken orally and administration once a day is adequate.

Normally start with 0.05 mg/day and increase by 0.05 mg every 2 weeks until the patient is taking a full maintenance dose of 0.15–0.2 mg/day. Continue this for the rest of the patient's life.

In younger patients, and in patients with a recent onset of hypothyroidism, e.g. after thyroidectomy, the starting dose is 0.1–0.15 mg/day.

In older patients or patients with ischaemic heart disease there is a risk of precipitating a myocardial infarction. Start with 0.025 mg/day or 0.025 mg on alternate days and increase only very gradually. Stop at the dose before the one where the chest pain becomes worse.

The full maintenance dose is normally decided on clinical grounds. Although it is usually unnecessary, plasma TSH estimation can help. A high plasma TSH implies that the patient is receiving a subnormal replacement dose.

Tri-iodothyroxine (liothyronine, T3 or Tertroxin)

This drug has the same action as thyroxine but has a more rapid effect. A dose of 0.04 mg is equal to 0.1 mg thyroxine. It is used where a particularly rapid effect is desired as in the treatment of hypothyroid coma. It is NOT the treatment of choice for long-term therapy.

Treatment in children

Treatment of hypothyroidism in children is also with thyroxine starting with a small dose of 0.025 mg daily and working up to the level of dosage required to suppress TSH to normal and maintain a thyroxine level in the normal or high normal range. This is usually of the order of 0.15 mg daily.

With treatment, the features of hypothyroidism disappear and the child begins to grow more rapidly than normal, i.e. catch-up growth occurs. The behaviour becomes more active and gradually school performance improves although, initially, some fall-off may occur as the child adjusts to a more normal tempo of living.

Treatment in infants

Infants also show an improvement in their development unless severe hypothyroidism has occurred *in utero* or during early life before the diagnosis is made, in which case there is permanent mental defect. It is of great importance to treat infants with hypothyroidism as early as possible to preserve whatever brain function has not been permanently damaged. Fortunately, where the hypothyroidism is partial, the brain may escape damage even at this crucial period.

NON-TOXIC GOITRE

The word goitre is derived from Latin 'guttur', meaning throat, and the term 'non-toxic goitre' is used in clinical medicine to refer to any enlargement of the thyroid gland excluding thyrotoxicosis and tumours.

Prevalence

Goitre still represents one of the world's commoner health problems and is one of the commonest endocrine problems facing the doctor.

In some parts of the world it is endemic – over 10% of the population have a goitre, e.g. Kashmir where up to 90% have a goitre. Elsewhere, it is sporadic, e.g. the incidence in east Brazil is reported to be only 1%.

In the UK, the incidence varies in different parts of the country and up to 5% of the population have been found in some areas to have significant thyroid enlargement. Many of these were small goitres and, overall, the figure is much less.

The sex incidence in non-toxic goitre is striking. When everybody in a community has a goitre, the sex ratio must necessarily approach unity. With decreasing prevalence, however, females appear to be progressively more often affected than males and in Britain the ratio is about 10 to 1.

Goitre occurs particularly commonly in women during pregnancy.

This is frequent in some areas of Africa where increase in thyroid size is so common during pregnancy that it is the custom to tie a straw around the girl's neck when she is married. When she develops a goitre, the string breaks and this is taken as an indication of pregnancy.

Causes

The four most important causes of non-toxic goitre are:

- iodine deficiency
- autoimmune thyroiditis (Hashimoto's thyroiditis)
- drugs which inhibit thyroid hormone formation
- dyshormonogenesis (inborn errors of thyroid hormone synthesis)

The basic mechanism in all of these cases is the same. The production of thyroid hormone is impaired and there is a tendency for the gland to secrete insufficient hormone for the body's needs. This leads to an increase in TSH production.

Iodine deficiency

Iodine deficiency is the commonest cause of non-toxic goitre. In the UK the dietary intake is only just sufficient and, although iodized salt is available, it is not routinely used. Avoidance of fish and dairy produce may result in iodine deficiency.

Autoimmune thyroiditis

This disease can occur at any age but is found typically in middle-aged females. The patient, while initially euthyroid, later may become hypothyroid and may present with a goitre or with hypothyroidism (CP-52). The gland is usually diffusely enlarged and firm in consistency. Occasionally the enlargement may be asymmetrical.

The diagnosis is confirmed by finding in the serum thyroid antibodies reacting with:

- thyroglobulin
- thyroid microsomes
- a soluble thyroid component

Histologically, the gland consists of rather small thyroid follicles infiltrated by lymphocytes and plasma cells. Sometimes there is germinal centre formation and an increase in the fibrous tissue stroma (CP-53).

Drug-induced goitre

Goitrogenic drugs include:

- iodide
- lithium carbonate
- phenylbutazone
- para-amino salicyclic acid

Patients first develop a goitre and later, if the drug is continued, hypothyroidism. Iodide is the most important cause and is found in a wide variety of proprietary preparations, particularly cough mixture and tonics.

A survey of patients attending an asthma clinic revealed a number of iodide-produced goitres. This cause of goitre can be missed unless a careful drug history is taken.

Naturally-occurring goitrogens in foodstuffs may play a part in goitre production in some parts of the world. In UK, however, this has not been proven as a significant factor.

Dyshormonogenesis

Here there is a deficiency in the synthesis of thyroid hormone due to a complete or partial enzyme defect. It may be inherited as a Mendelian recessive. At least five separate enzyme defects have been defined. All are rare. Features suggesting dyshormonogenesis as a possible cause in a patient with a goitre are:

- family history of goitre (CP-54)
- presence of a goitre in childhood

The enzyme defect may be a partial one and the patient's thyroid may have compensated for some time in response to TSH.

In Pendred's syndrome there is defective binding of iodine to tyrosine together with nerve deafness.

Radioiodine tests can help confirm the diagnosis and differentiate the different types of enzyme defects.

Management

The treatment of non-toxic goitre depends on the cause.

In iodine deficiency, iodized salt should be taken or the dietary iodine intake increased. In addition, thyroxine may be administered for 18 months to reduce TSH production.

In autoimmune thyroiditis and dyshormonogenesis the patient should be instructed to take thyroxine for life. In neither condition is the cause for the goitre likely to disappear but the goitre often decreases in size.

If the goitre is due to drug ingestion, then the drug should be stopped. Only if this is not possible is it necessary to prescribe thyroxine.

Surgery is required rarely and only:

- for cosmetic reasons
- where there are pressure effects from the goitre
- where there is doubt about the diagnosis and a carcinoma is suspected

SUBACUTE THYROIDITIS

Causes

This painful enlargement of the thyroid is usually attributed to a viral infection. It is sometimes called viral thyroiditis, granulomatous thyroiditis or de Quervain's thyroiditis. It is an uncommon cause of thyroid enlargement but is commonest in the 20–40 age group and has a female/male ratio of 4/1.

Clinical features

It can usually be easily recognized as it has three characteristic features (CP-55):

- preceding general malaise, pyrexia or upper respiratory infection

- usually a history of short duration

- pain and tenderness
  Pain from the thyroid radiates from the neck to the ear.

The disease varies in its severity. Transient hyperthyroidism can result from release of thyroglobulin and excessive amounts of thyroid hormone.

Investigations

The diagnosis is essentially clinical but:

- the ESR is markedly elevated

- serum thyroxine may be raised in the acute stage
  In the chronic phase of the disease it may be low.

- thyroid autoantibodies may be present
  They are rarely of high titre and disappear as the disease settles.

- viral antibodies may be helpful retrospectively in confirming the diagnosis

- radioiodine studies show that the iodine uptake is low during the acute phase but is high during recovery

Management

It is usually self-limiting over a period of a few weeks to a few months.

Mild analgesics often suffice to control symptoms. When these are severe, response to oral steroids is dramatic but they need to be tapered down over several weeks.

HYPERTHYROIDISM

Hyperthyroidism is a clinical condition resulting from overactivity of the thyroid gland and an excess of circulating thyroid hormone. Thyrotoxicosis is used as a synonym for hyperthyroidism; Graves' disease is often used as an eponym for one form.

In most patients the circulating levels of both thyroxine and tri-iodothyronine are increased. In a few patients, however, only an excess of tri-iodothyronine is secreted by the thyroid. This condition is called T3-toxicosis.

Few diseases have had more eponyms than hyperthyroidism. Although it was first described by Parry, an Edinburgh graduate, in 1786, the name of Graves is often attached to the disease following his description of three cases in 1835. On the continent, the disease is associated with Basedow who gave a description of it in 1840.

Causes

The term hyperthyroidism refers to a clinical state and is not itself a disease entity.

The following pathologies may be associated with hyperthyroidism:

- diffusely enlarged thyroid gland producing too much hormone (Graves' disease)
  This is the most common cause of hyperthyroidism. The plasma TSH is low and the gland is probably stimulated by an IgG autoantibody.

- a multinodular goitre producing excess hormone (Plummer's disease)
  This may be the result of the development of hyperthyroidism in a multinodular goitre.

- an autonomously functioning solitary thyroid nodule (toxic adenoma)
  The remainder of the thyroid gland is usually suppressed.

- excess TSH produced by a tumour in the pituitary or elsewhere.
  This is a rare cause of hyperthyroidism.

- ingestion of large doses of thyroid hormone (thyrotoxicosis factitia)
  Some patients surreptitiously take large quantities of thyroid hormone.

An autonomously functioning nodule can be suspected clinically when features of hyperthyroidism are found in a patient with a single thyroid nodule (CP-56, 57).

Excess TSH production is suggested by other features of the underlying pathology or by the finding of a high plasma TSH.

A careful history may reveal a history of thyroxine ingestion and, where suspected, it can be proved by finding a low uptake of radioiodine by the thyroid gland.

Graves' disease

Causes of the hyperthyroidism

Four aetiological factors have been identified:

- circulating immunoglobulins
- hereditary factors
  A higher incidence of goitre, hyperthyroidism or hypothyroidism is found in the relatives of patients.
- stress
  The precise relationship is not clear.
- iodide ingestion
  A rise in iodide intake increases the incidence of hyperthyroidism.

A high plasma TSH level was initially thought to be responsible for Graves' disease. It has since been demonstrated, however, that in this condition the pituitary gland is suppressed by the high circulating thyroid hormone level and that the plasma TSH level is, if anything, low.

In 1956 Adams and Purves observed that in many patients with hyperthyroidism there was a circulating IgG which became known as long-acting thyroid stimulator (LATS). More recently, other thyroid stimulating IgG molecules have been identified in patients with hyperthyroidism – thyroid stimulating immunoglobulins (TSI or TSAb).

A relationship between stress and hyperthyroidism has often been claimed but studies of this have produced conflicting results.

One of the first descriptions of hyperthyroidism was in a patient shortly after his unattended wheelchair proceeded out of control down a steep hill. More recently, an increased incidence of hyperthyroidism was found in the wives of members of armed forces serving in Vietnam (but not in the soldiers and airmen themselves). Evidence for iodide ingestion as a factor includes the finding of an 'epidemic' of hyperthyroidism in Tasmania when iodide was added to the bread.

Prevalence

Hyperthyroid Graves' disease is one of the commonest endocrine disorders. In Britain, the average general practitioner will have three hyperthyroid patients for every 2500 patients in his practice. It is found in both sexes and at all ages. However, it occurs most commonly in women between the ages of 30 and 50 and is six times more prevalent in women than in men. Hyperthyroidism occurs rarely in the newborn when it is usually due to transplacental transmission of thyroid stimulating antibody. This is usually only a transient phenomenon.

Clinical features

The clinical features of hyperthyroidism may be obvious. However, the picture is not always a clearcut one and patients may present with either:

- a single feature, e.g. atrial fibrillation
- an atypical presentation, e.g. weight gain instead of weight loss

Thyrotoxicosis may present in many different ways and, unless the possibility of hyperthyroidism is considered by the doctor, the diagnosis will be missed.

The signs and symptoms of thyrotoxicosis may be associated with:

- a diffuse increase in the size of the thyroid gland
- the effects on the body of the increased circulating hormone
- other features associated with Graves' disease and exophthalmos

Basedow, in one of the earliest descriptions of a patient with hyperthyroidism, described the story of a Madame F. about whom the rumour was spread: 'That this patient was crazy and was soon going to be taken to an asylum and in fact she had an unfriendly attitude towards the physician: she never had, however, and that I can assure you, any insane ideas: she never showed any abnormal signs and, if her astonishing carelessness over her truly sad condition seemed to be the result of her phlegmatic temperament, to the fastness of her speech, the uncertain holding of her body and her hands; the tendency to go about naked or very lightly dressed were undoubtedly symptoms of her heart disease and glandular disorder'.

Goitre

The patient may present with a goitre. Examples are shown in CP-58 and CP-60, 61. The gland is usually firm on palpation and the pyramidal lobe may be enlarged. A goitre is present in about 90% of patients with hyperthyroidism.

The gland is usually vascular and a bruit is often heard over it.

Effects of excess thyroid hormone

Thyroid hormones increase the body's metabolism. The resulting signs and symptoms are many. The most important symptoms diagnostically are:

- heat intolerance and excessive sweating
- nervousness and irritability
- weight loss and normal or increased appetite
  Occasionally the patient's appetite may overcompensate and increase in weight may result.

Other symptoms include:

- tiredness
- diarrhoea
- absent or scanty periods
- palpitations
- dyspnoea on exertion

The most important signs are:

- tachycardia or atrial fibrillation
- hot moist hands and skin
- finger tremor and hyperkinetic movements

Systolic hypertension may be present.

The patient may wear fewer clothes and may avoid warm situations. Arguments over bedclothes and electric blankets may result in matrimonial discord. On examination, the hands may be hot and moist.

Another common presenting complaint is nervousness and excitability. A common complaint among young mothers is that they find themselves increasingly irritable with their children. Examination of the patient may reveal a finger tremor and movements may be hyperkinetic.

Eye signs

The other most important feature of Graves's disease is involvement of the eye. Exophthalmos and lid retraction are most commonly found and there may be lid lag (CP-25, 27).

Ophthalmoplegia may also be present and the patient may complain of diplopia usually when looking in an upward and outward direction (CP-26).

The pathogenesis of the eye signs in hyperthyroidism is even less well understood than the hyperthyroidism.

Eye signs may develop for the first time when the patient with hyperthyroidism is treated or may be present in a patient in whom there is no history of thyroid disease. The eye signs may become serious and threaten the patient's vision (CP-28).

In a published account in 1835 Graves gives a good description of one sign: 'It was now observed that the eyes assumed an irregular appearance for the eye-backs were apparently enlarged. When the eyes were open the white sclerotic could be seen to a break of several lines all round the cornea'.

Other features of Graves' disease

These include:

- pretibial myxoedema (CP-11, 12)
- vitiligo (CP-59)
- muscle wasting (CP-23)

- thyroid acropachy (CP-14.1)
- onycholysis (CP-14)

Pretibial myxoedema is a less common feature of Graves' disease. The skin usually appears brown and indurated. On palpation it feels thickened. There is no pitting oedema.

The muscle wasting in hyperthyroidism usually affects the large proximal muscle groups. The patient may complain of weakness in the legs on climbing stairs.

Acropachy is an uncommon sign of Graves' disease which occurs particularly in association with pretibial myxoedema. Clinically, the clubbing may resemble hypertrophic pulmonary osteoarthropathy.

A feature of hyperthyroidism which has no therapeutic significance is onycholysis. In this condition the nail is separated from its bed leaving a greater margin of white at the tip of the nail. The patient may complain he has difficulty keeping his nails clean.

Thyroid crisis

Fortunately, thyroid crisis is now rarely seen. This is the result of earlier diagnosis and treatment. The patient develops:

- hyperthermia
- marked tachycardia
- cardiac failure
- severe mental and physical exhaustion

Investigations

All cases in which the diagnosis of hyperthyroidism is suspected should have it confirmed by:

- serum thyroxine
- serum tri-iodothyronine

Determination of the serum thyroxine level is the simplest and most valuable test. If combined with a measure of the degree of saturation of the binding sites in the serum, the free thyroxine index can be calculated. This is useful in situations where there are abnormalities in thyroid hormone-binding proteins.

Radioiodine uptake measurements are usually diagnostic but are usually unnecessary if the diagnosis is confirmed by a high thyroxine level.

The plasma tri-iodothyronine level is usually elevated and in 'T3-toxicosis' only the T3 is elevated.

If the diagnosis is in doubt after routine tests have been undertaken, the following will give useful additional information:

- tri-iodothyronine suppression test
- TRH test

Management

The aim of treatment is to decrease the amount of hormone produced by the thyroid. Three approaches have been adopted:

- drugs which decrease the synthesis of thyroid hormone
- surgery which reduces the amount of functioning thyroid tissue
- radioactive iodine
  This is taken up by the thyroid cells and, by causing cell damage, results in decreased activity.

While in some situations there would be broad agreement among physicians as to the best form of treatment, in others opinions as to the most appropriate treatment differ.

Antithyroid drugs

Experimentally it was found that administration of thiourea resulted in a goitre. This led to the discovery of a number of drugs which inhibited thyroid hormone synthesis. These antithyroid drugs act by blocking incorporation of iodine into the tyrosyl residues to form monoiodotyrosine and di-iodotyrosine (LD-49).

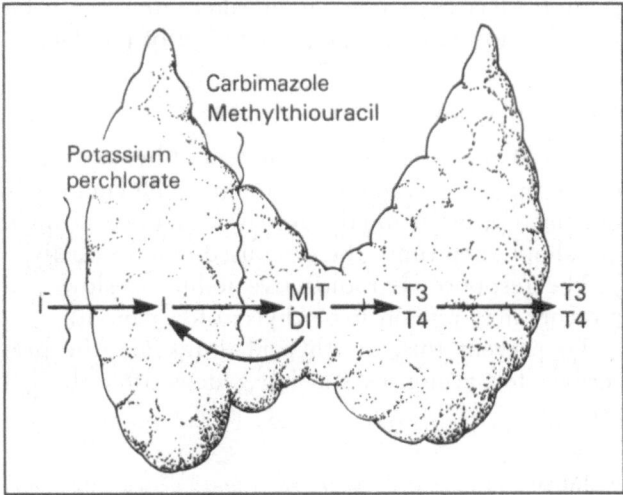

LD-49. *Blocking action of antithyroid drugs.*

Antithyroid drugs can be used:

- as the sole form of treatment for hyper-thyroidism
- to prepare the patient for thyroidectomy
- as an adjunct to radioactive iodine therapy

Initially the drugs are given in a high dose and the symptoms of hyperthyroidism take from 2 to 6 weeks to come under control. Thereafter, the dose is reduced to a maintenance one and continued for 12–18 months. Too high dosage results in hypothyroidism.

Carbimazole (Neo-mercazole)

This is the most widely used antithyroid drug in the UK. Methimazole is more commonly used in some countries. The initial dose is 30–60 mg/day in divided doses. The maintenance dose is 10–20 mg/day. The commonest side effect is skin rashes. Agranulocytosis is a rare but significant side effect.

Propylthiouracil

This is used where a patient has side-effects to carbimazole as cross-sensitivity between them is uncommon. The dose of propylthiouracil is ten times the carbimazole dose. Methylthiouracil is an alternative. These drugs are used less frequently in the UK but propylthiouracil is most popular in North America.

Iodide

This may be used for a limited period as an anti-thyroid drug. It has a more rapid action than the above drugs as, in addition to blocking synthesis of thyroid hormone, it prevents the release of hormone already formed and stored within the gland. It exerts an effect within 24 hours of its administration. Iodide is not suitable for long-term therapy as the thyroid escapes from the effect of iodide inhibition after about 2 weeks.

Adrenergic blockers

These include propranolol and may be used to control the clinical features of hyperthyroidism. They do not significantly affect the amount of circulating thyroid hormone but reduce the symptoms and signs of the disease.

Tri-iodothyronine and antithyroid drugs

Some physicians recommend that tri-iodothyronine

be given along with carbimazole in the treatment of hyperthyroidism. This has the advantage that it is not necessary to adjust the dose of the antithyroid drug so carefully. It also allows the clinician to assess the patient's thyroid suppressibility.

Relapse

There is some evidence that the following two categories of patients are more likely to relapse when drugs are stopped:

- patients with a high circulating level of thyroid-stimulating globulins
- patients whose thyroid uptake of radioactive iodine is not suppressed by tri-iodothyronine

All patients can be controlled by drugs. Failure to control a patient suggests:

- inadequate dosage prescribed
- irregular intake by patient
- initial diagnosis is incorrect

Complications

The problems associated with antithyroid drug therapy can be summarized:

- the patient must continue the treatment for many months
- rarely side-effects arise
  These include skin rashes and agranulocytosis.
- when drugs are stopped the hyperthyroidism may recur.

Surgery

The second approach to the treatment of hypothyroidism is surgery.

In skilled hands, thyroidectomy is now a relatively safe procedure. This has not always been so and, in the early days, the mortality was appreciable. A factor in this improvement has been control of the patient with antithyroid drugs prior to the operation (CP-60, 61).

Relapse

Relapse of hyperthyroidism following thyroidectomy commonly occurs years after treatment as does an increasing incidence of hypothyroidism. Long-term follow-up is, therefore, important.

Complications

The complications of thyroidectomy are:

- recurrence of hyperthyroidism (5–10%)
- onset of early hypothyroidism (5–10%)
- vocal cord damage (rare)
- hypoparathyroidism (rare)

Indications

the indications for surgery are:

- some contraindication to using ^{131}I therapy
- side-effects to drugs or unwillingness to take drugs
- a large goitre
  For cosmetic reasons or because of pressure effects surgery may be required.
- relapse after a full course of antithyroid drugs
- social factors

Thyroidectomy has obvious advantages as a method of treating patients with hyperthyroidism and these must be weighed against its disadvantages.

The final decision as to the choice of treatment will often be determined by the views or circumstances of the patient who, for domestic or social reasons, may prefer an operation to other forms of treatment. A girl may prefer a barely noticeable thyroidectomy scar to the presence of a goitre.

Patients should be prepared for thyroidectomy with antithyroid drugs so that they are euthyroid at the time of the operation. Potassium iodide 60 mg thrice daily is usually added for the 14 days before the operation. This ensures that the patient is euthyroid and also reduces the vascularity of the gland. Propranolol has been used as an alternative to carbimazole and iodide in the preparation of patients.

Radioactive iodine

The third approach to the treatment of hyperthyroidism is the use of radioactive iodine. The commonly used isotope is ^{131}I and the dose is normally 6–20 mCi, i.e. about 1000 times the dose used diagnostically.

The response to radioactive iodine is slow and patients may remain hyperthyroid for 3 months.

To prevent this, antithyroid drugs may be prescribed for 3 months starting 3 days after the ^{131}I therapy.

Relapse

If the first dose is insufficient, patients will need a

second dose. Recurrences of hyperthyroidism once the patient is euthyroid are, however, rare.

Complications

The only common complication of radioiodine therapy is hypothyroidism. The proportion of patients who become hypothyroid is high – as many as 50% at 8 years – and the proportion continues to rise by about 2–3% per year thereafter. It is important, therefore, to follow patients up wherever this is possible.

Indications

The indications for ^{131}I therapy are:

- patients over 40 years of age (20 years in some centres)
- patients with a limited life expectancy
- relapse after thyroidectomy
- relapse after a course of antithyroid drugs

Radioactive iodine is usually used as the first approach to therapy in patients over the age of 40 years. It is sometimes restricted to this age group, not because of the risk of cancer, but because the chance of developing hypothyroidism increases in the younger patient.

In patients with a coexisting medical condition, and with a reduced life expectancy, radioactive iodine is perhaps the simplest approach.

Radioactive iodine may be used in patients who relapse after a thyroidectomy or a course of drugs. Second thyroidectomies are not recommended.

TUMOURS OF THE THYROID GLAND

Tumours of the thyroid gland may be:

- simple
  adenoma
- malignant
  primary – carcinoma, lymphoma
  secondary – from breasts, lungs and kidneys

Prevalence

While thyroid adenomas are not uncommon, malig-nant tumours of the thyroid are rare and are responsible for fewer than 400 deaths per year in England and Wales. This contrasts with the mortality for malignant tumours of the lung which is about 70 times as great.

In oncology departments thyroid tumours provide a focus of interest which is out of proportion to their frequency. There is a great variety of natural history, malignancy and response to treatment.

Simple tumours

Adenomas or cysts may be found in the thyroid. Calcification may be present in the cysts and haemorrhage may occur in the adenoma. Most adenomas are follicular in type.

Malignant tumours

Four types of thyroid cancer occur:

- papillary ⎫
- follicular ⎭ – differentiated
- anaplastic or undifferentiated
- medullary – arising from parafollicular C-cells

Some features of the four types are shown in Table 50.

Table 50. Features of thyroid carcinomas

Type	Frequency (%)	Spread	10-year survival (%)
Papillary	60	Lymph nodes	80
Follicular	25	Blood stream	60
Anaplastic	10	Blood stream and local invasion	1
Medullary	5	Local	50

Differentiated carcinomas

Papillary and follicular carcinomas are the commonest and mixed papillary/follicular carcinomas may be found. Medullary carcinoma is rare.

The following clinical features should warn of the possibility that a lump in the neck is due to a thyroid cancer (CP-62, 63):

- a history of a recent increase in size

- a history of pain in the thyroid
- hard consistency of the gland
- a single nodule (especially in a male)
- cervical lymphadenopathy (CP-64)
- fixation of the gland
  This is usually a late sign

Cancer should always be considered in a patient with a single thyroid nodule. All such patients should be scanned following administration of an isotope. Tumours usually appear as a cold nodule. Hot nodules are almost never due to a thyroid cancer. Cysts and adenomas may also appear as cold nodules on a scan and indeed only about 5% of cold nodules turn out to be due to cancer.

Medullary carcinoma has only been recognized in the past decade. Its features may include:

- a family history (some cases only)
- diarrhoea
- associated phaeochromocytoma
- associated neuromas
- flushing attacks

Since these tumours invariably produce calcitonin, measurement of the level of this hormone in the blood provides us with a unique tumour marker.

Management

The four approaches are:

- thyroidectomy
- external radiation
- radioiodine therapy
- thyroxine

The precise regime will depend on the type of cancer and all patients should be treated in a centre with experience in this field. A total thyroidectomy by an experienced surgeon should be considered in all cases.

In anaplastic cancer, surgery is of little help except to confirm the diagnosis. External radiation is indicated for this tumour and lymphomas.

Radioiodine is given following total thyroidectomy in differentiated tumours.

Thyroid hormones should be given in all patients to suppress TSH as some tumours are hormone dependent.

PARATHYROID AND CALCIUM METABOLISM

The disorders which will be considered are:

- primary hyperparathyroidism
- secondary hyperparathyroidism
- tertiary hyperparathyroidism
- pseudohyperparathyroidism
- hypoparathyroidism
- pseudohypoparathyroidism
- renal stones
- osteomalacia and rickets
- osteoporosis
- osteogenesis imperfecta
- Paget's disease of bone

PRIMARY HYPERPARATHYROIDISM

Causes

Excessive secretion of parathyroid hormone (PTH) most commonly results from a parathyroid adenoma (77% of cases) but may be caused by hyperplasia of all four parathyroids (19%) or by parathyroid carcinoma (4%).

The characteristic abnormality is an increase in the plasma calcium level. Three factors contribute to this rise (LD-50). The most important are the increase in absorption of calcium by the intestine and the increase in calcium reabsorption in the renal tubules. In the most severe cases, increased calcium reabsorption from bone, involving both osteoclasts and osteocytes, is also important.

Clinical features

History

The presenting symptoms of hyperparathyroidism are summarized in Table 51.

Examination

The tumour causing hyperparathyroidism is gener-

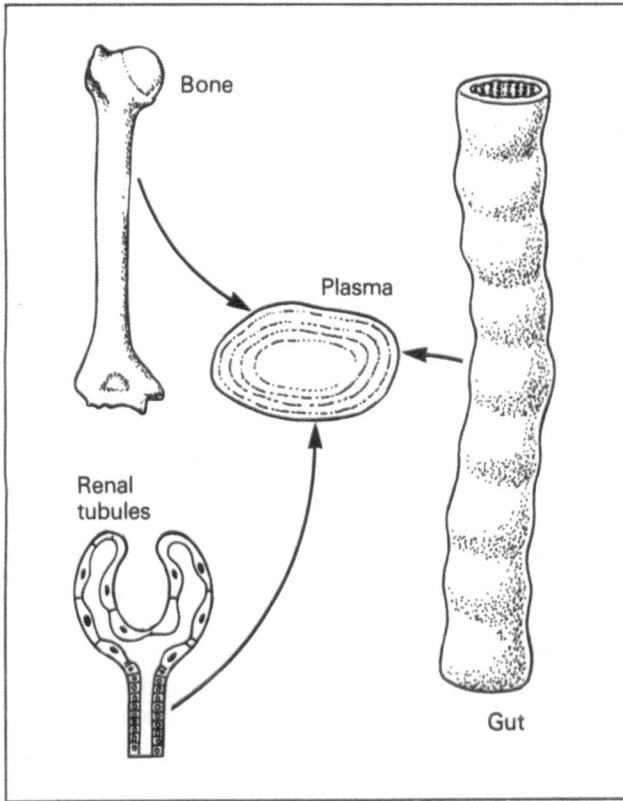

LD-50 Factors contributing to the rise in plasma calcium levels in hyperparathyroidism.

Table 51. Presenting symptoms of hyperparathyroidism

Symptoms	Incidence (%)	Comment
Renal stones	40	
Abdominal pain	20	Usually non-specific but sometimes caused by peptic ulcer or pancreatitis
Bone pain fractures	5	
Non-specific	15	These result from the hypercalcaemia and include anorexia and psychiatric disturbances
No symptoms	20	High plasma calcium is an incidental finding

ally small and deep-seated. It is scarcely ever palpable. A few patients with long-standing hypercalcaemia have calcification of the cornea (CP-65). This may be distinguished from an arcus by its gritty appearance, by the lack of a clear band between it and the scleral margin and by the fact that it is generally maximal medially and laterally.

Investigations

Laboratory

Almost all patients have a raised plasma calcium. The plasma phosphate is low in about half of the cases because there is excessive phosphate loss in the urine. The plasma alkaline phosphatase level is increased in about 10% of cases – generally those cases who also have radiological abnormalities.

Estimation of plasma PTH levels may be useful in diagnosis but, at present, the methods available cannot always distinguish between normal subjects and patients with hyperparathyroidism.

The urinary excretion of calcium is often, but not always, raised in hyperparathyroidism. For a given plasma calcium level the 24 hour urinary calcium is lower for hyperparathyroidism than for a patient whose hypercalcaemia has another cause.

Radiology

This is normal in about 90% of cases of hyperparathyroidism. In 10%, evidence of bone resorption may be seen particularly in the form of subperiosteal erosions in the phalanges (X-10).

Other areas where abnormalities may be seen include the:

- skull ('pepper-pot skull') (X-11)
- distal parts of the clavicles
- lamina dura (the cortical bone of the tooth sockets)

There is little point in looking for these changes if the hands are normal.

In a few cases bone cysts may develop (X-12).

Management

Most patients with hyperparathyroidism are treated surgically. The search for the abnormal parathyroid tissue can be difficult; occasionally an adenoma is not found in the neck but in the mediastinum.

In some patients surgery is contraindicated because of coincidental disease and in others the disorder is very mild itself. In an increasing number of such cases conservative management is providing a reasonable alternative to surgery. Patients with symptoms can be treated with oral phosphate to reduce the plasma calcium and those without symptoms only need regular follow-up.

In phosphate therapy the dose of phosphate is

adjusted to bring the plasma calcium into the range 2.6–2.8 mmol/l (10.4–11.2 mg/dl).

Patients treated surgically also need regular follow up.

Acute hyperparathyroidism

Acute hyperparathyroidism ('parathyroid crisis') is a rare but dangerous disorder. The plasma calcium rises rapidly, the patient becomes:

- dehydrated
- confused
- finally comatose

The disorder has a high mortality unless treated urgently by rehydration and either surgery or intravenous phosphate therapy.

SECONDARY HYPERPARATHYROIDISM

Secondary hyperparathyroidism is the parathyroid overactivity which occurs in response to the longstanding hypocalcaemia of, for example, renal failure or osteomalacia. There may be radiological changes identical to those of hyperparathyroidism (but the plasma calcium is, of course, low) (X-13).

TERTIARY HYPERPARATHYROIDISM

This is a term used to describe a patient with secondary hyperparathyroidism in whom an autonomous adenoma has developed within a hyperplastic gland. In this rare condition, the plasma calcium is high.

PSEUDOHYPERPARATHYROIDISM

This is a term used to describe a patient with hypercalcaemia (and often hypophosphataemia) due to the ectopic production of PTH or a PTH-like substance by a tumour. The commonest primary tumours to behave in this way are those of the:

- bronchus
- kidney

HYPOPARATHYROIDISM

Hypoparathyroidism is an uncommon disorder.

Causes

The commonest cause is damage to the glands or their blood supply during surgery to the neck. At one time, this was a not uncommon sequel to partial thyroidectomy for hyperthyroidism but surgeons now take special care to avoid parathyroid damage. Hypoparathyroidism remains a frequent complication of the extensive neck surgery needed in patients with carcinoma of the thyroid or larynx.

Rarer causes of hypoparathyroidism are an autoimmune disorder and congenital absence of parathyroid tissue.

Clinical features

Most symptoms are those of hypocalcaemia. This may be dramatic as in the onset of tetany (CP-66) or 'carpopedal spasm'.

Tetany results from a muscular spasm provoked by a low level of ionized calcium in the plasma. This can be caused either by hypocalcaemia (from, for example, hypoparathyroidism or vitamin D deficiency) or by alkalosis due to overbreathing or vomiting.

Other symptoms of hypoparathyroidism and hypocalcaemia from other causes include paraesthesiae and psychiatric disturbances such as depression and irritability.

A few patients with hypoparathyroidism have cataracts, loss of hair, abnormalities in the finger nails (CP-13) or hypoplasia of the enamel of the teeth.

In severe cases, tetany occurs spontaneously and the hand takes up the position shown in CP-66. The feet may also be affected.

In Trousseau's test, frank tetany may be provoked in patients with hypocalcaemia by applying a sphygmomanometer cuff to the upper arm and inflating it above the systolic pressure for 2 minutes.

In Chvostek's test, the facial nerve is tapped in front of the ear (LD-51). In patients with latent tetany, the corner of the mouth twitches upwards.

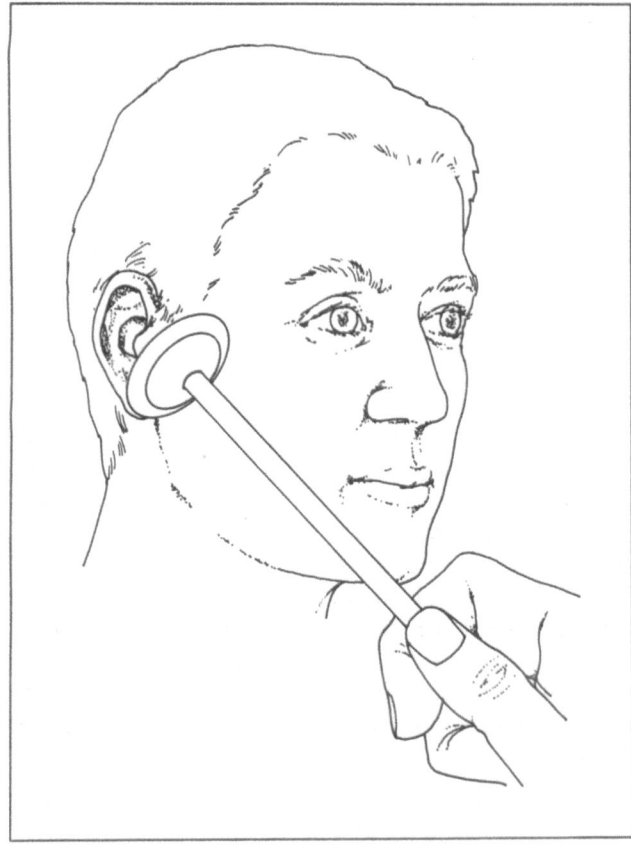

LD-51. *Chvostek's test.*

Investigations

Laboratory

Since PTH production is impaired, the plasma calcium level is reduced. The plasma inorganic phosphate is generally raised while the plasma alkaline phosphatase is normal. Methods for PTH assay currently available are unhelpful in the diagnosis of hypoparathyroidism.

In a patient with hypocalcaemia, the simple plasma assays of inorganic phosphate and alkaline phosphatase are of great value in diagnosis since, in vitamin D deficiency, plasma inorganic phosphate is often low and plasma alkaline phosphatase is raised.

Radiology

There are no characteristic radiological abnormalities in the bones in hypoparathyroidism. Patients with longstanding hypoparathyroidism may develop calcification of the basal ganglia (X-14).

Management

Treatment is usually with vitamin D in large doses by mouth. Most patients require between 0.5 mg and 2 mg daily and the dose should be adjusted so that the plasma calcium is in the range 2.0–2.2 mmol/l (8.0–8.8 mg/dl). The patient must be followed up indefinitely since unexplained variations in sensitivity to vitamin D may occur spontaneously and vitamin D intoxication is a dangerous disorder.

In patients in whom rapid relief of symptoms is important, treatment is with 1α-hydroxycholecalciferol ('one-alpha') with which normocalcaemia may be achieved within a few days. Patients receiving 'one-alpha' need very frequent follow-up to prevent intoxication so that vitamin D is generally preferable for long-term control.

PSEUDOHYPOPARATHYROIDISM

Pseudohypoparathyroidism is a rare inherited disorder in which the chemical abnormalities of hypoparathyroidism (low calcium and high phosphate) are not caused by lack of PTH. The cause appears to be an insensitivity by the tissues (notably the kidney) to circulating PTH probably due to a defect in adenyl cyclase. These patients may have curious skeletal abnormalities, notably short metacarpals (CP-67).

RENAL STONES

Causes

Calcium and phosphate stones

These stones account for 90% of all renal stones. They are important as a presenting feature of hyperparathyroidism or other disorders characterized by hypercalcaemia. However, only about 10% of patients who form recurrent stones have hypercalcaemia. Another 20% have a normal plasma calcium but an excessively high urinary excretion of calcium which may be caused by:

- immobilization
- malignant disease
- vitamin D excess
- Cushing's syndrome
- idiopathic hypercalciuria

Idiopathic hypercalciuria is a poorly understood disorder in which calcium absorption from the intestine is excessive. Other factors which contribute to the formation of stones containing calcium and phosphate include:

- dehydration
- urinary infection
- urinary tract abnormalities

Uric acid stones

Uric acid stones – 8% of all stones – may form in patients with an excessive urinary excretion of uric acid due to:

- gout
- myeloproliferative diseases
- uricosuric drugs

In 80% of cases none of these are present.

These patients form stones because their urine has a generally very low pH and uric acid is much less soluble in acid urine than in neutral or alkaline urine.

Cystine stones

Cystine stones – 2% of all urinary stones – result from cystinuria, an inborn error affecting the renal tubular reabsorption of cystine and three other amino acids. The urine has a high urinary cystine content.

Investigations

The first step is to check the plasma calcium and measure the 24 hour urinary excretion of calcium on an ambulant patient. At the same time, plasma and urinary levels of uric acid can be measured and qualitative tests for cystine content of urine can be carried out. If a stone becomes available, it should be analysed.

Management

Appropriate treatment should be given for any urinary tract infection, anatomical abnormality or disorder of calcium metabolism.

Patients with excessive uric acid secretion should receive allopurinol or similar therapy.

Patients with recurrent uric acid stones should be given appropriate alkali to render the urine alkaline and prevent further stone formation.

All patients need to be encouraged to have a high intake of fluids particularly in the evening and at night.

OSTEOMALACIA AND RICKETS

Causes

Osteomalacia (in adults) and rickets (in children) are the names given to the bone disorder of vitamin D deficiency. Rickets differs from osteomalacia mainly in that the epiphyses show the most striking abnormalities.

Vitamin D deficiency is, in the world as a whole, a very common disease. It occurs readily whenever patients have both an inadequate diet and limited exposure to sunlight. In Britain this combination occurs particularly in elderly housebound women and Asian immigrants of all ages. In addition, patients with intestinal disorders may fail to absorb their dietary vitamin D. The commonest causes are:

- coeliac disease
- partial gastrectomy

The biological appearances of normal and osteomalacic bone are shown in CP-68, 69.

Clinical features

History

As in hypocalcaemia due to other causes, tetany does occur in some cases, particularly Asian immigrants and in patients with coeliac disease.

Most adults and adolescents with vitamin D deficiency complain of bone pain and often muscular weakness as well.

In very young infants with rickets tetany may take the form of laryngeal stridor.

Examination

Physical signs of rickets may include knock-knees (CP-70) or bow-legs. There may be obvious enlargement of the wrists or a 'rickety rosary' caused by expansion of the costochondral junctions (CP-71).

Most adults with osteomalacia have bone tenderness which can be elicited, for example, by springing the ribs or squeezing the radius and ulna together.

Investigations

Laboratory

The plasma calcium is low or in the low–normal range; the plasma inorganic phosphate level is low or normal. The plasma alkaline phosphatase is raised in almost all adults with osteomalacia.

In children with rickets the plasma alkaline phosphatase level is greatly raised but, since normal growing children have a higher alkaline phosphatase than adults, this may not be easy to interpret.

Radiology

Radiology is not always helpful in osteomalacia since less than 50% of cases show any radiological abnormality. The characteristic abnormality is a pseudofracture or 'Looser's zone' (X-15) which is most commonly seen in the pelvis, ribs, upper femur or scapula.

The characteristic radiological abnormality in rickets is widening of the epiphysis which also appears ragged (X-17). The abnormalities may be absent in a rachitic child who is not growing. Occasionally, pseudofractures are seen in childhood rickets.

In general, clinical chemistry is more helpful than radiology for the diagnosis of the adult with vitamin D deficiency while the reverse is true of childhood rickets.

Management

In simple nutritional vitamin D deficiency, the disorder is corrected by giving vitamin D by mouth in a dose of 2000 units (50 μg) daily.

Patients with vitamin D deficiency due to malabsorption need vitamin D by injection. The osteomalacia of coeliac disease improves rapidly when the underlying disorder is treated with gluten-free diet.

Patients who are likely to default from follow-up should be given a single, very large dose (600 000 units or 15 mg intramuscularly). This vitamin D is stored in the tissues and lasts for 6 months or more.

Patients who do not respond to the doses of vitamin D described should be reinvestigated for rare rickets-like disorders such as vitamin D dependent rickets and familial hypophosphataemic rickets.

OSTEOPOROSIS

Osteoporosis is defined as any condition in which the amount of bone is reduced but the bone itself is normal in composition. In this respect, osteoporosis may be contrasted with other disorders of bone.

Osteoporosis is a very common cause of symptoms in the elderly.

Causes

In the great majority of cases osteoporosis has no identifiable cause but results from progressive loss of bone which is a feature of normal ageing in women and, to a lesser extent, men (LD-52).

In a small number of cases osteoporosis has a cause such as:

- immobilization
- Cushing's syndrome
- steroid therapy
- hypogonadism

The bone loss in women is known as 'involutional osteoporosis' or 'postmenopausal osteoporosis'. All women are affected but only a minority have symptoms.

Clinical features

Patients with thin bones may have no symptoms at all or may present as a result of backache due to crush

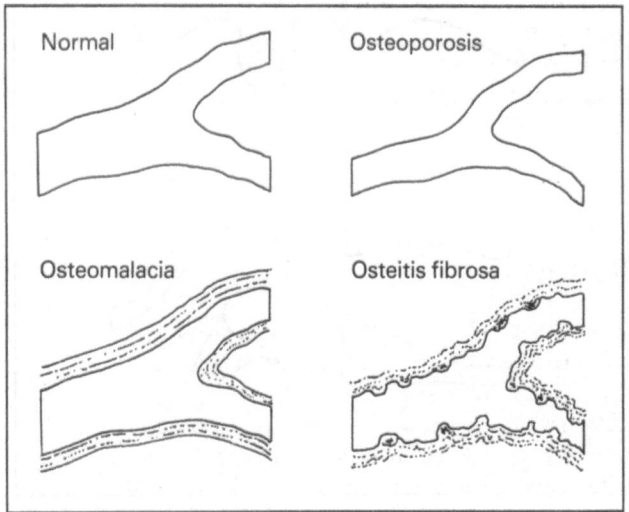

LD-52. Osteoporosis.

fractures of the vertebrae, or other fractures. The commonest fractures are:

- fractures of the femoral neck
- Colles fracture at the wrist

While fractures with minimal trauma may be the result of osteoporosis, it is important to be aware of other causes such as:

- secondary neoplasm in the bone
- Paget's disease
- osteomalacia
- osteogenesis imperfecta
- hyperthyroidism
- fibrous dysplasia of bone

The backache of osteoporosis is generally episodic. Each episode of pain represents a crush fracture and the symptoms generally improve in 4–6 weeks. The patient loses height and develops an increased kyphosis but the loss of height is generally self-limiting; in many patients loss of height stops after a few years (LD-53).

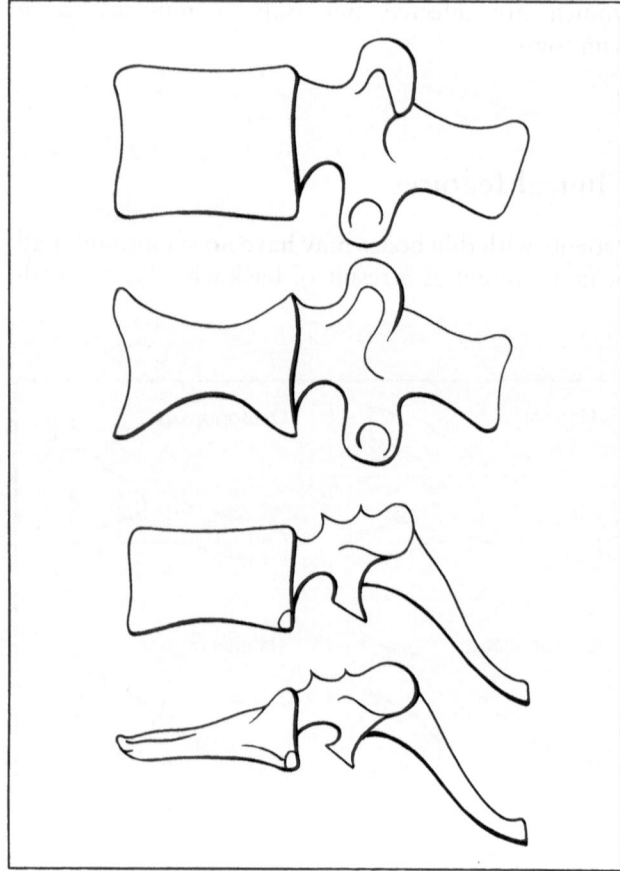

LD-53. *Radiological progression of osteoporosis.*

Investigations

Laboratory

In osteoporosis plasma levels of calcium, phosphate and alkaline phosphatase are generally normal, although the alkaline phosphatase may be raised after a fracture.

Radiology

The assessment of bone mineral from ordinary radiographs is very unreliable. It has been estimated that 50% of the bone mass must be lost before radiographs can be regarded as abnormal.

Better evidence of osteoporosis is obtained if the patient shows evidence of structural failure of the skeleton such as crush fractures of the vertebrae (X-9).

The appearance of bone on an X-ray depends on many factors including the amount of overlying soft tissue, the physical factors of the X-ray beam and the quality of the film development.

Special techniques such as photonabsorptiometry are available for the assessment of bone mass in certain centres.

Management

No form of treatment has yet been shown to improve the bone mass in osteoporosis. The rate of bone loss in postmenopausal patients can be reduced by cyclical oestrogen therapy but how widely this treatment should be advised is uncertain since the incidence of side-effects in long-term use is unknown.

Patients with backache can be reassured that the symptoms will go away and that loss of height will cease. They may need analgesics. After fractures, immobilization should be kept to a minimum to avoid making the osteoporosis worse.

OSTEOGENESIS IMPERFECTA

Causes

Osteogenesis imperfecta is the most common inherited disorder of the bone. In Western countries its incidence is approximately 1 in 20 000 but it is found in all races. The term osteogenesis imperfecta (brittle bones) probably includes several distinct disorders which have as yet been poorly differentiated from each other. The underlying disorder is in the collagen component of bone and not the mineral.

Clinical features

The main feature of the disorder is the fractures which occur with little or no identifiable trauma. Other clinical features may be:

- blue or grey sclerae in two thirds of the patients
- deafness in early adult life
- hyperextensibility of joints
- excessive sweating
- spontaneous bruising

There is a family history of the disorder in about half the cases.

The severity of the disorder can vary greatly from patient to patient. The most severe cases are born with multiple fractures sustained in intrauterine life and have very poor growth thereafter. Milder cases may have infrequent fractures and normal growth. Almost all the patients have normal intelligence.

Investigations

Laboratory

Plasma levels of calcium, phosphate and alkaline phosphatase are generally normal in this condition. No biochemical investigations useful in the diagnosis of this condition are yet known.

Radiology

In severe cases X-rays may show fractures in various stages of healing at the time of birth. In milder cases, the bone may appear normal at the time of the first few fractures; the X-ray changes seen later probably reflect the many fractures and the immobilization used in their treatment.

Management

There is no effective drug therapy for this disorder. Individual fractures are treated in a conventional manner and heal at a normal rate. In some cases the fracture rate in particular bones may be greatly reduced by the use of intramedullary rods (LD-54). The more severe cases may need powered wheelchairs and other specialized aids.

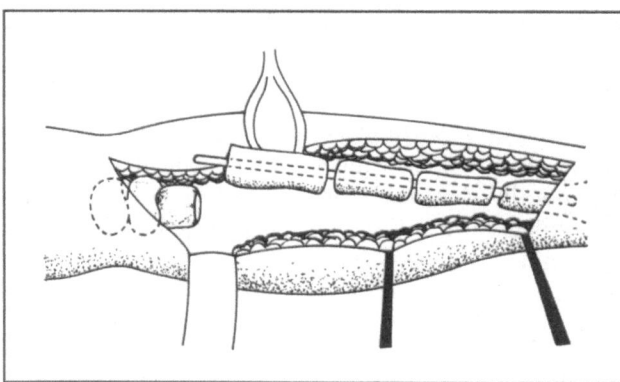

LD-54. *Treatment of osteogenesis imperfecta with intramedullary rods.*

PAGET'S DISEASE OF BONE

Paget's disease is characterized by localized areas of increased skeletal remodelling.

Causes

Paget's disease of bone is common in middle-aged or elderly people and, until recently, its cause was unknown. Recent studies have demonstrated inclusion bodies in the osteoclasts of affected bone suggestive of a slow virus. Affected bones have successive cycles in which bone is resorbed by osteoclasts after which it is replaced by abnormal new bone with a characteristic histological appearance.

Sir James Paget, Bt., 1814–1899, described the precancerous disease of the nipple in 1874 and the disease of bone in 1877.

Clinical features

At most only 10% of patients with Paget's disease of bone have any symptoms. The main ones are:

- bone pain
- deformity of affected limb-bones
- deformities of the skull
- other neurological disorders
- high output cardiac failure (extremely rare)
- sarcoma formation (rare)

Investigations

Laboratory

The striking abnormality is that the plasma alkaline phosphatase is greatly increased. The plasma calcium is generally normal unless the patient has been immobilized for any reason when raised values may be found.

Radiology

In early Paget's disease the characteristic abnormality is an area of osteolysis. In the skull such areas are known as 'osteoporosis circumscripta'.

Later successive waves of bone destruction and replacement leave a coarse trabecular pattern and generally increased density compared with unaffected bone. One or more bones may be affected, common ones being:

- the skull
- vertebrae
- pelvis
- femora
- tibia
- clavicles

Management

Calcitonin may be used in the treatment of Paget's disease. It must be given by intramuscular injection and is very expensive. Great care is needed in the selection of patients for treatment.

Not all patients respond to calcitonin and some become resistant after a while. Analgesics are often needed and, in severe cases with secondary osteoarthritis, hip replacement may be required.

Selection for calcitonin therapy

Calcitonin therapy should be considered for those patients:

- with severe bone pain not caused by osteoarthritis
  In doubtful cases a three month trial may be helpful.
- with progressive neurological complications such as deafness
- with high-output cardiac failure

- requiring orthopaedic surgery
  Paget's disease may make surgery such as hip replacement very difficult.

TESTIS

The disorders which will be described are:

- hypogonadism
- infertility
- testicular tumours
- mechanical disorders of the testes

The functions of the adult testis are:

- spermatogenesis which occurs in the seminiferous tubules
- secretion of testosterone by the Leydig cells

Control of these functions is mediated through the gonadotrophins FSH and LH from the anterior pituitary gland. These hormones stimulate spermatogenesis and testosterone production respectively. Abnormalities of testicular function may thus be due to:

- primary testicular disease

or be:

- secondary to gonadotrophin deficiency

Patients with testicular disease present with hypogonadism due to androgen deficiency or infertility due to failure of the germinal cells or with both of these features.

HYPOGONADISM

Causes

The causes of hypogonadism are:

(1) Hypergonadotrophic hypogonadism (primary testicular failure):

- chromosomal abnormalities
  Klinefelter's syndrome with XXY sex chromosome pattern
- anorchidism
  In some male patients the testes fail to develop.
- cryptorchidism
  Undescended testes are more likely to result in

infertility due to tubular dysfunction than true hypogonadism.

- castration

(2) Hypogonadotrophic hypogonadism (gonadotrophin failure):

- panhypopituitarism
  Hypogonadism may be part of a generalized deficiency of anterior pituitary hormones.

- isolated gonadotrophin deficiency
  When familial and associated with anosmia, colour blindness and facial developmental abnormalities such as hair lip this is called Kallman's syndrome.

- delayed puberty
  Most boys will show signs of sexual maturation by mid-teens. However in some patients puberty is delayed until as late as 18 years of age (p. 107).

Clinical features

The most important symptoms diagnostically are:

- loss of libido
- impotence
- infertility
- decreased frequency of shaving
- delay in establishing potency and secondary sex characteristics

On examination the male hypogonadal patient may exhibit:

- a eunuchoidal body habitus (long limbs with delay in closure of epiphyses)
- a high pitched voice
- lack of adult male hair distribution (failure of scalp hair recession, sparse facial, pubic and axillary hair)

- small genitalia
- muscle hypotonia and decreased bulk
- gynaecomastia

Investigations

These are:

- estimation of radiological bone age
  This provides an index of skeletal maturation and is useful diagnostically in 'delayed puberty'.

- chromosomal analysis for sex chromatin and karyotyping

- plasma testosterone
  This will be reduced in primary and secondary testicular disease.

- Plasma gonadotrophins (LH and FSH). These may be:
  elevated in primary testicular failure
  reduced in gonadotrophin failure with secondary testicular failure

In patients with reduced plasma gonadotrophin levels failure to produce a rise in LH and FSH by administration of luteinizing hormone release hormone (LHRH) helps to confirm gonadotrophin insufficiency.

Human chorionic gonadotrophin (HCG) administration may differentiate primary and secondary testicular failure – in patients with primary failure a response would not be expected (Table 52).

Management

Patients with primary testicular failure require treatment with one of the testosterone analogues. This can be administered orally, sublingually or by depot injection.

Table 52. Differentiation of primary and secondary testicular failure

	Plasma testosterone	Plasma LH	Plasma FSH	Testosterone response to HCG	LH, FSH response to LHRH
Primary testicular failure	↓	↑ (or normal)	↑ (or normal)	↓	↑
Secondary testicular failure (gonadotrophin deficiency)	↓	↓	↓	normal	↓

In secondary testicular failure due to gonadotrophin deficiency, the androgen deficiency can be treated by testosterone administration. However, this therapy will not induce spermatogenesis and the hormonal defect is more appropriately treated with chorionic gonadotrophin (HCG) until full masculinization has occurred.

INFERTILITY

Causes

The cause of primary tubular dysfunction is frequently not identifiable but occasionally a recognizable aetiology is found:

- gonorrhoea
- postpubertal mumps orchitis
- non-specific 'viral' orchitis
- cytotoxic drugs
- irradiation

Clinical features

Patients presenting with infertility rarely show evidence of androgen deficiency but usually have selective seminiferous tubular dysfunction. Only rarely will clinical examination reveal obvious causal abnormalities:

- undescended testes
- varicocoele

Investigations

How should infertility be investigated in a male?

- plasma gonadotrophins
  These, particularly FSH, will be elevated in seminiferous tubular failure. In secondary tubular failure due to gonadotrophin deficiency, the LH and FSH levels may be decreased.
- plasma prolactin
  Impotence and infertility may be the consequence of hyperprolactinaemia (see Hyperprolactinaemia, p. 113)

- seminal analysis
  The volume, the sperm count and the percentage of mobile normal forms are valuable.
- testicular biopsy
  This may give some indication of the degree of spermatogenetic arrest and be a useful indicator of the possible outcome of therapy. Absence of germinal epithelium indicates that therapy is not likely to be of any value. A normal biopsy may suggest mechanical abnormalities such as vas and epididymis blockage.

Management

Treatment depends on the degree of spermatogenic arrest and whether an identifiable cause is apparent.

Patients with varying degrees of spermatogenic arrest due to idiopathic tubular dysfunction may respond to preparations containing high concentrations of FSH. These are normally isolated from postmenopausal urine (HMG human menopausal gonadotrophin). Response can be assessed by repeated testicular biopsy.

Testicular failure due to selective gonadotrophin deficiency is treated by HCG administration. Infertility associated with hyperprolactinaemia will usually respond to bromocriptine therapy.

TESTICULAR TUMOURS

The majority of testicular tumours are malignant; however, they account for <1% of all malignant tumours.

They are:

- seminomas
  These are malignant tumours arising from germinal epithelium.
- teratomas
  These arise from all three germinal layers and may contain islands of cartilage and bone; some contain trophoblastic tissue (testicular chorion carcinoma)
- interstitial cell tumours
  These are usually benign.

The major endocrine abnormalities associated with such tumours include:

- feminizing syndromes
  These are due to gonadotrophin and oestrogenic hormone production by chorion carcinomas particularly.

- sexual prematurity
  This follows androgen secretion by interstitial cell tumours.

MECHANICAL DISORDERS OF THE TESTIS

These will be considered under three headings:

- problems of descent
- pain in the testis
- swellings of the testis

Problems of descent

Embryology

The testis develops from the genital fold in the region of the kidney. The mesonephros contributes to development of the vasa efferentia and the Wolffian duct to the epididymis and vas deferens (LD-55, 56).

The gubernaculum is a tissue connecting the lower fold of the developing testis to the skin of the primitive scrotum. It is differential growth which results in the gubernaculum drawing the testis down into the scrotum where it normally arrives during the last month of intrauterine life.

Incomplete descent of the testis

This is a term to be applied when the descent of the testis has been arrested along its normal course of descent. Hormonal factors, probably mostly HCG, control the effect of the gubernaculum in drawing the testis into the scrotum.

Clinical features

Testicular descent may not be complete at birth, especially in premature babies. Some descend spontaneously in the early months of life. Incomplete descent should not be confused with physiological retraction of the testis into the superficial inguinal pouch under the influence of the cremaster muscle. This occurs particularly under the influence of cold as,

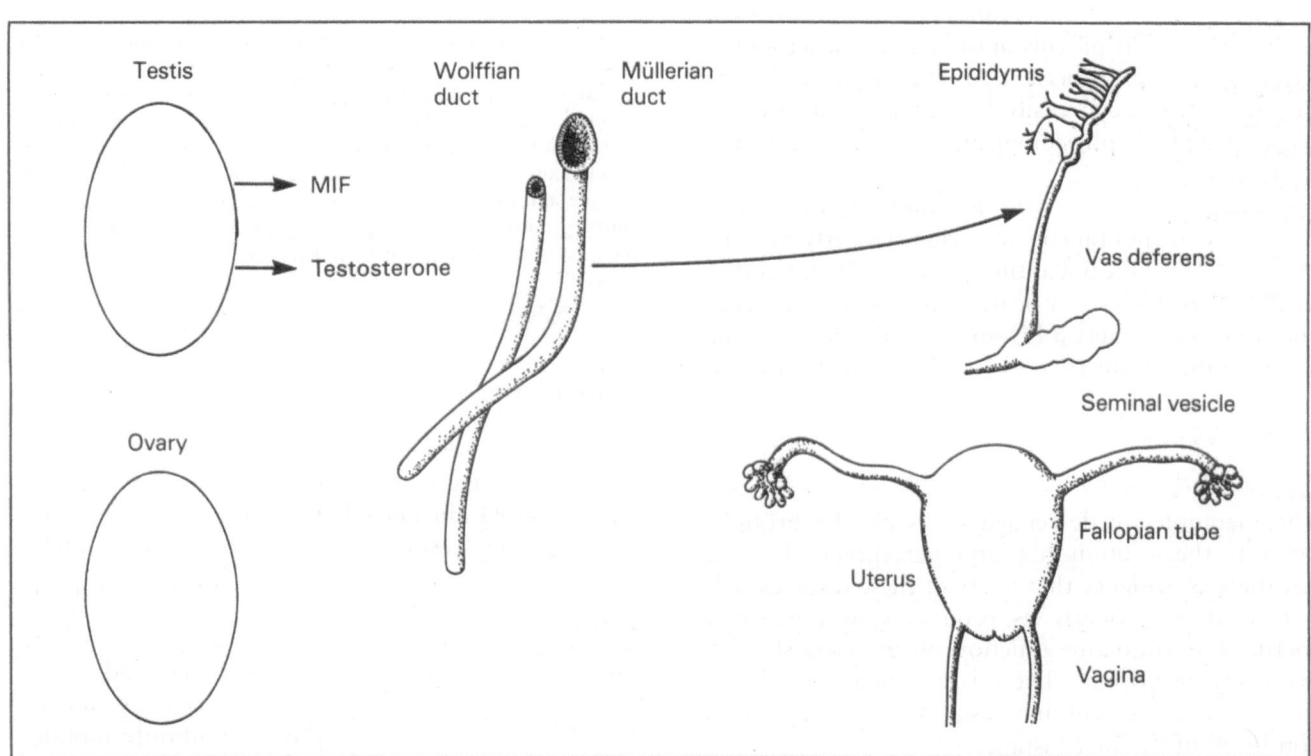

LD-55. *Differentiation of the internal genitalia in response to the dual endocrine secretion of the fetal testis.*

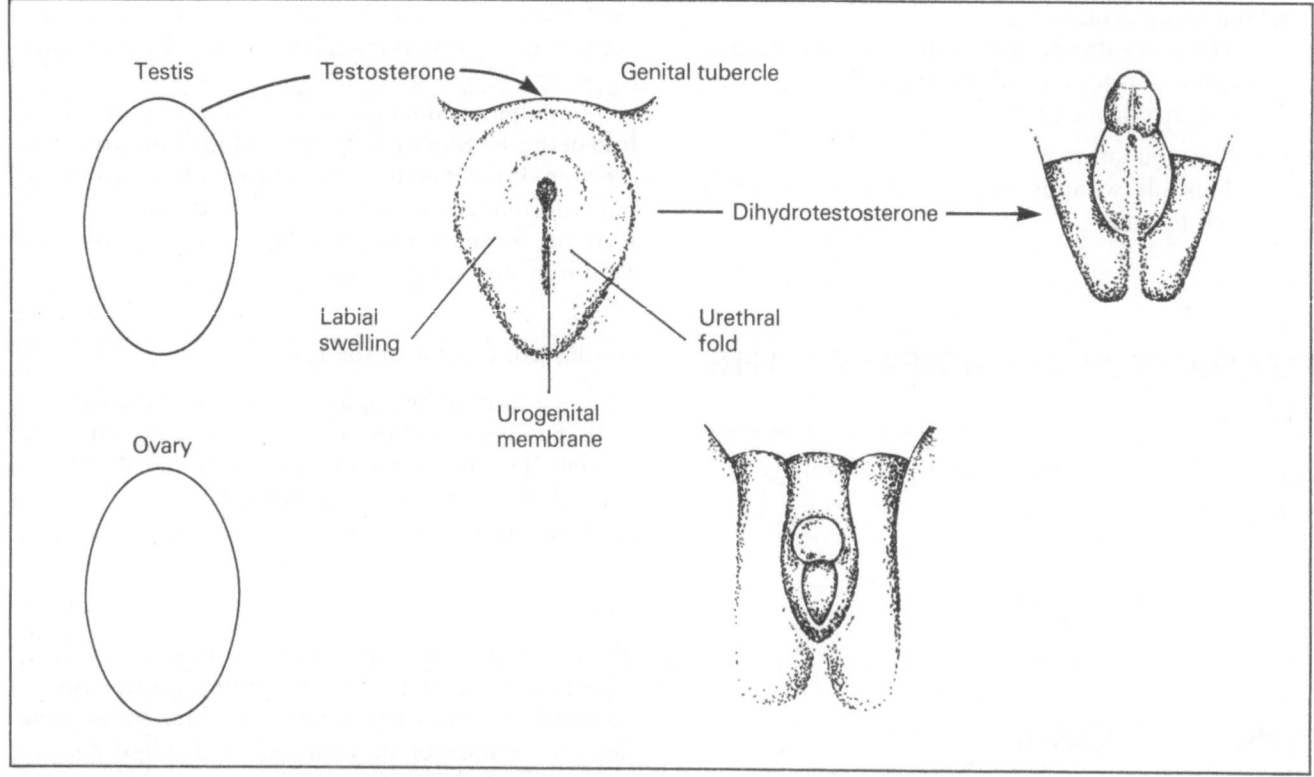

LD-56. *Differentiation of external genitalia.*

for example, when a child is undressed for purposes of examination. The parents must be asked whether the testicles have ever been seen within the scrotum. Retractile testes can easily be manipulated into the scrotum. This is a physiological state and no treatment is indicated.

Testicles which are truly incompletely descended may descend spontaneously during the early months or even years of life but as time passes the likelihood of this declines. HCG may be used to encourage descent and appears to exert a favourable influence in some cases. Longer-term effects of such treatment are not known.

Management

The incompletely descended testis can be brought down to the scrotum in a large percentage of cases, but there is evidence that many of these testicles will not function properly as producers of functional sperm. The endocrine function of the testis should, however, be normal. There is a definite possibility that failure of descent may result because the testis is abnormal in the first instance.

Where incomplete descent is bilateral there is a high incidence of subsequent infertility in spite of early and adequate orchidopexy and normal endocrine function. Orchidopexy should always be attempted because the incompletely descended testis is prone to trauma and there is a higher incidence of tumours in such organs.

Maldescent of ectopic testes is to be distinguished from incomplete descent. The gubernaculum has its major attachment to the skin of the primitive scrotum but there are minor attachments at the root of the penis, the perineum, the femoral triangle and the superficial inguinal region. A testis which has followed one of these alternative pathways will never come to lie in the scrotum spontaneously and surgery should not be unduly delayed.

Pain in the testis

This is an important symptom but examination may not be too easy in the presence of severe pain. A careful history must be taken. Trauma occurs in some cases and may result from games or from assault.

Torsion

Pain may be severe where the testis has undergone torsion often as a result of persistence of the primitive mesorchium which leaves the testis unduly mobile. Pain is often felt first as abdominal colic referred from the testis as a consequence of its intra-abdominal origin.

Management

Fixation is made locally to the scrotum to prevent recurrence and the opposite side should also be fixed since the predisposition (persistent mesorchium) is normally bilateral.

Acute epididymo-orchitis

This is usually due to a coliform infection and is another important cause of acute pain. It can be difficult to distinguish from torsion. If there is a history of urinary tract infection, or pus is present in the urine, or acute prostatitis can be diagnosed on rectal examination, then acute epididymo-orchitis can be diagnosed.

If doubt exists as to the diagnosis, the testis must be explored early so that a torsion can be diagnosed and undone while the testis is still viable.

Management

Epididymo-orchitis is treated with antibiotics. The urinary tract should be investigated bacteriologically and by intravenous pyelography. Since acute urogenital infections may supervene upon tuberculous disease, three overnight samples should be cultured for *Mycobacterium tuberculosis*.

Swellings of the testis

Swellings of the testis are both common and important. First it is important to determine that the swelling is of the testis or associated with the testis and not due to some other cause such as a scrotal hernia. A scrotal hernia can be differentiated because it:

- may be reducible
- should transmit a cough impulse
- is not possible to get the fingers above the swelling

Causes

Swellings in the scrotal sac are:

- non-testicular:
  hydrocoele
  spermatocoele
- testicular:
  granulomas
  tumours (seminoma, teratoma)

Hydrocoele

This is characteristically a painless and slowly enlarging swelling anterior to the testis. It is brilliantly transilluminent and it is easy for the examining fingers to get above it. It should be aspirated then excised.

Spermatocoele

This is also characteristically painless but the testis is most usually anterior to the swelling. It is not so brilliantly transilluminent as a hydrocoele. Treatment is the same as for hydrocoele.

Tuberculosis

This is the most important granulomatous disease, syphilis being now quite rare. Tuberculosis may be easily diagnosed if there is a sinus, otherwise the diagnosis requires full clinical examination, intravenous pyelography and the submission of three overnight urine specimens for culture. Treatment consists of chemotherapy and excision of the affected tissue.

Tumours of the testis

These are practically all malignant. Ninety-nine percent are accounted for about equally between seminoma and teratoma. Tumours of interstitial tissue account for the remaining 1% and these may be functioning endocrine tumours.

Seminoma

Seminoma is commonest in men in the second and third decades. There is painless enlargement of one testis. Spread is central by lymphatics to the para-aortic nodes which may be palpable or their presence may be determined by displacement of the ureter laterally on intravenous pyelography.

Treatment is by removal of the testis and cord from the internal ring, the cord being clamped before the tumour is handled. Radical radiotherapy has led to considerable improvement in survival.

Teratoma

Teratoma arises from cells with total potential, and elements of ectoderm, mesoderm and endoderm may be seen. One is normally predominating and is the source of the malignant change. A number contain chorionic tissue and plasma and urine levels of HCG may help to establish the diagnosis. These can also be useful in following progress after treatment. Treatment is surgery plus radiotherapy.

OVARY AND BREAST

Here the topics to be discussed are:

- anovulation
- ovarian tumours
- virilism
- disorders of gender differentiation
- disorders of the breast

ANOVULATION

Causes

Many causes of infertility are not amenable to treatment. Notable among these are:

- intersex problems
- congenital abnormalities
- male infertility
- mechanical causes such as blocked tubes

The outlook is much better when there is a failure of ovulation. When the failure is at the ovarian level, e.g. menopause praecox, surgical or radiological castration, it cannot be treated.

Although a few women with failure of ovulation have regular cycles, anovulation is commonly associated with secondary amenorrhoea or oligomenorrhoea. Rarely they may present as primary amenorrhoea. Even those with primary amenorrhoea usually produce enough ovarian hormones for secondary sex characters to develop completely.

Some causes of infertility for which logical treatments do exist are associated with a very low success rate. Among these are:

- the repair of blocked fallopian tubes
- the psychiatric treatment of psychosexual problems
- attempts to suppress immunological immobilization of sperm

Primary amenorrhoea or a marked failure of secondary sex characteristics carries a poor prognosis in infertility.

Investigations

Failure of ovulation is easy to establish. There is:

- no rise in basal body temperature
- no secretory endometrium
- a plasma progesterone below 20 nmol/l in the latter part of the menstrual cycle

It may arise from many causes and the clinician must exclude:

- polycystic ovarian disease (Stein–Leventhal syndrome)
- hyperprolactinaemia
- anorexia nervosa
- Cushing's syndrome

Often no precise cause can be ascribed, but an attempt should be made to assess the functional state of the ovary and the pituitary by measurement of:

- oestradiol, LH, FSH on 7th, 14th and 21st days after menstruation.
- progesterone on 14th and 21st days
- prolactin, thyroxine and androstenedione on 7th day

Once it is established that ovulation is rare or infrequent, specific causes should be considered. It is not sensible to undertake a battery of tests designed to pick up all possible causes. One should be guided by the clinical presentation:

- hirsutism in Stein–Leventhal syndrome and adrenal hyperactivity
- emaciation in anorexia nervosa
- visual disturbance in pituitary tumours

Some general tests are helpful because they point to specific lines of treatment.

- elevated levels of prolactin should be confirmed by measurements on several occasions.
- low gonadotrophin levels throughout the cycle should be investigated by studying the response to gonadotrophin releasing hormone (LHRH).

Management

The primary treatment for anovulation in cases where there is no specific cause is the use of antioestrogens such as:

- clomiphene
- tamoxifen

Only if three courses at increasing dosage fail to produce ovulation should more drastic forms of treatment be considered. The usual regime of anti-oestrogen therapy is to administer it for 4 days and expect ovulation 10 days after the start of treatment. Whether ovulation is induced can be determined by a single plasma progesterone assay done 21 days after treatment. It will also give some indication of the quality of luteal function.

In patients in whom a deficient secretion of gonadotrophin is demonstrated, FSH (in the form of human menopausal gonadotrophin, HMG) is injected daily. The ovarian response is monitored by measurements of urinary or plasma oestrogens. When oestrogen levels attain a satisfactory level, ovulation is induced by one injection of LH (in the form of HCG).

OVARIAN TUMOURS

Although tumours in the sense of an enlargement of the ovary are common, it is not usual for such swellings to have endocrine consequences.

Often such tumours are not neoplasms but excessive growth of some ovarian component as occurs with simple follicular cysts or corpus luteum cysts.

The endocrine characteristics of the true neoplasms are related to the ovarian component from which they arise. They may be:

- may be feminizing
- may be masculinizing
- have no endocrine consequence
  This is the rule when the tumour springs from undifferentiated germ cells, dysgerminomas in the female and seminomas in the male.

Feminizing tumours

These develop from either granulosa or theca cells.

Usually granulosa cell tumours arise, not from intrafollicular granulosa, but from stray cells in the medulla or interstitial tissue of the ovary. They owe their feminizing effect to the secretion of oestradiol.

Occasionally, thecal tumours produce progesterone, not oestradiol, and cause a secretory change in the endometrium.

Granulosa and theca cell tumours commonly occur in the third or fourth decade and are apt to disrupt the normal menstrual cycle. When, rarely, they occur earlier, they may cause precocious puberty.

Masculinizing tumours

These, particularly arrhenoblastomas, arise from undifferentiated mesenchyme in the hilus of the ovary where such embryonic survivals contain Leydig (testosterone secreting) cells. Sometimes masculinizing tumours arise not from gonadal tissue but from adrenal rests in the ovary.

When endocrinologically active ovarian tumours arise during menstrual life, they are seldom recognized until they are big enough to give rise to symptoms as space-occupying lesions. The menstrual irregularity, hirsutism or secondary virilization to which they might give rise more commonly have non-neoplastic causes. They feature, however, more often in precocious puberty and in this condition an ovarian or testicular tumour should always be considered.

VIRILISM

Clinical features

Excessive hair growth in women is termed 'hirsutism'. In the majority of these patients this feature is the only clinical abnormality and only infrequently is it associated with a recognizable hormonal abnormality.

Occasionally, however, hirsutism is part of the condition known as virilism. These patients also have in varying degree:

- amenorrhoea or disturbance of menstrual pattern
- recession of frontal hair line and male pattern hair distribution
- deepening of the voice
- enlargement of the clitoris
- masculine body habitus with reduction in breast size

These features are always associated with increased androgen production.

Causes

The causes may be classified as:

- ovarian
- adrenal
- drugs

Ovarian

The normal ovary produces only small amounts of androgen.

Polycystic ovary syndrome (Stein–Leventhal)

In this condition the pattern of hormone secretion is changed so that androstendione – a normal intermediate of oestrogen synthesis – may be produced in excess. This may then be converted peripherally to testosterone or dihydrotestosterone – both potent androgens.

Masculinizing ovarian tumours

Although rare, these may occur at any age, from childhood to menopause. A rapid onset and severe symptoms suggest this pathology.

Adrenal

Congenital adrenal hyperplasia and *Cushing's syndrome* may be associated with virilization.

Drugs

Some drugs may cause virilization in females including:

- androgens – prescribed for advanced breast cancer and for treatment of menopausal symptoms
- anabolic steroids – prescribed to promote 'muscle bulk' in athletes
- glucocorticoids
- phenytoin

Investigations

The finding of elevated serum testosterone, dihydrotestosterone and androstenedione indicates increased androgen secretion. To distinguish between ovarian and adrenal disease, however, it is necessary to proceed to dynamic function tests. Providing no autonomously functioning adrenal or ovarian tumour is present:

- in *adrenal disease* ACTH will stimulate and increase androgen metabolites and betamethasone will suppress and decrease androgen metabolites
- in *ovarian disease* HCG will stimulate and increase androgen metabolites and oestrogenic steroids will suppress and decrease androgen metabolites

DISORDERS OF GENDER DIFFERENTIATION

Sexual differentiation begins during the 8th week of fetal life. The primitive genital tract, from which the internal sexual organs develop, is in the form of hollow tubes, known as the Müllerian and Wolffian ducts. They are associated with primitive gonads and a primitive common primordium which will develop later into the perineum. If the chromosomal constitution of the individual is female, the gonads become ovaries and, if male, the gonads become testes (LD-55).

In females, the Wolffian duct atrophies and the Müllerian duct differentiates into the Fallopian tubes, uterus and upper vagina. The perineum becomes female, with a clitoris, separate urethral opening, vaginal opening and formation of the lower vagina and labia majora and minora. These processes are not dependent upon the secretion of oestrogens as they occur in Turner's syndrome, in which the ovaries fail to develop (LD-56).

In males, the Müllerian duct atrophies in response to the secretion of Müllerian inhibiting hormone (MIH) secreted by the testes. The Wolffian duct, under the influence of testosterone, differentiates into an epididymis, vas deferens and seminal vesicle on each side and the prostrate. Under the influence of testosterone and dihydrotestosterone, the perineum becomes male with a penis, penile urethra and the formation of a scrotum fused across the midline containing the testes which have descended in response to testosterone.

Clinical features

The rare disorders of gender differentiation are of three main types:

- a female child, with normal sex chromosomes
  There is virilization due to excess adrenal androgen secretion, e.g. 21-hydroxylase deficiency. This is by far the commonest type (LD-11)
- a male child with normal sex chromosomes
  There is incomplete virilization, either due to an enzyme defect on the pathway of testosterone or dihydrotestosterone synthesis, or due to failure of the cells of the genital organs to respond to testosterone. This failure of cells to respond is a consequence of absence or marked reduction in the number of androgen receptors in cells.

- abnormalities of chromosomal origin leading to intersex situations

    There is some female and some male development of the gonads, genital tract and perineum. These conditions are exceedingly rare.

Details of 21-hydroxylase deficiency are given in the section on the adrenal gland, see p. 128. In affected girls, in the absence of MIF the Müllerian duct develops normally, producing female internal organs. The adrenal androgens are not powerful enough to stimulate the Wolffian duct but they cause virilization of the perineum with enlargement of the clitoris. The urethral opening usually remains separate but, in severe forms of the deficiency, the urethra may open at the tip of the clitoris. The vaginal opening may be absent and there may be poor development of the lower vagina and of the labia majora and minora. These patients should be brought up as girls.

In incompletely virilized males, in the presence of MIF, the Müllerian duct atrophies and there are no female internal organs. Wolffian duct development and masculinization of the perineum may be deficient and, in extreme situations, the external appearance of the child may be that of a normal female, as in testicular feminization.

The appearance of the perineum is of vital importance when the decision is made to bring the child up as male or female. Incompletely virilized male children and children with some female and some male development often have to be reared as girls.

Investigations

The investigative approach to 21-hydroxylase deficiency is covered elsewhere (see p. 128).

THE BREAST

Clinical features

Developmental abnormalities

Complete *absence* of a breast or both breasts occurs but is very rare.

Absence of a nipple or the presence of accessory nipples is more common. An accessory nipple may be mistaken for a pigmented naevus.

Abnormalities may arise at puberty, if development is *asymmetric*. This worries both the girl and her parents, but they can be reassured that symmetry will normally be expected to be restored in time and reassurance is all that is usually necessary.

Abnormalities in size and shape

Failure of breasts to develop normally at puberty occurs in:

- ovarian failure
- hypopituitarism
- some hypothalamic conditions

Asymmetric development at puberty, as mentioned above, is not uncommon. Tumours, if large, may cause obvious asymmetry.

Mastoplasia (enlargement of the breast to excessive size) does occur rarely in Graves's disease and Cushing's syndrome.

Gynaecomastia is enlargement of the male breast due to increased glandular tissue and is considered on p. 84.

Retraction of the nipple

This is an important clinical sign. It may be of long standing, often from puberty. Retraction is usually due to some minor congenital abnormality interfering with normal growth and development. It predisposes to:

- problems with feeding during lactation
- entry of infection into the breast

Minor degrees can be treated by the patient drawing out the nipple daily until it remains everted. The resistant case can be improved by minor surgery.

Recent retraction of the nipple is a serious sign often indicating an underlying carcinoma.

Pain in the breast (mastodynia)

Some degree of discomfort in the breasts is common in association with menstruation and may be regarded as almost physiological. Painful breasts are usually worst premenstrually and the symptom is rare after the menopause. The breasts may be normal to palpation, though tender, or have the shotty feeling of generalized chronic mastitis.

Management

It can be difficult to relieve this condition. Minor degrees may respond to reassurance that the pain does not indicate serious breast disease. A good supporting brassière should be worn, if need be, at night.

On the assumption that pain is consequent upon hormonal changes, cyclical hormone preparations can be tried for a few cycles and may give relief.

Occasionally pain is a feature in women already

using the contraceptive pill and it may be worthwhile discontinuing it for a few cycles. Alternatively, pain may respond to use of a diuretic. Resistant cases may respond to treatment with bromocriptine.

Discharge from the nipple

This is often very worrying to the patient. Discharges from the nipple may be:

- milky
- serous
- coloured (yellow, green, brown)
- bloodstained

Milky and serous discharges are considered under Galactorrhoea (Section III, p. 85).

Yellow, green or brown discharge usually indicates underlying chronic mastitis with duct ectasia. There is little that can be done by way of treatment but it is essential to exclude serious (malignant) disease of the breast.

In resistant cases it can be managed by excision of the underlying duct-containing portion of the breast with conservation of the nipple.

Bloodstained discharge should be regarded as pathognomonic of intraduct papilloma or carcinoma until proved otherwise. It may be possible to identify a single duct or at least the quadrant from which the blood is coming and an intelligent patient may be of great help here. The appropriate sector can be explored through a circumareolar incision, the lesion identified and excised for histopathology.

It is not always this easy, however, and it may prove very difficult to identify the duct or sector from which the blood is coming. If it is a papilloma the underlying duct-containing breast can be excised with nipple conservation.

Lumps in the breast

Few conditions cause more distress to a woman than the finding of a lump in her breast.

The possible causes of a lump in the breast form a very long list indeed, but in practice the vast majority are due to a small number of conditions:

- fibroadenosis (chronic mastitis)
- cyst
- fibroadenoma
- breast cancer

No age group is exempt from cancer but below the age of 40 the statistical chances of a lump being simple are much greater than over the age of 40.

Two conditions:

- haematomas
- traumatic fat necrosis

may result from trauma. The latter is a granulomatous condition. It is important because it may mimic carcinoma of the breast.

It is not uncommon for a woman complaining of a lump in her breast to relate its development to some traumatic episode. There is rarely any valid association. Usually the incident has directed attention to a change already present in the breast.

Infection of the breast

Some degree of breast development and even milk secretion is normal in the newborn, due to the hormonal environment *in utero*. Infection with *Staphylococcus aureus* can result and unskilled surgery can lead to abnormalities of development of the female breast at puberty.

Most infections of the breast occur in association with lactation. Infection, usually with *Staphylococcus aureus*, may pass through a cracked nipple from the infant to the mother. The engorged lactating breast has poor resistence and clinical infection supervenes.

If not aborted with antibiotics, breast abscess results. Treatment of the breast abscess is by incision and drainage or by excision and closure.

Chronic granulomatous infections such as tuberculosis, syphilis and actinomycosis are known but are now rare in Western society.

Fibroadenosis (chronic mastitis)

This may be regarded as an exaggeration of the normal physiological changes accompanying the hormonal changes of the menstrual cycle and pregnancy.

Formation of cysts and fibrosis may produce what is clinically a lump in the breast. There may be no other changes in the breast but often the shotty changes of generalized fibroadenosis can be detected in one or both breasts. The lump may be tender but there is no tethering to skin or fascia.

Axillary glands may be enlarged and tender due to the absorption by the lymphatics or the cyst content. The patient may complain of a yellowish or greenish discharge from the nipple.

Fibroadenosis confers no immunity to carcinoma and, since both the conditions are common, they may coexist. All suspicious lumps should be biopsied by

- aspiration
- needle
- excision

Management

If carcinoma is excluded, treatment is by reassurance and the wearing of an adequate supporting brassière. Painful lumps may need excision.

It is not uncommon to see such patients repeatedly with different lumps in either breast. Each new lump should be treated on its merits and the possibility of carcinoma borne constantly in mind.

Very rarely pain is so severe as to warrant subcutaneous mastectomy usually together with a prosthetic implant.

Cysts in the breast

The clinically solitary cyst may present as a painless lump. None of the features pathognomonic of carcinoma is present. Aspiration and cytological examination can help to rule out cancer. The cyst may be cured by the aspiration, but frequently it recurs and may require to be excised.

Fibroadenomas

Fibroadenomas are usually small and very freely mobile lumps in the breast of younger patients. There is a rare 'giant' equivalent.

Treatment is by excision. The clinical differential diagnosis from carcinoma is not often difficult.

Cancer of the breast

No age group is exempt, but there is a peak incidence between 40 and 50 years. The incidence is relatively lower in parous women, but carcinoma of the breast is a common disease and these factors should not influence the diagnosis in the individual case.

Clinical presentation

Characteristically it is painless and found by accident or nowadays by self-examination or at a screening clinic.

Sixty per cent of carcinomas arise in the upper and outer quadrant of the breast. There may be tethering to the skin or to the deep fascia and later skin may become invaded. Metastases to the axillary nodes occur in time. Other features may include recent nipple retraction.

Diagnosis

The diagnosis may be confirmed clinically by evidence of local invasion. Thermography, ultrasonography and mammography may be of value in elucidating the nature of lumps in the breast, and have been used in screening populations for detection of early breast cancers.

Aspiration, needle biopsy or excision biopsy of all suspicious breast lumps are vitally important to achieve a histopathological diagnosis. Scintiscanning of liver and skeleton, radiological skeletal survey and laparoscopy are essential in detecting spread of carcinoma and in determining the most appropriate treatment.

Management

Surgically there is no general agreement as to what methods are best but there is an overall similarity of results, whether treatment is by:

- simple mastectomy and radiotherapy
- radical mastectomy

Survival depends on the degree of the spread of disease at the time of diagnosis and treatment.

A proportion of carcinomas of the breast are *hormone stimulated*. The disease may respond well to alteration of the hormonal environment. In the age group up to just beyond the menopause, the tumour may be stimulated by female sex hormones and a proportion do respond well to removal of the ovaries and adrenals, or to hypophysectomy. This may be conveniently carried out by local implantation of yttrium 90 rather than open operation. It is possible that the use of antioestrogen drugs (such as tamoxifen) will replace these surgical procedures.

Older patients may do well on oestrogen replacement. When hormone therapy is being considered, case selection is difficult since only about 20% of tumours respond really well. Assay of hormone receptors in the primary tumour is a guide to which tumours are likely to respond to hormone manipulation.

It is usual to talk in terms of 'disease free interval' rather than 'cure'. Almost certainly the outcome is dependent on a complex biological equation between the inherent invasiveness of the tumour and, of course, ability to resist. These are important but ill-understood concepts.

Paget's disease of the nipple

This is an eczematous-like condition usually affecting only one nipple, unlike true eczema which normally affects both. It is due to an underlying carcinoma of the breast with spread through the main ducts to the nipple and areola.

It is important to distinguish it from eczema. Paget's disease is a manifestation of carcinoma of the breast and should be treated as such.

The male breast

Virtually all of the pathological processes seen in the female breast can be seen in the male breast but they are very much more rare.

Carcinoma of the male breast accounts for much less than 1% of all cases of cancer of the breast. It is treated along lines similar to those used in the management of carcinoma of the female breast but the prognosis is worse.

Gynaecomastia

Gynaecomastia is enlargement of the male breast due to an increase in its glandular component. It should be distinguished from enlargement of the breast consequent upon simple obesity but, in practice, it is not always easy to differentiate the two (Section III, p. 84) (CP-20).

Causes

The many causes of gynaecomastia are outlined in Table 53. Only the physiological ones are common.

Investigations

See Section III, p. 85.

CARBOHYDRATE METABOLISM

The three principal topics to be covered are:

- diabetes mellitus
- hypoglycaemia
- simple obesity

DIABETES MELLITUS

Clinical features of diabetes were described in early Greek literature, and even in Egyptian writings many centuries earlier. However, the most vivid of early descriptions is attributed to Aretaeus the Cappadocian writing in the second century AD:

Diabetes is a wonderful affection, not very frequent among men, being a melting down of flesh and limbs into urine . . . life is short, disgusting and painful, thirst unquenchable, death is inevitable.

Certainly the symptoms of established overt diabetes of severe thirst and polyuria are unmistakable and always associated with glycosuria. A blood glucose estimation will confirm the diagnosis. However, in the grey area of impaired glucose tolerance there may be only intermittent or no glycosuria because the plasma glucose level rises above the renal threshold infrequently.

Epidemiological evidence in whites and most other races has suggested that there is no sharp distinction between diabetics and non-diabetics. The label is used to divide conveniently the smooth distribution curve of plasma glucose values at various times after a glucose load.

Prevalence

Accurate figures for the prevalence of diabetes cannot be given in the absence of a generally accepted definition and of notification of the disease. Nevertheless, about 1% of the population of Western countries are known diabetics and surveys based on glycosuria have usually disclosed a similar number of undiagnosed diabetics.

It is commoner in the elderly and obese, the

Table 53. Causes of gynaecomastia

Physiological	Exogenous hormones	Drugs	Endocrine	Other
Neonatal	Oestrogens	Reserpine	Testes	Cirrhosis
Pubertal	Chorionic gonadotrophin	Digitalis	Klinefelter's	Bronchogenic carcinoma
Old age		Spironolactone	atrophy	Starvation
			tumours	
			Adrenal cortex	
			hyperplasia	
			tumours	
			Graves's disease	
			Pituitary	
			acromegaly	
			tumours	

prevalence in children being only about o.1%. A survey carried out in Birmingham (England) in 1961 showed that newly diagnosed diabetes was slightly commoner in men than women up to the age of 40 years but thereafter the situation gradually changed so that, by the age of 60, almost twice as many women as men were being diagnosed. More recent evidence suggests that there is no longer this excess of female over male new cases between the ages of 40 and 60 years, and in the Far East, the incidence in men tends to exceed that in women.

Much of the postmenopausal increase in the incidence of diabetes in women can be attributed to the late effects of parity for the incidence is actually slightly less in nulliparous women than in men. Although multiparity increases the chance of future diabetes, it does not make it appear earlier.

There is no doubt that obesity increases the risk of future diabetes but this cannot be held responsible for the differential sex incidence because, in Western countries, diabetic women outnumber men even among the non-obese. The menopause has been held responsible for some of the extra diabetes in women after the age of 45 but, if it is, the mechanism is not understood and there is usually no obvious change in the severity of established diabetes at the time of the menopause.

Classification

There are two main types of diabetes and these are important to recognize because their treatment and outlook are different. They are:

- Insulin dependent diabetes (IDD) or Type I diabetes or juvenile onset diabetes
- Non-insulin dependent diabetes (NIDD) or Type II diabetes or maturity onset diabetes

The use of the term 'insulin dependent diabetes' rather than 'juvenile onset diabetes' takes account of the group of patients who present in later adult life and are ketosis-prone from the outset, and also the interesting maturity onset diabetes of youth (MODY). In the latter there is a dominant type of inheritance of the diabetes which presents in childhood or early adult life. MODY patients are usually mild diabetics who are not ketosis-prone and often will not require insulin therapy.

Clinical features

These are summarized in Table 54.

Table 54. Clinical features of insulin dependent and non-insulin dependent diabetes

	Type I Insulin dependent, juvenile onset	Type II Non-insulin dependent maturity onset
Age	Children and young adults	Middle-aged and other
Sex	Male = Female	Male < female
Proportional frequency	About one third	About two thirds
Onset	Acute or subacute	Gradual
Symptoms	Present	Often absent
Nutrition	Usually normal or thin	Usually obese
Weight loss	Marked	Slow or absent
Ketosis	Common	Absent or slight
Response to insulin	Sensitive	Relatively insensitive
Plasma insulin	Absent or low	Normal or raised
Response to oral antidiabetic drugs	Absent or slight	Present

Insulin dependent diabetes (IDD)

Here the symptoms may be dramatic enough to make the diagnosis virtually certain, for only uncontrolled hyperthyroidism can cause a similar rate of weight loss in spite of a normal or increased appetite. In children and young adults, previously undiagnosed diabetic ketosis may present as severe abdominal pain and vomiting with associated fever and leukocytosis. Younger children may already be in ketoacidotic coma when first diagnosed.

Careful inquiry into symptoms before the onset of those due to raised blood sugar levels may often reveal others such as episodes of:

- unusual hunger
- weakness
- sweating
- tremor
- irritability

These suggest that phases of hypoglycaemia were then occurring. This is in keeping with the known delay in insulin release in response to a carbohydrate load found in very early stages of diabetes.

Non-insulin dependent diabetes (NIDD)

In contrast, the symptoms of non-insulin dependent diabetes are seldom dramatic and frequently misinterpreted. About 20% of such diabetics in Britain have had continuous symptoms for at least a year before the diagnosis is made. A fat, middle-aged woman is unlikely to complain if she loses weight with unexpected ease and she may easily attribute a slow progression of:

- lassitude
- urinary frequency
- failure of vision

to the natural process of aging. She may be too embarrassed to complain of her pruritus vulvae and may seek relief with local treatment. Regrettably these patients will often present with established complications of diabetes rather than symptomatology related to hyperglycaemia.

Defining diabetes

The 2 hour post-breakfast blood glucose is the best way of confirming diabetes. A value in excess of 10 mmol/l effectively confirms the diagnosis and a value of less than 7 mmol/l refutes it.

However, when the level falls into the grey area between these limits the standard oral glucose tolerance test should be performed.

The test should be carried out in standardized conditions and the results interpreted in relation to age of the patient and whether or not the patient was severely stressed at the time of the procedure.

The test is carried out in the morning after an overnight fast. The patient should have a normal diet for at least 3 days prior to the test. 75 g of glucose is given orally in the form of a sugary drink.

Recent British Diabetic Association criteria have defined diabetes on glucose tolerance testing as a fasting venous blood glucose level in excess of 7 mmol/l and a level at 2 hours greater than 10 mmol/l. Levels lying below these limits but which are clearly abnormal were previously defined as chemical diabetes but are now termed 'impaired glucose tolerance' (LD-57).

Staging of diabetes is schematized in Table 55.

It is important to take into account both the age of the patient and the site of blood sampling in interpreting the test. Glucose tolerance gradually deteriorates with age. Capillary blood or venous plasma tends to measure 1.0 mmol/l higher than an equivalent venous blood sample.

Aetiology

Diabetes is a syndrome rather than a single disease and many causative factors can be recognized. In only

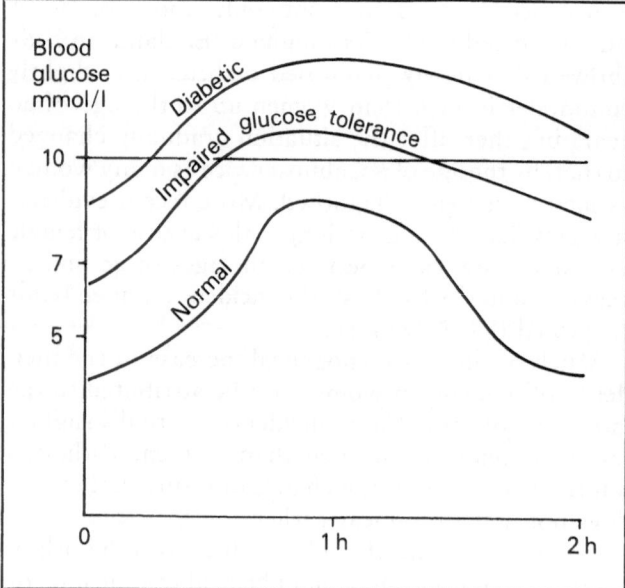

LD-57. *Normal and abnormal glucose tolerance curves. showing the defined range of glucose intolerance and unequivocal diabetes.*

rare instances, such as when the pancreas is largely destroyed or removed, is it possible to define a single cause of the consequent diabetes. The following factors appear to be important:

- age
- sex
- obesity
- heredity
- immune deficiency
- virus infections
- injury and stress

Obesity

Obesity is often associated with high circulating levels of insulin to which the fat person is relatively insensitive possibly because insulin receptors on cell surfaces are reduced in numbers and sensitivity. There is an associated hyperplasia of the beta cells in the pancreatic islets until, after many years, they show degenerative changes and irreversible diabetes results.

Obesity is a feature of the NIDD rather than the IDD type of diabetes. Most of those with the former condition over 45 years of age are on average about 15% above their expected weight at the time of the diagnosis of diabetes and by then the diabetes has usually produced some weight loss. The average degree of obesity is approximately twice that found in

Table 55. Staging of diabetes

Stage	Type of patient
Potential	A person with a normal glucose tolerance but having one or more stigmata associated with the disease: strong family history in first degree relative(s) past obstetric history of a heavy-for-dates baby or stillborn baby with hyperplasia of the islets of Langerhans chronic obesity
Latent	A person with a normal glucose tolerance test who is known to have had temporary asymptomatic or classical diabetes at some time in the past during: pregnancy infection or other stress a previous obese phase A person with a normal glucose tolerance test but with an excessive hyperglycaemia response to a glucocorticoid
Impaired glucose tolerance (previously known as chemical diabetes)	A person in whom glucose tolerance testing is abnormal but is in the borderline range not reaching frank diabetes
Clinical	A person with abnormal glucose tolerance and diabetic symptoms or complications
Prediabetic	This term is best used only in retrospect to describe the time before diabetes was diagnosed

the non-diabetic background population. In recording the history of a new diabetic, it is worth noting the maximum weight and its timing as well as his or her actual weight.

Heredity

Genetic factors play a definite role in the cause of both insulin dependent and non-insulin dependent diabetes but not a dominant one. Several types of inheritance are concerned. These statements are based on:

- twin studies
- the frequency of a family history of diabetes
- histocompatibility antigens

The genetic factors in diabetes appear to be multifactorial and relatively weak. It is never possible to predict with certainty that a person will develop diabetes, however close the family history of the condition. The exceptions are in some of the rare

diabetes-associated syndromes and in uniovular twins with maturity onset diabetes. The risk of children of a diabetic developing diabetes is not high. Thus, less than 1% of the children of a diabetic parent will develop diabetes before adult life and, when both parents are diabetic, only about 5% of the progeny can be found to be diabetic at the time of testing.

Twin studies

Identical twins (monozygotic) who develop insulin dependent diabetes over the age of 45 years are 100% concordant, i.e. both twins are affected. Concordance for insulin dependent diabetes in children, however, is low.

Family history

Insulin dependent diabetics under the age of 16 years will have a family history of diabetes in just over 20% of cases. In these patients there is a marked peaking of presentation around the age of 11 years and to a lesser extent at 5 years. In addition there is a marked seasonal variation of presentation during the winter and spring months in such children.

These findings tend to point to the possible role of the environmental factors of viral infection and the hormonal changes at puberty as important aetiological factors in onset of diabetes in the young. Genetic factors are less likely to be important in such patients whereas in non-insulin dependent diabetics inheritance may be a more significant factor.

Maturity onset diabetes of youth

There are a number of rare strongly inherited syndromes of which diabetes is but one component, but these are quite distinct from the ordinary insulin dependent and non-insulin dependent types of diabetes. Maturity onset diabetes of the young (MODY) is a not-so-uncommon type with a dominant type of inheritance.

Histocompatibility and immunological factors

Insulin dependent, but not non-insulin dependent, diabetes has been found to be positively associated with certain human leukocyte antigens (HLA) genes otherwise called histocompatability antigens. These are antigens B8, B15 and B18, DW3, DW4, DRW3 and DRW4, the effects of which are additive when more than one occurs in the same individual.

The presence of these genes does no more than render that person susceptible to some other agent, e.g. a virus. The genes are distributed differently in various parts of the world and may account at least in part for the scarcity of insulin dependent or juvenile onset diabetes in some countries in the world such as Japan. HLA B8 is positively associated with other endocrine deficiency diseases as well as pernicious anaemia which has long been known to be commoner in diabetics.

As part of the immune mechanism of insulin dependent diabetes, many newly diagnosed young diabetics have antibodies to pancreatic islet cells. These persist for a few years but seem to persist indefinitely in diabetics carrying the HLA B8 tissue type.

Thus two immune processes may be involved in the development of insulin dependent diabetes – the genetically determined susceptibility to a virus and the damaged beta cells setting up an autoimmune process which may complete their destruction.

Virus infection

Insulin dependent, but not non-insulin dependent, diabetes has a seasonal incidence with peaks in winter and the spring and relatively few new cases in the summer, mimicking the pattern of virus infections. Furthermore, as previously suggested there are peaks in the incidence of diabetes in children in the UK at about 5 and 11 years of age. This coincides with the time of first going to school and then changing to a secondary school with consequent increased exposure to viruses. It is common to find a history of vague flu-like illness within a month of the onset of diabetic symptoms in children. In spite of such epidemiological evidence, and suspicion having fallen on the Coxsackie B4 and mumps strains of viruses, no satisfactory laboratory confirmation has been forthcoming.

Injury and stress

It is excessively rare for a direct injury to damage the pancreas sufficiently to produce diabetes but any severe stress may well be sufficient to precipitate the onset of symptoms of diabetes in a patient who previously had impaired glucose tolerance.

Chronic complications of diabetes

Although called 'complications', there are a number of changes which occur in longstanding diabetes which may in fact be inherent features of the disease itself. These include:

- retinopathy
- nephropathy
- neuropathy

All of these may reflect the results of an underlying and specific diabetic microangiopathy.

In addition, diabetics are at greater risk than the normal population for:

- atherosclerosis and all its cardiovascular and cerebrovascular effects
- infections, especially of skin and urinary tract
- hyperlipidaemias

Diabetic microangiopathy

Underlying the retinopathy and the renal changes in diabetes is a specific process affecting capillaries and other small blood vessels.

The initial change – sometimes pre-dating the onset of clinical diabetes – is a thickening of the basement membrane. This shows up as PAS-positive lesions affecting the capillaries and precapillaries. Variations of vessel diameter occur. These changes occur throughout the body and also contribute to:

- gangrene
- neuropathy
- skin changes

Diabetic retinopathy

Since the late 1960s, diabetic eye disease has been the single most common cause of blindness in England and Wales.

Although diabetes affects the eye in several ways including:

- cataracts
- glaucoma
- palsies of the external ocular muscles

the hallmark of diabetic eye disease is the specific retinopathy which occurs.

The classical features (CP-73) of this condition include:

- microaneurysms (CP-73.2, 73.4)
- haemorrhages (CP-73.2)
- exudates (CP-73.1)
- new vessel formation (CP-73.3)

Microaneurysms

The most specific and probably the earliest vascular

changes that occur in diabetes involve the small vessels, so-called microangiopathy, and the eye offers a unique opportunity for examining these *in vivo*.

When inspecting the normal eye through an ophthalmoscope, the patient's lens is used to give a 15–20 times magnification of the retina and, at this magnification, many of the medium or larger sized capillary aneurysms can be seen.

Haemorrhages

These occur early in the course of diabetic retinopathy and are usually rounded, the smaller ones being hard to differentiate from capillary aneurysms (CP-73.2). Flame-shaped haemorrhages are deeper and lie in the nerve fibre layer. Large haemorrhages may develop from preretinal new vessels and intrude to lie immediately behind the vitreous. As they are slowly reabsorbed, they tend to be replaced by fibrous tissue. Intraretinal haemorrhages are absorbed more quickly. Bleeding into the vitreous is more serious and, although the first few will usually clear within a matter of weeks, later ones are absorbed more slowly and less completely. Vitreous haemorrhages are commonly followed by gross deterioration of vision within the next year or so.

Hard exudates

These are distinctive manifestations of diabetic retinopathy and are very variable in size (CP-73.1). They are especially common near the macula on its temporal side. Sometimes they form circinate patterns. With the passage of months or years, some disappear as new ones appear but, in general, they extend or coalesce. When the neighbouring retinal vessels are occluded by photocoagulation, hard exudates tend to disappear, suggesting that they are due to a leakage of oedema fluid and lipid material from abnormal capillaries.

New vessel formation

Formation of new vessels is characteristic of proliferative retinopathy. Before this stage is reached, there is often a period of only background retinopathy.

Background retinopathy

Microneurysms and haemorrhages occur in any area. There may be some scattered hard exudates and arteries and veins may be slightly dilated. It may be seen at any stage of diabetes at any age and may show little tendency to progress for many years. Signs of impending progression are intraretinal dilated capillaries, multiple cottonwool spots and irregularities in the large veins.

Maculopathy

In background retinopathy with maculopathy, visual impairment is due mainly to macular oedema. It usually occurs in insulin independent diabetics and sometimes very early in the history of diabetes. In general, the visual prognosis at 5 years is worse the later the age of onset of diabetes. Maculopathy is in fact a major cause of blind registration over the age of 60 years.

Proliferative retinopathy

This is a manifestation of longstanding diabetes which is usually of juvenile onset type. It is marked by new vessels developing in, and in front of, the retina and followed by the growth of strands or webs of fibrous tissue (CP-73.4, 73.3). While new vessels can be destroyed by photocoagulation, the fibrous changes are irreversible. As the new vessels grow laterally and forwards into the vitreous, a fibrous tissue support develops. Later, the fibrotic process obliterates the new vessels so that there is little risk of further bleeding but, by then, the fibrous tissue has effectively destroyed useful vision.

Proliferative retinopathy is usually slow in its evolution but there is also a more rapidly progressive and florid form seen mainly in patients under 35 years old with poorly controlled diabetes.

Factors affecting evolution of the retinopathy

In juvenile onset diabetes little retinopathy is seen in the first 5 years but the prevalence rises steeply between 5 and 15 years and less so after 20 years' duration of diabetes. There is a shorter history of diabetes before the onset of retinopathy in patients developing diabetes late on in life. Irrespective of age, there is a high frequency of retinopathy after 15–20 years of diabetes.

Good diabetic control is widely considered to be a major factor in delaying the onset and progress of retinopathy although control is notoriously difficult to measure. There is now evidence that good control during the first 5 years of known diabetes is particularly important in delaying the progress of retinopathy.

Diabetic nephropathy

In addition to the small vessel changes, a specific

glomerular sclerosis is found in diabetes. It takes the form of nodular thickening of the intercapillary spaces or a less specific diffuse thickening of the basement membrane.

These changes are very common in longstanding insulin dependent diabetics and are the main cause of death in that group of patients.

The disease usually presents with increasing proteinuria with the later development of the nephrotic syndrome. At this stage, the prognosis is very poor and most patients die from azotaemia within 5 years.

Diabetic neuropathy

A number of neurological features are associated with longstanding diabetes. Although most are thought to be due to a metabolic effect on neural cells, some of the syndromes involving only one nerve are thought to be microvascular in origin.

Peripheral neuropathy

Diabetic peripheral neuropathy usually affects the lower extremities and is mainly sensory causing paraesthesia and loss of most modalities of sensation. Motor and mixed variations also occur. In the absence of pain and pressure sensation, there is a risk of pressure ulcers on the feet (CP-74). In the motor variant, rapid onset of muscular weakness may cause incapacitating paralysis. The weakness usually affects the quadriceps muscles of the thigh (CP-75). Not infrequently, the extraocular eye muscles are involved. Occasionally, neuropathic joint changes can occur and can lead to a presentation similar to Charcot's joint.

Mononeuropathies

These are not uncommon and are usually rapid in onset. There is often complete recovery over a period of months.

Autonomic neuropathy

This causes impotence and neurogenic bladder problems which, along with the increased tendency to infection, often lead to chronic urinary tract infection.

Postural hypotension, gastric distension and nocturnal diarrhoea are other manifestations of the effect of diabetes on the autonomic nervous system.

Diabetic gangrene

Foot ulcers, infections and gangrene of the lower extremities are common in longstanding diabetics of both the juvenile onset and maturity onset types (CP-76).

They are due to a combination of factors including:

- an increased tendency to accelerated atherosclerosis
- an increased tendency to infections
- microangiopathy
- neuropathy
- trauma

Early attention to foot problems and good foot care are very important in preventing the progression to gangrene.

Acute complications of diabetes

These include (see Table 41, p. 98):

- ketoacidosis
- hyperosmolar (aketotic or non-ketotic) coma
- lactic acidosis
- hypoglycaemia

Ketoacidosis

This is due to acute insulin lack and produces a severe state of rapid tissue breakdown in which high levels of growth hormone, glucocorticoids, catecholamines and glucagon play a part. The pathophysiology is represented in diagrammatic form in LD-58.

From this it can be seen that:

- hyperglycaemia is responsible for dehydration
- ketonuria leads to the loss of cations
- cellular breakdown leads to the loss of nitrogen and phosphate

Vomiting, as well as the osmotic diuresis, contributes to the loss of chloride. The fall in volume of the extracellular fluid produces:

- hypovolaemia
- hypotension
- impaired renal function

so that blood concentrations of potassium and sodium may be above normal in spite of a gross total body deficit.

In an adult with severe ketoacidosis the loss of water averages about 6 litres, sodium 500 mmol, chloride 400 mmol and potassium 350 mmol.

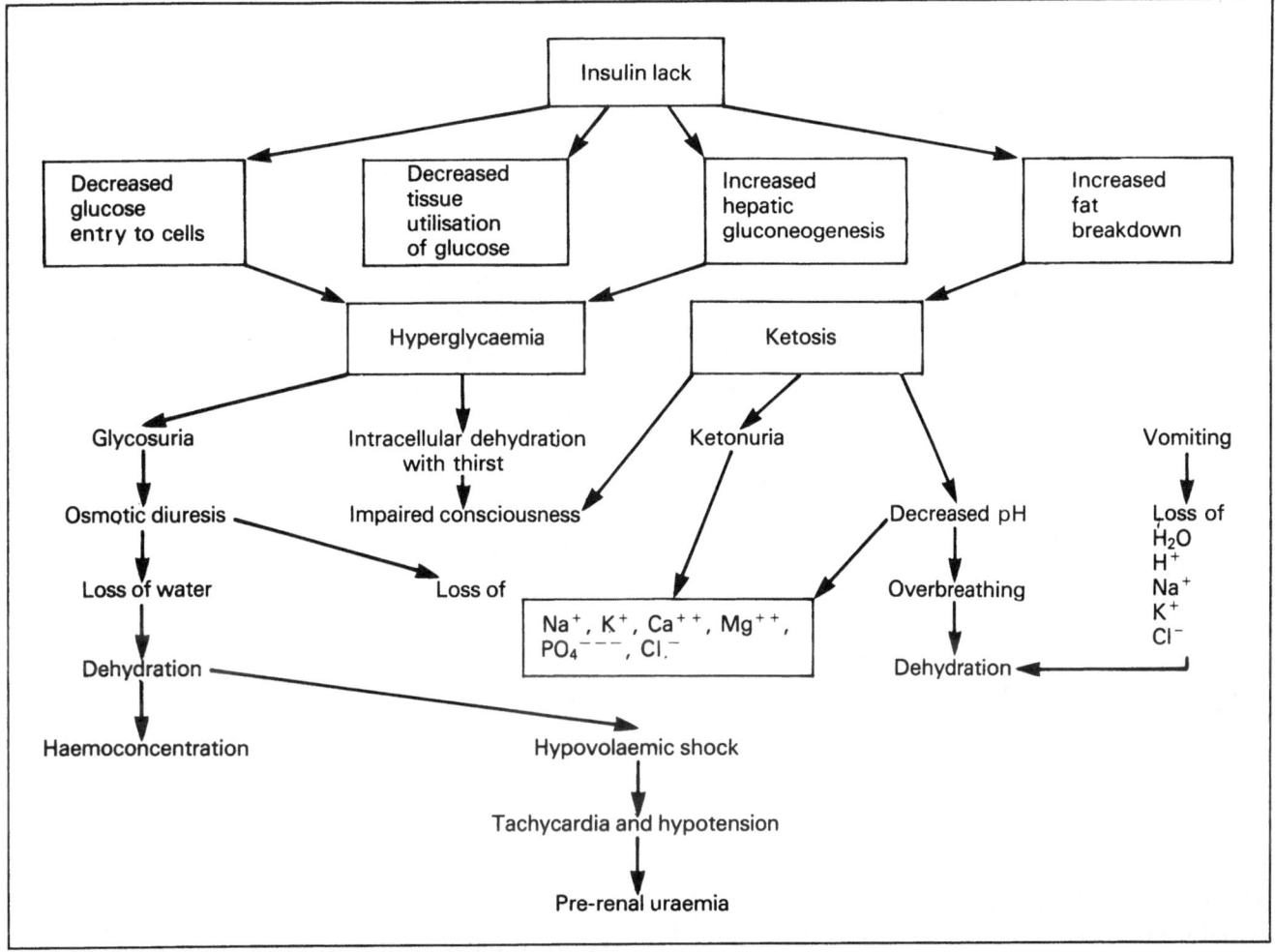

LD-58. *Pathophysiology of ketoacidosis.*

Although the restoration of water and extracellular fluid takes only 2 or 3 days, the intracellular repair process is not usually complete for 10 days or more.

Before the insulin era, most diabetics died in ketoacidotic coma, but the proportion is now less than 1% in most developed countries and even these few cases are in theory preventable.

Insulin dependent diabetics are much more prone to ketoacidosis than obese, maturity onset patients because they typically lack endogenous insulin.

Predisposing factors for ketoacidosis

These are:

- omission or reduction of insulin
- undiagnosed diabetes
- intercurrent illness, especially acute infections

About 25% of all cases of ketoacidosis are due to undiagnosed diabetes with a failure to recognize the significance of symptoms which have been developing for weeks or even months.

Acute infections, especially pyogenic ones, sharply increase the requirement for insulin. Insulin dependent diabetics should test their urine for ketones during an infection, and increase the dose of insulin in steps of 20–25% if the tests are consistently positive.

Not all episodes of severe ketoacidosis can be explained but it is known that acute emotional stress can initiate the process and, in a few young women, menstruation is associated with a variation in insulin requirement.

Clinical features

Severe ketoacidosis develops slowly with warning symptoms of the hyperglycaemia with which it is associated:

- thirst
- polyuria

- lassitude
- dehydration
- vomiting
  often with colicky abdominal pain

By this time the respirations will be deep and rather fast as if the patient had been running (Kussmaul respiration) and the sweet smell of ketones can be detected on the breath.

The patient looks drowsy and has marked signs of dehydration in the skin and tongue with a sunken appearance of the eyes. The pulse is usually fast and the blood pressure may be low but, in spite of this, the face may be flushed. The abdomen may be diffusely tender with few or no bowel sounds and often a succussion splash can be obtained due to the paralytic ileus.

It is important to look for signs of any infection and to exclude other possible precipitating causes such as myocardial infarction.

The urine shows 2% sugar (glucose) and ketones to Acetest, Ketostix or to ferric chloride (Gerhardt's test: this is less sensitive but more significant when positive). The urine may also be infected.

Biochemical findings

Although the blood glucose is always raised, the actual value depends on factors such as the timing of the last meal or the occurrence of vomiting so that the glucose level does not necessarily reflect the severity of the disturbance. The range is about 16–84 mmol/l (approximately 290–1500 mg/dl).

Blood ketone levels are consistently raised to about 100 times normal or more. A rough estimate can be obtained by dropping serum on an Acetest tablet or Ketostix strip and, with significant ketosis, the result will be positive even when the serum is diluted eight or ten times.

The plasma bicarbonate is low and is usually less than 14 mmol/l. A more direct measure of the severity of the acidosis is the plasma pH which may fall below 7.0.

Electrolyte concentrations do not reflect the extent of the deficits, for the water loss may be relatively greater (with acidotic breathing, vomiting or diarrhoea) or less (if the patient has continued to drink to make up for some of the polyuria).

The initial serum potassium is seldom below normal due to its rapid transfer from the cells to the extracellular fluid. The blood urea is usually moderately raised (e.g. 7–25 mmol/l, i.e. 42–150 mg/dl) due

to dehydration, decreased glomerular filtration and rapid tissue breakdown.

There is usually leukocytosis even in the absence of an infection.

Diagnosis

The diagnosis of diabetic ketoacidosis is not difficult. The history of:

- primary diabetic symptoms
- dehydration
- overbreathing
- the slow onset of drowsiness

is characteristic. It is confirmed by finding:

- heavy glycosuria
- ketonuria

The main differential diagnoses are:

- uraemic acidosis
- salicylate poisoning
- severe pneumonia
- subarachnoid haemorrhage with glycosuria and ketonuria

Hyperosmolar or aketotic diabetic state

This occurs nearly as frequently as ketoacidosis but more often in elderly, maturity onset, undiagnosed diabetics. Ketosis does not develop probably because they still have sufficient endogenous insulin. These patients are usually managed later quite successfully with diet and oral hypoglycaemic agents.

The blood glucose values need to be very high for sufficient cerebral dehydration to be produced to induce coma and are usually over 58 mmol/l (approximately 1004 mg/dl). The haemoconcentration also causes the serum electrolytes and urea to rise so that the total osmolality is usually more than 80 mosmol/l above the normal figure of approximately 285 mosmol/l.

The urine contains no more than a trace of ketones and the breathing is not acidotic. The serum bicarbonate is above 16 mmol/l.

The prognosis for this type of coma is less favourable than for diabetic ketoacidosis partly because older patients are involved, because of delay in diagnosis, and because of the severe degree of dehydration and associated hyperviscosity effects.

Lactic acidosis

This is by no means limited to diabetics, but has been associated with diabetes mainly because of the property of the biguanide, phenformin, to bring it on, especially in individuals with a significant degree of renal or hepatic failure. Alcohol may also be a further factor for it raises NADH levels which inhibit lactate dehydrogenase. Phenformin is significantly more likely to produce lactic acidosis than metformin probably because it is concentrated in the liver which is the main site for removal of lactate.

Clinical features

The patient has acidotic breathing but is not necessarily dehydrated. The onset occurs over a matter of hours and the patient is usually restless and may suffer from abdominal pain and vomiting before becoming drowsy and then comatose.

The urine contains a variable amount of glucose and few or no ketones but the most suggestive indirect biochemical finding is a high anion gap in the serum electrolytes, $(Na^+ + K^+) - (Cl^- + HCO_3^-)$, the normal gap being less than 16 mmol/l. If:

- uraemia
- ketonaemia
- salicylate poisoning

are not present, the extra unmeasured anions may be assumed to be mainly lactate.

The mortality of lactic acidosis is high, approaching 50%, especially in association with phenformin, and is even higher if the patient is also in a state of shock.

If lactic acidosis is associated with some degree of ketoacidosis, as is often the case, the ratio of 3-hydroxybutyrate to acetoacetate will be higher than in typical ketoacidosis. Since Rothera's test (e.g. Acetest or Ketostix) detects only acetoacetate, it is an insensitive detector of ketoaemia in association with hyperlactataemia. Blood lactate varies according to the severity of the condition but may reach 20–30 mmol/l, being normally less than 1 mmol/l.

Hypoglycaemic coma

This is not a complication of diabetes but rather of overtreatment with insulin or less often with one of the sulphonylurea drugs. The level at which symptoms occur varies and tends to be higher in older patients but is usually less than 2.2 mmol/l (40 mg/dl) true glucose. It is useful to confirm the clinical diagnosis with a plasma glucose estimation.

Hypoglycaemia due to insulin comes on at a time depending on the peak action of that particular insulin, e.g. about 4 hours after soluble insulin or about 8 hours after lente or isophane insulin or in the early hours of the morning after protamine zinc insulin or Ultratard. Similarly, long-acting chlorpropamide is likely to produce nocturnal hypoglycaemia whereas shorter-acting glipizide or glibenclamide may do so within 3–6 hours of administration.

In an insulin dependent diabetic the causes of hypoglycaemia will be one or more of the following:

- too little to eat
- food taken too late
- unusual, fairly heavy or prolonged exertion
- a mistake with insulin dose
- deliberate overdose
- injecting a long-acting insulin inadvertently into a small vein

Hypoglycaemia may follow a sudden change from a conventional insulin to a highly purified porcine insulin to which the patient is more sensitive.

Management

The management of diabetes mellitus in all its presentations is described in Section V, p. 199.

HYPOGLYCAEMIA

Causes

Hypoglycaemia may be related to:

- insulin overdosage
- overtreatment with hypoglycaemic drugs
- reactive hypoglycaemia
- insulin-secreting pancreatic tumours
- extrapancreatic tumours
- failure of gluconeogenesis
- infants of diabetic mothers.

Insulin overdose

The commonest cause of hypoglycaemic symptoms is relative overdosage with injected insulin. Overdosage effects can occur, not only by mistake in measurement, but also in the method of administration. A long acting insulin injected inadvertently into a small blood vessel has a rapid action. Thus, the piston of the syringe should always be withdrawn slightly after insertion of the needle to confirm that its tip is not in a small blood vessel before the insulin is injected.

Even a usually correct dose of insulin can cause hypoglycaemia if a meal has been missed or delayed or if there has been prolonged and fairly strenuous exercise.

Rare cases of hypoglycaemia artefacta, where insulin has been surreptitiously self-administered to simulate the effects of an insulin-secreting tumour, can now be recognized since such cases have an absence of immune reacting 'C' peptide in their blood whereas it is present in excess when insulin is of endogenous origin.

Overtreatment with hypoglycaemic drugs

The next most frequent cause of hypoglycaemia is overtreatment with a sulphonylurea oral hypoglycaemic agent.

Hypoglycaemia due to a sulphonylurea may occur in a few hours after an inappropriately large dose of one of the newer such agents which are more potent on a weight-for-weight basis. Alternatively, chlorpropamide, which is metabolized slowly, may produce severe and long-lasting hypoglycaemia even in a modest dose, especially if there is renal failure.

A number of drugs can increase the hypoglycaemic potency of the sulphonyureas. These include:

- phenylbutazone
- oxyphenbutazone
- sulphonamides (sulphaphenazole or sulphamethoxazole – one of the two constituents of Septrin and Bactrim)
- clofibrate
- chloramphenicol
- dicoumarol
- salicylates (in large doses)

The biguanides do not produce hypoglycaemia on their own but they can facilitate hypoglycaemia from a sulphonylurea or from insulin.

Reactive hypoglycaemia

This is a fairly common sequel to a partial gastrectonomy or gastroenterostomy. It is also commonly seen in the early stages of diabetes.

Pancreatic tumours

Insulin-secreting (islet cell) adenomas and some carcinomas of the pancreatic beta cells characteristically produce symptoms after an overnight fast, in contrast to reactive hypoglycaemia which usually occurs 2–3 hours after a meal containing carbohydrate and/or protein.

Extrapancreatic tumours

Extrapancreatic large, rapidly-growing tumours, e.g. hepatic and retroperitoneal lesions, may produce hypoglycaemia. The mechanism is not well understood, but may relate to secretion of peptides with insulin-like activity or to the avid utilization of glucose by a large tumour mass.

Failure of gluconeogenesis

Failure of gluconeogenesis, such as occurs with adrenal cortical failure or severe hypopituitarism, can lead to hypoglycaemia after starvation, as can severe liver failure or an enzymatic defect in glycogen mobilization (von Gierke's disease).

Failure of the pituitary or adrenal glands also removes the main hormonal antagonists to the hypoglycaemic action of insulin.

Infants of diabetic mothers

Infants of diabetic mothers tend to have hyperplasia of their pancreatic islets and are prone to hypoglycaemia in the first 2 days or so of life (CP-77).

Other causes

Babies of below average birth weight may also suffer neonatal hypoglycaemia probably due to low levels of liver glycogen or immaturity of one or more enzymes needed for glycogenolysis.

Babies of a few months to 2 years of age may develop idiopathic hypoglycaemia which disappears as they grow older.

Among rare causes of hypoglycaemia are:

- leucine sensitivity (noted first in familial hypoglycaemia of childhood)
- various adulterated alcoholic drinks
- certain toadstools such as *Amanita phalloides*

Clinical features

Symptoms are likely to occur when the capillary plasma glucose concentration falls below about 2.2 mmol/l (40 mg/dl). The brain is mainly involved, especially the function of the higher centres since it, except after prolonged starvation, is dependent almost wholly on glucose for its energy. In the presence of starvation, the brain can derive an increasing proportion of its energy from the metabolism of ketones.

In general, the clinical manifestations can be divided into those dependent on inadequate fuel for the brain (neuroglycopenic) and those associated with the secondary release of adrenaline and noradrenaline (sympathomimetic).

Older individuals tend to tolerate low levels of blood sugar less well than the young and often have a higher threshold of the blood sugar for the appearance of symptoms.

Sympathomimetic symptoms

The sympathomimetic manifestations usually precede those of glycopenia and include:

- apprehension
- tremor
- sweating
- palpitations
- pallor
- widely dilated pupils
- physical aggression
- epileptiform seizures (in severe cases)

The sympathomimetic features of hypoglycaemia are seen when the blood sugar is already starting to rise as a result of glycogenolysis in the liver. They may be the first indication to the subject that he or she is hypoglycaemic. Drugs, such as the less specific β-blockers, can be dangerous to diabetics by blocking glycogenolysis and their warning adrenergic symptoms of hypoglycaemia.

Neuroglycopenic symptoms

These may reveal the inadequacies of the arterial circulation by producing a transient hemiparesis or an episode of angina pectoris.

Other symptoms include:

- diplopia or inability to focus the eyes
- ataxia
- dysarthria
- inappropriate changes in mood
  varying from depression and inactivity to euphoria and mania

The general impression is similar to that of the uninhibited behaviour of alcoholic inebriation. There may be partial or full amnesia of the event.

Investigations

If the diagnosis of acute, severe hypoglycaemia is suspected clinically, it should always be confirmed retrospectively by drawing blood for glucose prior to administering glucagon or intravenous glucose.

The diagnosis of chronic mild-to-moderate hypoglycaemia in the non-diabetic can be very difficult since the symptoms may mimic neuroticism and other psychiatric conditions. A key point in suggesting that hypoglycaemia is the cause of the patient's symptoms is the relationship of onset of the attacks to food intake (or lack of it) and to exercise. Even so, that the patient's attacks are due to hypoglycaemia must always be confirmed by finding a low blood glucose level at a time when the patient is symptomatic.

Reactive hypoglycaemia is not seen in the fasting state but occurs 2–3 hours after a meal. It can be confirmed by a prolonged (e.g. 4 hour) glucose tolerance test.

Although hypoglycaemia due to pancreatic tumours is rare, it is important to make the diagnosis since 95% are curable adenomas. This form of hypoglycaemia is always related to fasting (and may be precipitated by exercise). The diagnosis is confirmed by serial measurement of blood glucose and insulin (if available) levels during a prolonged fast.

Recently a test has been developed on the basis that the normal suppression of endogenous serum insulin levels following the injection of a small amount of fish insulin (which does not cross-react with human insulin in the assay) is not seen in patients with autonomous insulin-secreting tumours.

Management

The management of acute hypoglycaemia consists of intravenous glucose (usually as a 50% solution) in the severe cases or provision of rapidly assimilable carbohydrate orally in mild cases.

In diabetics, hypoglycaemia is a signal to adjust the insulin or oral hypoglycaemic agent dosage.

For further details see Section V, p. 213.

In most other cases this consists of treatment of the offending cause:

- surgical excision of tumours
- dietary management of reactive hypoglycaemia avoiding large carbohydrate meals and substituting frequent small meals
- steroid replacement therapy in hypoadrenalism

Where removal of the offending tumour is not possible, e.g. inoperable islet cell carcinomas, the hypoglycaemia may be treated by diazoxide. This is given orally in doses ranging from 50 mg twice daily to 800 mg daily in divided doses. Since it only prevents the release (and not synthesis) of insulin, it is ineffective in those tumours which secrete insulin without storing it.

Streptozotocin is a chemotherapeutic agent with specific efficacy against islet cell carcinomas.

SIMPLE OBESITY

This is defined as excessive accumulation of fat in the storage areas of the body. This definition is attractively simple but lacks objectivity regarding how to define 'excessive'. Two different standards can be and are used:

- average weight
  This is recorded for sex, height and age in a particular part of the world.
- optimum or ideal weight
  This takes the weight range shown by actuaries of life insurance companies to carry the best expectancy of long life (Tables 34, 35).

Average weights of populations are affected largely by the availability of fattening foods and thus tend to rise after periods of deprivation imposed by war or famine, whereas standards of optimum weight for longevity vary much less and form a more rational basis for the aim of treatment.

Prevalence

Obesity is by far the most common and serious nutritional disease in the affluent parts of the world.

Depending on the criteria for diagnosis, from a third to a half of the population of Britain is obese (Office of Health Economics, 1969) and in the USA, about 36% of men and 42% of women are classified as overweight.

The problem of human obesity in the westernized nations follows many centuries of man's history when famine was the nutritional threat. Individuals with a good capacity to lay down fat at times of adequate nutrition continued to do so mainly because of decreasing needs for exercise and the ready availability of manufactured foods of concentrated energy content. They are now at risk. Furthermore, their increased mortality seldom shortens their reproductive life so that natural selection plays little or no part in limiting the increase of obesity in the community.

Causes

The basis of obesity is an imbalance, usually of long duration, between the levels of energy intake and of expenditure but there is no doubt that some individuals gain fat with greater ease than others (CP-4). There is seldom a single cause of obesity and it is often difficult to assess the individual role of the multiple causes in any particular case.

Factors influencing the development of obesity include:

- economic factors
- eating pattern
- psychological factors
- genetic factors
- environment
- lack of exercise

Economic factors

An important economic influence is the uneven distribution of food resources over the surface of the earth.

Eating pattern

This, quite apart from energy intake, influences weight. One big meal a day induces the laying down of more fat than the spaced ingestion of the same amount of food at intervals in the day.

Psychological factors

There are important psychological aspects of simple obesity for there is no doubt that many obese people eat too much mainly as a compensation for frustration and boredom. In addition, the presence of obesity isolates the individual and makes it less easy for him or her to join the usual social and sporting activities of their associates.

Genetic factors

Genetic influences are important, not only in the production of obesity, but in its distribution in the body. This is shown best in studies of identical twins living apart because, in a family group, it is very difficult to dissociate genetic from environmental factors.

Environment

Social pressures may encourage the development of obesity in some countries just as they discourage it in others. In regions of the world where hunger is common, obesity may be considered a desirable manifestation of affluence.

Social or family conventions may lead to overfeeding the infant leading to an increase in the number of fat cells (hyperplastic obesity) or an adult of normal proportions may accumulate fat at a time when individual adipocytes enlarge rather than divide (hypertrophic obesity).

Lack of exercise

It has been shown that obese people take significantly less exercise than those of normal weight. Social class also influences the prevalence of obesity which tends to be higher in women in social classes (UK definition) IV and V than in I and II but, in men, the trend is in the opposite direction. Many women first become obese during pregnancy and, if they do not lactate, such fat stores tend to persist.

Metabolic features

There are various metabolic and hormonal changes found in the obese when compared to those of normal weight. These include increased levels of:

- fasting glucose
- free fatty acids
- cholesterol
- triglycerides
- insulin

Glucose tolerance tends to fall and insulin sensi-

tivity decreases with the development of obesity. After several days of fasting, there is less ketonaemia in the obese than in those of average weight. There is no evidence that any of these metabolic changes are actually causing obesity because they all revert to normal when dietary weight loss has occurred.

Complications

The importance of obesity depends on its influence on mortality and morbidity. Most of the information is derived from life insurance statistics and is to this extent selected. These show in both the USA and the UK that obesity is associated with increased rates of mortality and that the mortality rate varies more or less in proportion to the degree of obesity. In fact, the best survival figures are obtained by men at 20 lb (9 kg) less than average weight and by women at 30 lb (13.6 kg) less than average weight, the rise being steeper for men than women and steeper for younger than older men.

Among the obese, even without the additional risk factors of:

- hypertension
- family history of cardiovascular disease

mortality from cardiovascular and renal disease is 50% above normal. There is also an increase in deaths due to conditions of the biliary tract but the most dramatic increase is a fourfold one in deaths associated with diabetes. There is evidence that a substantial reduction of the weight of obese people is alone sufficient to diminish the greater death rate usually associated with obesity.

The effects of obesity on morbidity make a formidable list (Table 56).

Management

Not all obese people eat excessive volumes of food but, to gain weight and remain overweight, they must be eating a diet having too high an energy content. Some obese people can be considered addicts of concentrated carbohydrate and, like other forms of addiction, this is hard to treat. Any attempt to treat obesity is doomed to failure unless the patient is strongly motivated by a conviction that dietary control with its improved prospects outweighs the more immediate satisfaction of self-indulgence. It is

Table 56. Diseases associated with obesity

Class	Condition
Cardiovascular	Hypertension
	Myocardial infarction
	Cardiac failure
	Strokes
	Varicose veins
	Phlebothrombosis and embolism
Metabolic	Diabetes mellitus
	Hypercholesterolaemia
	Gallstones
Skeletal	Osteoarthritis
	Accident proneness
Respiratory	Respiratory failure
Gastrointestinal	Herniae
	Hiatus hernia
Skin	Intertrigo
	Fungus infections

always worth asking why the patient decided to seek medical advice about the treatment of obesity and then consider with them other important reasons for treating it.

Diet

Since obesity is caused by the ingestion of more food energy than is needed, the obvious aim in treatment is to change this excess into a deficit of calories. Such reducing diets are usually planned to contain about 4.184 MJ (1000 kcal). The normal recommended allowance for adults doing light work is 11.51 MJ (2750 kcal) for men and 9.41 MJ (2250 kcal) for women.

Fat is such a concentrated source of energy (9 kcal/g) that it is clear it must be used only sparingly in any weight-reducing diet but, if carbohydrate is strictly limited, unrestricted amounts of food containing only fat and protein can be allowed since these appear to be self-limiting when carbohydrate is low.

As it is always rather more interesting to learn about foods that can be eaten rather than those which cannot, the opportunity should be seized at the beginning of any dietary treatment to instruct about the use of low-calorie food such as:

- clear soups
- green vegetables
- raw fruits

Switching to such items can maintain a satisfying bulk to the meals while retaining their potential to allow weight loss.

Since may obese people find it hard to give up sugar, they should be told that, if they insist on masking the taste of various foods and drinks by the common factor of sweetness, saccharine should be used but, better still, they should teach themselves to appreciate the natural taste of foods.

It is more effective for those on a reducing diet to take small meals at intervals of a few hours rather than infrequent larger meals.

A more drastic form of dietary management – fasting – with the free provision of non-calorific drinks has been used. An obese subject so treated feels severe hunger only for about the first day and then settles down to a spell of unexpected wellbeing, associated with ketonuria and mild ketonaemia, not amounting to metabolic acidosis. Starvation has an initial diuretic effect usually of 2 or 3 litres in the first day or so, so that the initial weight loss is a morale booster, but later settles to 2–3 kg a week. Ingestion of carbohydrate, such as for a glucose tolerance test, rapidly primes the tissues to retain water again. The rate of weight loss decreases after starvation is continued for about 10 days due to a fall in metabolic rate and more efficient conservation of body stores. The rate of nitrogen excretion also diminishes and the blood urea falls. The weight loss comes unfortunately more from lean body mass than from fat. This is a good reason not to prolong such fasting for more than 8 days or so, before permitting some protein intake. Total fast should only be carried out in hospital.

Exercise

The role of exercise is important since it permits the maintenance of muscle mass while adiposity is decreasing and also tends to restore the natural relationship between energy intake and output seen in physically active animals and men.

Drugs

Although there is a great expenditure on a variety of anorectic drugs, their place in the treatment of obesity should be secondary and minor, if they are used at all.

Anorectic drugs may produce important side-effects dependent on the actual drug being used. They are of three main sorts:

(1) Stimulant effects on the central nervous system

- nervousness
- irritability
- insomnia
- decreased sense of fatigue

- euphoria

  These sensations may lead to addiction and, for this reason, amphetamine has been largely withdrawn from use.

(2) Sympathomimetic effects

- dry mouth
- tachycardia
- palpitations
- sweating
- blurred vision
- dizziness
- rise in blood pressure

(3) Dyspeptic effects

- nausea
- vomiting
- constipation

  Such symptoms may be equally common on placebo treatment in control trials.

Psychotherapy

In general, the failure rate of all medical regimes to reduce obesity is high but some patients do better when they get the additional incentive of competition with others in a group, as in Weightwatchers or in TOPS (Take Off Pounds Sensibly). Group therapy in hospital or general practitioner health centres may also have advantages over treatment in isolation.

Other treatments

The role of such drastic measures as ileocolic bypass surgery (surgically creating a malabsorption state), dental splinting (so that no solid food can be consumed) and destruction of hypothalamic feeding centres is not yet clear but is almost certainly extremely limited.

SHORT STATURE

This is much commoner than tall stature as a presenting feature requiring investigation.

Causes

These are:

- constitutional

  without skeletal and pubertal delay

  with skeletal and pubertal delay

- chronic systemic disorders
- abnormalities of chromosomes or genes
- intrauterine growth retardation
- bone and cartilage disorders
- psychological and social factors
- endocrine disorders
- iatrogenic

Constitutional

Without skeletal and pubertal delay

In this condition, the child's height is below the 3rd centile but not usually below the -3 standard deviation line, and it falls within the range of height predicted for the children from the midparental height. As the bone age is virtually normal and puberty occurs at approximately the normal time, the short stature seen in childhood persists into adult life. Investigation of such patients reveals no organic cause for the short stature.

With skeletal and pubertal delay

The height in this condition is also usually between the 3rd centile and the -3 standard deviation line on the chart. However, the bone age is considerably retarded of the order of 2–3 years or more, and the onset of puberty is correspondingly late. Growth continues for several years longer than in the average population, and the adult height falls near the lower end of the normal range.

Clinically, these individuals resemble patients with partial growth hormone deficiency and the distinction is made by finding a normal growth hormone response to the exercise test or insulin hypoglycaemia, or other growth hormone stimulation test.

It is unwise to embark on therapy with testosterone or oestrogens in an effort to produce a growth spurt, as this, by causing earlier closure of the epiphyses, would diminish the final height attained. Time should be taken to reassure the adolescents that they will be taller relative to their peers as adults. However, they still have to adjust to the fact that for several years they will remain short in stature and that, in addition, their sexual maturation will be delayed.

Chronic systemic disorders

Normal growth is a very accurate reflection of the health of a child. Any chronic systemic disease will interfere with growth to a greater or lesser extent.

Many disorders are diagnosed on the basis of specific symptoms and signs related to the system involved, for example, congenital heart disease, chronic asthma, cystic fibrosis, repeated urinary tract infections, chronic malignant disease etc.

However, from time to time, a child presents with short stature and, on investigation, a chronic systemic disease is found to be responsible. Of the conditions likely to be brought to notice in this way, examples are coeliac disease, presenting for the first time in later childhood, Crohn's disease and chronic renal failure (CP-78).

Abnormalities of chromosomes or genes

Children suffering from Down's syndrome due to trisomy of the number 21 chromosome or translocation of a number 21 chromosome on to another acrocentric chromosome are of short stature, but well-known clinical stigmata draw attention to the children in the neonatal period or early in life.

Similarly, in typical Turner's syndrome, where the chromosomal pattern is 45,XO, the characteristic physical features – for example, webbing of the neck, shield-shaped chest, congenital heart disease, an increased carrying angle at the elbow and, in a proportion of cases, mental defect – lead to the diagnosis being made in infancy. However, where mosaicism is present, these features may be modified. Where there is only partial deletion of one X chromosome, or ring forms of the chromosome exist, the girl may present in the first instance only with short stature. Delay in puberty will be noted later and patients may present at a gynaecological clinic on account of this if the short stature has not led to earlier referral to hospital. It is therefore very important to do a chromosomal analysis in all girls presenting with short stature or delayed puberty.

There are many inborn errors of metabolism due to an abnormality of a single gene associated with short stature. Most of them have other presenting features which enable the diagnosis to be suspected and confirmed by further investigation. Many are inherited as autosomal recessive conditions so that genetic counselling is of value.

Intrauterine growth retardation

Congenital defects arise in the first 3 months of pregnancy. They may be due to genetic causes or to causes which cannot be ascertained, or to seriously inadequate intrauterine conditions. They may kill the fetus or lead to intrauterine growth retardation so that the infant is born light-for-dates. Many congenital defects should be picked up at the routine neonatal examination.

When adverse factors affecting the fetus arise from the 4th month of pregnancy onwards, the only outcome of clinical importance may be intrauterine growth retardation. Severe instances of this will be detectable clinically by repeated evaluation of uterine size from measurements of the distance between the pubic symphysis and the height of the fundus.

When ultrasound measurements of the transverse diameter of the fetal skull are available, it is possible to pick up lesser degrees of intrauterine growth retardation and to distinguish between protracted failure of growth from the 3rd or 4th month of pregnancy and the commoner situation, when failure of growth takes place in the last month or two of pregnancy.

A recent follow-up study of babies, whose intrauterine growth was monitored in this way, has shown that prolonged slow growth *in utero* is likely to be followed by an inadequate velocity of growth after birth and, in addition, some degree of brain damage is common. In contrast, infants suffering from intrauterine growth retardation late in pregnancy often catch up in every respect and become indistinguishable from normal children.

It may be possible to improve the conditions for the fetus by careful care of the mother during her pregnancy. The following potentially adverse factors should be considered:

- poor nutrition with regard to the intake of:
  calories
  protein
  vitamins

- moderate or heavy smoking

- alcohol consumption

- drug addiction

- drug therapy, e.g. for epilepsy

However, some conditions are difficult to influence:

- pre-eclamptic toxaemia

- eclampsia

- inadequate vascular supply to the placenta

- intrauterine infections

Intrauterine growth retardation leads to permanent short stature if catch-up growth is inadequate. Usually the intelligence is normal and the bone age is at most only slightly delayed so that puberty occurs around the normal time.

In the Silver–Russell syndrome there is asymmetrical development of the face, limbs or trunk and the patients are thin with low skinfold measurements.

In addition, there are many rare syndromes of maldevelopment associated with intrauterine growth retardation and subsequent short stature. Some are hereditary and, if there is a known mode of inheritance, it will be important to give the family genetic counselling.

Bone and cartilage disorders

These disorders are very numerous although individual examples are rare. Some are due to inborn errors of metabolism with known modes of inheritance but many are of unknown aetiology. The short stature is usually because of subnormal growth of the legs and therefore the sitting height is increased in relation to the total height, as shown on the special charts available. Short stature due to subnormal growth of the spine occurs very rarely.

Children suspected of having such disorders should have an X-ray taken of the skull, spine, hips and one arm and hand, and one leg. The radiological abnormalities will usually lead to a diagnosis. Among the commoner disorders are:

- achondroplasia
- hypochondroplasia
- dyschondroplasia (type Schmid)
- the skeletal abnormalities associated with the mucopolysaccharidoses

Achondroplasia

This causes a severe degree of short stature because of defective proliferation of the cartilage cells of the growth plates. The bones formed in membrane are normal. The condition is inherited as an autosomal dominant but, because of the reduced reproductive capacity of affected individuals, most of the patients in the population have arisen as a result of a new mutation.

The clinical features of shortness of the limbs and shortness of the bones of the hands and feet associated with skull changes due to underdevelopment of the base, which is developed from cartilage, and full development of the vault of the skull, which is developed from membrane, may lead to the diagnosis in the neonatal period and certainly most cases are diagnosed in early childhood. The intelligence is usually normal.

Radiologically there is flaring and irregularity of contour at the metaphyses. The epiphyses appear irregularly.

Hypochondroplasia

This is a less severe defect of cartilage proliferation and is also of autosomal dominant inheritance. In this condition the skull and facies are relatively normal and the short stature is less obvious than in achondroplasia. Nevertheless, there is some shortening of the long bones and some flaring of the metaphyses. Such children are quite likely to present on account of short stature and are diagnosed on the basis of relatively short limbs and the radiological findings. *Short stature due to shortness of the trunk* is rare, the commonest type being spondylo-epiphyseal dysplasia of X-linked recessive inheritance.

Dyschondroplasia

The commonest form of metaphyseal dyschondroplasia is the type Schmid. It is often accompanied by bowing of the legs.

Radiologically there is splaying and irregularity of the metaphyses with deficient ossification. The appearances are similar to those of rickets which should be excluded.

Mucopolysaccharide-associated skeletal disorders

Those which may cause short stature are found, for example, in the syndromes of Hurler, Hunter and Morquio.

An abnormal mucopolysaccharide may be detected in the urine and the specific enzyme defect established by studies of cultured fibroblasts.

Other bone disorders

The congenital form of osteogenesis imperfecta is associated with extreme short stature but this condition is likely to be diagnosed at birth.

Multiple enchondromatosis (Ollier's disease) is another condition which may be associated with short stature.

Management

Nothing can be done to improve the growth of children with disorders of bone or cartilage. Nevertheless, it is important to try to reach a diagnosis so that the child may be spared unnecessary investigations for short stature. In addition, genetic counselling may be of value to the family.

Psychological and social factors

Psychological and social factors are frequently present together, hence the term 'psychosocial short stature'. Undernutrition or malnutrition of varying degree may also play a part in the aetiology. The children suffer from the psychological ill-effects of being unwanted and lacking in parental love and interest.

Typically, they are unhappy and sleep badly and their eating and drinking habits are disturbed. At times they eat and drink very little and at other times they have bouts of voracious eating and drinking when they steal food of any description and drink whatever they can find – even water from the lavatory bowl. Such bizarre behaviour makes the diagnosis relatively easy but the psychological disorder may be much more subtle.

The parents are often inadequate and they may be mildly mentally retarded. They do not usually seek advice early on account of the poor growth of the child so that the condition goes on for several years undetected, leading to short stature of considerable degree. However, it should be borne in mind that psychological causes may lead to poor growth in a child belonging to a high social class family. As such parents are often on the defensive when the history is taken, it may take some time to reach a diagnosis.

Differentiation of children suffering from psychosocial short stature from those with growth hormone deficiency of an organic nature may be difficult as the psychosocial group may show a poor growth hormone response to stimulation tests. Every effort should be made to improve the child's family circumstances either by help given directly to the parents at home or by removing the child to a foster home or a residential childrens' home. As the child responds to happier circumstances, catch-up growth can be expected, demonstrated by careful monitoring of the height velocity. Once this occurs, the tests for growth hormone response will no longer show deficiency.

Other manifestations of psychological disturbance, such as anorexia nervosa or severe food refusal, may lead to a reduction in height velocity. Growth resumes at a normal rate when treatment is successful. In these conditions, help is usually sought before they have continued for long enough for the poor growth to lead to short stature as defined.

Endocrine disorders

The causes of short stature from endocrine disorders are:

- growth hormone deficiency
- disorders of puberty:
  delayed
  precocious
- disorders of the adrenal cortex:
  Cushing's syndrome
  adrenal hyperplasia
- hypothyroidism
- poorly controlled diabetes mellitus

Growth hormone deficiency

Short stature due to growth hormone deficiency has been described earlier in this section (p. 111).

Disorders of puberty

If puberty is delayed (see p. 117) the growth spurt normally associated with it is also delayed. Short stature and delayed puberty often occur together.

In precocious puberty (see p. 119), short stature in adult life is a likely consequence due to early cessation of growth when the epiphyses close.

Disorders of the adrenal cortex

In childhood and adolescence Cushing's syndrome, which is a rare condition, causes short stature because of excess cortisol production. It leads to inhibition of growth hormone secretion, inhibition of somatomedin production in the liver, which is essential to growth hormone action, and direct inhibition of the multiplying cartilage cells of the growth plates in bone.

The features described in adults (p. 125) lead to the condition being suspected clinically.

Details of adrenal hyperplasia are given on p. 129. Careful monitoring of treatment with hydrocortisone, to preserve a normal bone age, should prevent the occurrence of short stature due to the inadequately suppressed production of adrenal androgens which leads to early closure of the epiphyses.

Hypothyroidism

The thyroid hormones speed up metabolism and, if they are deficient, there is slowing of cell function throughout all the body systems. The result is marked retardation of growth and bone age in infants and children with moderate or severe thyroid deficiency (see p. 134).

The clinical features are described on p. 134. In milder forms of hypothyroidism, e.g. due to TSH deficiency or early autoimmune disease of the thyroid gland, growth is affected, but possibly not enough to produce short stature, i.e. a height below the 3rd centile for age and sex. The parents may notice that the child is not outgrowing his clothes. Height velocity studies will reveal accelerated growth after treatment with thyroxine.

Diabetes mellitus

It has been reported that children with diabetes mellitus may not grow satisfactorily and may show delayed puberty. It is likely that such children had poorly-controlled diabetes leading to a degree of undernutrition. Well-controlled diabetic children, on the other hand, grow normally and enter puberty at the normal time.

Iatrogenic

The use of any corticosteroid preparation for therapy carries the risk of growth suppression during childhood and adolescence. The mechanisms are the same as those described under Cushing's disease and indeed other features of this disease also occur as side-effects of corticosteroid therapy. It has been proved that intermittent treatment on alternate days with, for example, oral prednisolone, has a less detrimental effect on growth than continuous daily therapy, although growth impairment occurs even with intermittent therapy if large doses are required and/or if treatment is prolonged. Topical steroids applied to the skin, e.g for eczema, or used by inhalation, e.g. for asthma, or even used as eyedrops for allergic conjunctivitis, may also interfere with normal growth. The height velocity of all children and adolescents receiving corticosteroid therapy should be carefully monitored if courses exceeding 3 months' duration are envisaged. Once therapy is stopped, catch-up growth occurs provided the patients are well.

Therapy with testosterone or its derivatives was once recommended to produce a growth spurt in boys with short stature. However, it leads to earlier closure of the epiphyses and a shorter adult height than would otherwise have been achieved.

Therapy with cytotoxic drugs, e.g. for leukaemia or tumours, may contribute to poor growth by their generalized toxic effect.

Deep X-ray therapy to the brain leading to involvement of the hypothalamus may impair growth hormone secretion and hence cause a reduction in height velocity. Deep X-ray therapy to the spine may directly interfere with its growth.

Conclusion

The investigation of children and adolescents with short stature is time consuming. However, the effort is very well worth while as, for a substantial number, appropriate treatment will be successful in improving the growth and general health of the patient.

TALL STATURE

For psychological reasons girls are more likely to seek advice on account of tall stature than boys. Investigation is indicated for either girls or boys whose heights lie above the 97th centile.

The history, family history and social history should be recorded as is done for short stature and the results of the examination plotted on a height chart. In the systemic examination particular attention should be paid to the central nervous, cardiovascular and locomotor systems, and the possibility of metabolic or endocrine causes should be borne in mind.

Causes

These are:

- constitutional
     without skeletal and pubertal advance
     with skeletal and pubertal advance
- disorders of the central nervous system
- abnormalities of chromosomes or genes
- those associated with obesity
- endocrine disorders

Constitutional

This is the commonest cause of tall stature and reflects tall stature on the part of the parents. It may be associated with a relatively advanced bone age and early puberty, resulting in the tall stature being more pronounced during childhood and adolescence than it is in adult life.

Disorders of the central nervous system

Occasionally tall stature due to the secretion of excess growth hormone from an anterior pituitary adenoma occurs in older children and adolescents. When it affects infants and young children secondary to brain damage it is called cerebral gigantism (Sotos' syndrome) (CP-79). Further details have been given above on p. 113.

Abnormalities of chromosomes or genes

Healthy males with a chromosome constitution of XYY have an adult height above the 50th centile and possibly above the 97th centile. The following genetically determined syndromes should be considered in children with tall stature:

- Marfan's syndrome
- homocystinuria

Marfan's syndrome

Marfan's syndrome is a connective tissue disorder with musculoskeletal, ocular and cardiovascular manifestations. Sporadic mutants are recognized. It is transmitted by autosomal dominant inheritance.

The limbs are abnormally long and thin and the patients have noticeably long fingers and toes. Intelligence is normal. Musculoskeletal disorders may occur, e.g. general hypotonicity and a tendency to kyphosis and scoliosis. Ocular manifestations may include dislocation of the lens of the eye. Aortic regurgitation may be present and dissection of an aortic aneurysm is a recognized association.

The metacarpal index is a useful diagnostic aid. On an X-ray of the hand the combined lengths of the four metacarpal bones relative to the summation of their widths is found to be greater than normal.

Homocystinuria

This is an autosomal recessive condition which may be associated with clinical features resembling those of Marfan's syndrome. However, about half of these patients are mentally retarded.

The diagnosis is made by finding excess homocystine in the urine. It is confirmed by enzyme assays on cultured fibroblasts.

Although the condition is very rare it is amenable to treatment and should always be excluded before a definitive diagnosis of Marfan's syndrome is made.

Causes associated with obesity

The height of obese children is usually above the 50th centile and in some it may even be above the 97th centile. Such tall fat children usually have an advanced bone age and early puberty. Whether this advanced development is due to overnutrition is a matter for conjecture, due to a mechanism the reverse of that well recognized as causing short stature and delay in puberty in children suffering from undernutrition.

Endocrine disorders

The thyroid gland

Hyperthyroidism is rare in childhood, but, when it occurs, it may be associated with tall stature and advanced bone age. There is a high thyroxine level in the serum.

The gonadal axis

Precocious puberty leads to an early spurt in growth and resultant tall stature. It should be possible to diagnose the condition on clinical grounds by finding breast development in girls and enlargement of the penis and testes in boys, followed by the appearance of pubic hair before the age of 9 years in girls and 10 years in boys. The condition is confirmed by finding high LH, FSH and testosterone or oestradiol levels in the serum. As the child gets older, however, the epiphyses fuse early and short stature will be present in adult life. Early treatment of boys with cyproterone acetate may lessen the risk of final short stature.

Delayed puberty, because of delayed closure of the epiphyses, allows extra years of growth so long as growth hormone remains active, and eunuchs tend to be tall.

The adrenal cortex

In untreated or undertreated adrenal hyperplasia, adrenal androgens are synthesized in excess and result in a similar growth pattern to that described for precocious puberty, i.e. tall stature in childhood and short stature in adult life, unless the situation is corrected by adequate dosage of hydrocortisone.

Growth hormone excess

In acromegaly, excess growth hormone is produced by a tumour of the anterior pituitary gland and, if such a tumour arises during childhood and adolescence the excess growth hormone leads to tall stature or 'gigantism'. The possible contribution of growth hormone excess in the causation of cerebral gigantism has been discussed (p. 113) under the heading Cerebral gigantism (Sotos' syndrome).

Investigations and management

The relevant investigations have been described elsewhere.

The management of boys and particularly girls with constitutional tall stature must include adequate explanation of the condition to the parents. It is possible to predict the adult height with enough accuracy to be able to reassure the majority, for the commonest time for them to present is just after the peak of the puberty growth spurt, which has often occurred rather earlier in them than in their friends. It is not easy to be certain whether oestrogen therapy for girls, which should ideally begin in early puberty at the latest, does in fact lead to a significant reduction in final height, although growth inhibition of a few centimetres has been claimed. This approach might be considered if the height prediction for the girl exceeded 1.8 metres (6 feet).

MISCELLANEOUS CONDITIONS

These include:

- carcinoid syndrome
- ectopic hormone production by non-endocrine tumours
- multiple endocrine adenomatosis

CARCINOID SYNDROME

Carcinoid tumours are tumours of argentaffin cells with very characteristic histological features which occur most commonly in the stomach, intestine and bronchus. In some 3% of patients with a carcinoid tumour, mainly those in which metastatic spread has already taken place, a very striking clinical syndrome may occur:

- attacks of flushing
- attacks of diarrhoea
- development of cardiac valvular lesions

Causes

The basic cell of which the tumours appear to consist are chromaffin cells of the gut. The commonest site is the ileum but carcinoid tumours have arisen in:

- pancreas
- stomach
- colon
- rectum

as well as in non-gastrointestinal sites such as:

- lung
- ovary
- thyroid

The carcinoid tumour produces excessive amounts of serotonin (5-hydroxytryptamine 5-HT) and other amines. Some also produce peptides such as kallikrein which is responsible for the production of bradykinin within the plasma. It is thought that bradykinin is responsible for the flushing attacks. It is uncertain whether the role of 5-HT (serotonin) is in the production of the clinical features but it may have some role in causing diarrhoea.

Clinical features

The syndrome is commoner in males and can occur at any age. The most characteristic features of this syndrome is the carcinoid flush (90% of all cases). Initially this is episodic: the skin, particularly of the face and upper part of the trunk, becomes generally red or cyanotic. The flushes may be spontaneous or provoked by excitement, alcohol or straining at stool. Some patients develop a permanent erythema with numerous superficial dilated blood vessels.

Many patients have excessive gastrointestinal mobility with abdominal pains and diarrhoea (50% of cases). Some patients with longstanding carcinoid develop a valvular heart disease consisting predominantly of right-sided lesions, notably pulmonary and tricuspid stenosis (50% of cases). The mechanism of production of the cardiac lesions is unclear.

Less common features are:

- asthma
- oedema
- peptic ulcers

Investigations

The screening test for the carcinoid syndrome involves the estimation of the urinary excretion of 5-hydroxy-indole-acetic acid (5-HIAA). Normally, less than 8 mg is excreted daily in the urine; in the carcinoid syndrome the excretion may be many times greater. Radiology of the gastrointestinal tract and liver scan are necessary to localize the tumour.

Management

Once a carcinoid syndrome has developed, it is likely

that the tumour has metastasized, usually to the liver. Despite this, symptoms may be alleviated by surgical removal of the primary tumour and as much metastatic tissue as possible.

Various drugs may also be effective in relieving the flushing attacks:

- methyldopa
- adrenergic blocking agents

or the diarrhoea:

- methysergide
  This drug has the disadvantage of causing retroperitoneal fibrosis in some cases.

ECTOPIC HORMONE PRODUCTION BY NON-ENDOCRINE TUMOURS

It has become increasingly recognized that some non-endocrine tumours are capable of producing sufficient quantities of hormones to produce endocrine syndromes. The clinical and biochemical manifestations of such hormonal-producing tumours include:

- hypercalcaemia
- hypoglycaemia
- hyponatraemia
- hypokalaemia
- neuromuscular abnormalities
- polycythaemia
- hyperpigmentation

Syndromes which have been identified as being due to ectopic production of hormones include:

- ectopic ACTH/MSH syndrome
- inappropriate ADH syndrome
- hyperthyroidism

and a number of other very rare syndromes.

It is known that all tumours produce, apparently randomly, a variety of peptide chains most of which have no biological activity.

Some tumours produce biologically active peptides in sufficient quantities to act as a marker but the occasional tumour produces sufficient quantities of hormonally active peptides or glycoproteins to produce the clinically evident ectopic hormone syndromes.

Without exception, the known ectopic hormone syndromes involve production of peptides or proteins.

The steps required to produce steroid, catecholamine or thyroid hormones are less easily mimicked by the random genetic activity of tumours.

Hypercalcaemia

About 10% of patients with cancer have hypercalcaemia. In most, the apparent cause is dissolution of bone by metastases. It has been estimated that the destruction of 1 gram of bone by tumour metastases releases 100 mg of calcium.

In the other patients, elaboration by the tumour of one or more of three humoral substances:

- parathyroid hormone
- prostaglandin E_2
- osteoclast stimulating factor (OSF)

is the cause of the raised serum calcium. The commonest tumours producing these substances are:

- squamous carcinoma of lung
- renal tumours
- tumours of ovary and uterus
- various tumours of the gastrointestinal tract

Management

This involves removal of the offending tumour where possible or the palliative control of the hypercalcaemia with:

- steroids
- oral phosphate

Some success in reducing blood calcium by treatment of the prostaglandin-induced hypercalcaemia with inhibitors of prostaglandin synthesis (aspirin and indomethacin) has been reported.

Hypoglycaemia

Rarely is hypoglycaemia due to non-pancreatic malignant tumours but a wide variety of tumours have been associated with this condition. These include:

- mesotheliomas (the commonest cause)
- hepatic tumours
- adrenal tumours
- adrenal carcinomas

The mechanism of hypoglycaemia induction probably differs between the various types of tumours. Some may elaborate peptide hormones with insulin-like activity. Large tumours may induce hypoglycaemia by avid metabolism of glucose.

Management

This is by operative removal which may be feasible in low-grade malignancy. Where inoperable tumours exist, frequent carbohydrate feeding and diazoxide therapy are indicated.

Polycythaemia

The most common tumour associated with erythrocytosis has been renal carcinoma. Other tumours include:

- cerebellar haemangiomas
- renal adenomas and cysts
- uterine fibroids

The most common mechanism is elaboration of erythropoeitin or similar substance by the tumour.

Management

This consists of removal of the offending tumour or palliative phlebotomies when the tumour is inoperable.

Ectopic ACTH syndrome

This is one of the commoner ectopic hormonal production syndromes. Most lung cancers appear to produce a pro-ACTH polypeptide ('big' ACTH), which is biologically inert. The syndrome only occurs in the few patients who also produce the enzyme necessary for the conversion of 'big' ACTH to the active ACTH 39 amino acid peptide.

β-LPH (lipotropin) is produced by all of these tumours and is responsible for the hyperpigmentation.

The commonest tumours associated with the clinical syndrome are:

- oat cell carcinoma of the lung
- other lung cancers
- thymic tumours

- pancreatic tumours

The syndrome usually develops so rapidly that the full-blown classical picture of Cushing's syndrome is not seen.

The usual clinical picture includes:

- hypokalaemia
- diabetes
- hyperpigmentation
- muscle weakness
- hypertension

The diagnosis is made by demonstrating elevated plasma cortisol and ACTH levels – the plasma ACTH may be exceedingly high.

Management

This includes:

- removal of the tumour if possible
- metyrapone, aminoglutethimide
  These are steroid enzyme blockers.
- o,p'-DDD
  This is an adrenolytic drug used if surgical removal of the tumour is not possible. It is extremely toxic and is frequently not tolerated by patients.
- trilostane
  This is a new competitive enzyme inhibitor

Inappropriate ADH syndrome

A variety of tumours:

- carcinomas of lung (squamous, anaplastic and oat cell)
- carcinoma of prostate
- carcinoma of adrenal cortex
- Hodgkin's disease

produce a syndrome due to sustained and inappropriate secretion of ADH (vasopressin).

Possibly as many as 4% of patients with carcinomas of the lung secrete ADH and are liable to develop the syndrome if they drink too much fluid.

The syndrome also occurs in a number of other non-malignant conditions – here the mechanism is unclear.

The patients present with:

- drowsiness

- inappropriate behaviour
- coma
- seizures

and have severe hyponatraemia.

The diagnosis is confirmed by finding:

- hyponatraemia
- hypervolaemia
- urinary osmolality greater than that of plasma
- high urinary sodium

Management

In all cases this is by severe fluid restriction and/or the use of drugs which inhibit renal response to ADH, e.g.:

- lithium carbonate
- demethylchlortetracycline

Other syndromes

Other syndromes have been reported with a variety of tumours but are very rare. They include:

- hyperthyroidism
  A TSH-like substance is produced mainly by trophoblastic tumours.
- gynaecomastia
  HCG is produced by a wide variety of tumours.
- watery diarrhoea with hypokalaemia and achlorhydria (Verner–Morrison syndrome)
  This results from excess gastrointestinal hormones mainly produced by pancreatic tumours.

MULTIPLE ENDOCRINE ADENOMATOSIS (MEA)

Two syndromes of multiple endocrine neoplasia are described; hyperparathyroidism may be a component of either.

Type I

In the most common, Type I (formerly known as the pluriglandular syndrome), the following abnormalities may be found:

- hyperparathyroidism (usually chief-cell hyperplasia)
- pituitary tumours (acidophil or chromophobe adenomas)
- pancreatic tumours (gastrin-producing and insulin-producing)
- adrenal cortical adenomas (rarely)
- thyroid tumours (rarely)

This disorder may be familial and inherited in an autosomal dominant manner.

Type II

Multiple endocrine neoplasia Type II (formerly known as Sipple's syndrome) may also be familial. Patients have one or more of the following:

- medullary carcinoma of the thyroid
- phaeochromocytoma
- neurofibromatosis
- hyperparathyroidism
- Cushing's syndrome (rarely)

SECTION V

The Principles of General Management and Treatment

SURGERY

There are two ways in which operation can alleviate disorders:

- Surgery may be used to treat some abnormality of endocrine function, for example, the removal of a parathyroid adenoma in hyperparathyroidism.

- Some abnormalities, not of function but of structure, may be treated surgically – for example, a simple goitre.

On occasions, the solution to an endocrine problem is supplied by surgical operation. An example is the removal of a phaeochromocytoma where the importance of close co-operation between physician, anaesthesiologist and surgeon in preoperative preparation and perioperative management cannot be over emphasized.

In other situations, surgery by itself cannot be regarded as curative but as an important part of a planned programme of therapy by surgeon, radiotherapist and oncologist. Clearly, the treatment of thyroid lymphoma, or cancer of the breast, is of this nature.

Occasionally, surgery has an important part to play in palliation even although cure cannot be expected. Relief of respiratory obstruction in anaplastic carcinoma of the thyroid is an example where much may be done to relieve distress.

The optimum surgical approach for each pathology and each endocrine organ has been summarized in Table 57. Fuller details are given under the appropriate disorder.

RADIOTHERAPY

Radiotherapy is used in endocrine disorders either to treat tumours or to modify function. It may be applied externally as supervoltage X-ray therapy or internally in the form of radioisotopes. In general, radiotherapy:

- plays a large role in the management of thyroid disorders

- has moderate usefulness in some pituitary and gonadal disorders

- is of little use in managing disorders of the adrenals, parathyroid and endocrine pancreas

Radiotherapy for thyroid disorders

The thyroid is uniquely susceptible to attack by radioisotopes in that it is the only tissue in the body which concentrates a specific element – iodine. Normally, this concentration of iodine by thyroid cells is of the order of twentyfold but in disease, particularly hyperthyroidism, this can rise to a hundredfold.

Since the thyroid cell cannot distinguish between stable iodine (^{127}I) and its radioactive isotopes, these can be administered therapeutically in order to damage thyroid cells selectively.

The isotope most commonly used for treatment is ^{131}I. The majority of the radiation damage caused by this isotope is by the beta-rays which have very short pathways and thus cause no damage to surrounding tissues.

Hyperthyroidism

Hyperthyroidism may be treated by:

- surgery

- antithyroid drugs

- radio-iodine

Surgery is especially indicated where there is a solitary autonomous nodule or where there are pressure symptoms from a large goitre. Radio-iodine is especially indicated in elderly patients with cardiovascular complications, though in such cases they should be made euthyroid by antithyroid drugs before giving ^{131}I.

Radio-iodine is rarely used in adolescents unless all other methods have failed, and is absolutely contraindicated in pregnancy. Uncomplicated hyperthyroidism in adults may be treated with equal success by radio-iodine or antithyroid drugs though many centres do not routinely use radio-iodine in patients under 35, but prefer partial thyroidectomy in antithyroid drug failures.

Radio-iodine is given in a single dose, the dose depending on the

- uptake of isotope by the thyroid

- rate of turnover

- mass of the thyroid

- severity of the disease

- previous treatment

The effects do not appear much before 2 weeks and are not maximal until about 12 weeks following the treatment dose.

Antithyroid drugs or β-adrenergic blocking agents such as propranolol may be used to control the patient's symptoms while awaiting the full effect of radioactive iodine. β-adrenergic blockers should

Table 57. Summary of surgical approach to endocrine therapy

Endocrine organ	Pathology	Surgical approach
Hypothalamus		Not amenable to surgical treatment; only radiotherapy is available
Pituitary	Adenoma	Transphenoidal approach if the adenoma is confined to the pituitary fossa Craniotomy if there is visual field involvement or suprasellar extension
	Carcinoma	Much less common. Craniotomy required and radiotherapy

Thyroid	Thyrotoxicosis	Thyroidectomy for:
		    failure of drug therapy
		    patient preference
		    large goitre
		    social/occupational needs

	Simple goitre including mediastinal thyroid	Thyroidectomy for:
		    cosmetic effect
		    pressure symptoms on trachea,
		      oesophagus and nerves

	Lump in thyroid	Removal of cysts and adenomas for diagnostic purposes
	Carcinoma	Papillary – total thyroidectomy (controversy exists, see p. 144) Follicular – total thyroidectomy Anaplastic – decompression as an adjunct to radiotherapy Medullary – total thyroidectomy
	Lymphoma	Primarily diagnostic biopsy – definitive treatment with radiotherapy and drugs
Thyroglossal tract	Cysts Fistulae	Complete removal of the thyroglossal tract

Pancreas	Endocrine tumours:
	    gastrinoma
	    insulinoma
		Excision – pancreatectomy if malignant

	Hyperplasia: e.g. of islet cells	Partial pancreatectomy
Breast	Cysts Lumps	Excision for diagnostic purposes
	Carcinoma	Mastectomy (simple/radical) with or without radiotherapy
Carotid body tumours	Chemodectoma	Excision with preservation or grafting of carotid vessels
Gut	Argentaffinoma Vipoma etc.	Excision
Parathyroid	Primary	Removal of adenomas Subtotal removal of glands for hyperplasia
	Secondary	Subtotal removal of glands
	Tertiary	Removal of adenomas Subtotal removal of glands for hyperplasia
Adrenal cortex	Cushing's disease	Bilateral adrenalectomy in preference to hypophysectomy where there are no signs of pituitary adenoma
	Primary adenoma usually unilateral	Adrenalectomy
Adrenal medulla	Phaeochromocytoma	Adrenalectomy (10% are bilateral and 10% are malignant)
Testes	Incomplete and maldescent	Orchidopexy
	Tumours	Orchidectomy and radiotherapy

	Testicular destruction:
	    acute orchitis
	    tuberculosis
		Orchidectomy and/or epididymectomy

Ovary	Cysts	Ovariotomy or ovariectomy
	Tumours	Ovariectomy
	Stein–Leventhal syndrome	Wedge excision of ovaries

always be used when very large doses of radio-iodine are being administered.

Radio-iodine therapy may be repeated. There are no immediate complications and the sole long-term morbidity is the development of postradiation hypothyroidism in a substantial proportion of patients. Patients treated with ^{131}I for hyperthyroidism should be checked regularly for developing hypothyroidism for the rest of their lives.

Thyroid tumours

Differentiated carcinoma

The definitive treatment of these tumours is surgery but, in operable cases, and in patients with residual or metastatic disease, supervoltage X-ray therapy and/or radio-iodine are useful. Once all normal thyroid tissue has been removed or destroyed, 20–30% of well-differentiated carcinomas (either papillary or follicular) will pick up radio-iodine and may thus be destroyed by it.

This method can be used to treat widespread functioning metastases sometimes with long term control. Supervoltage therapy should be given to inoperable or recurrent primary tumours and may be combined with radio-iodine.

Radioactive iodine probably has no role in the management of the other types of thyroid tumours.

Anaplastic carcinoma

This highly malignant tumour is treated by thyroidectomy if operable but this must always be followed by supervoltage X-ray therapy, otherwise local recurrence is inevitable. Inoperable cases should have supervoltage X-ray therapy.

Medullary carcinoma

This rare tumour arises from the parafollicular cells. It should be treated by thyroidectomy and, if removal appears complete, radiotherapy is unnecessary. Inoperable cases should have supervoltage therapy.

Lymphoma

This condition is exceedingly radiosensitive and the treatment of choice is supervoltage therapy. Surgery is confined to biopsy for histological confirmation.

With the exception of thyroid lymphoma, which responds well to cytotoxic agents, the role of chemotherapy in the management of thyroid carcinoma is as yet undefined. Adriamycin has been used with some palliative success in patients with metastatic papillary or follicular carcinomas.

Radiotherapy for pituitary disorders

Radiotherapy has an important role to play in the management of pituitary tumours. The commonest tumours of the pituitary gland are the benign adenomas:

- eosinophil adenoma – associated with gigantism and acromegaly
- basophil adenoma – associated with Cushing's disease
- chromophobe adenoma – once believed to be non-functioning but commonly secreting prolactin

They may be treated by surgery or radiotherapy.

All three types of adenomas, when confined to the sella turcica, respond satisfactorily to carefully planned, and accurately directed, supervoltage therapy and there are few complications.

Although symptomatic improvement occurs fairly rapidly, the full effect of radiotherapy in terms of the return of circulating hormones to normal levels may not be seen for several years after treatment. The usual form of radiotherapy is supervoltage but implantation of needles containing radioactive substances (e.g. radioactive gold and yttrium) which destroy nearby tissues, has been used.

Craniopharyngioma

This is a rare, benign tumour which is not of pituitary origin but arises in proximity to the sella turcica. Surgical removal is very difficult because of its intimate relationship to vital structures.

Carefully planned high dose supervoltage therapy will give good results with little morbidity.

Cytotoxic therapy has no place in the treatment of pituitary tumours.

Radiotherapy for testicular tumours

Initial treatment is by high inguinal orchidectomy to remove the primary and to establish an exact histological diagnosis. In all patients, a lymphangiogram must be done to detect possible involvement of the inguinal lymph nodes but, irrespective of the result of this examination, operation must be followed by supervoltage radiotherapy to these lymph nodes and to the primary site.

The remaining testis may or may not be included in the treatment volume. (There is a small incidence of second tumours on the opposite side.) Irradiation of the remaining testis will not affect the patient's sexual potency but will render him sterile. However, it is possible to collect semen prior to treatment and store it for future artificial insemination.

The prognosis depends on the histology and the extent of the disease at diagnosis but the cure rates for early tumours are high when treated in this way, especially in seminoma where they may exceed 90%. Even advanced seminomas may be cured by radiotherapy.

Cytotoxic drugs are playing an increasing part in the management of advanced testicular tumours though only palliatively since there is, as yet, little evidence of their value as adjuvant therapy in early cases.

Radiotherapy for ovarian tumours

The place of radiotherapy depends on the histological type of tumour. It has no place in the management of benign tumours. The common serous and mucous adenocarcinomas are only moderately radiosensitive and, if complete surgical removal is possible, postoperative radiotherapy is not indicated. In inoperable tumours, and in patients with residual disease, there is good evidence that supervoltage radiotherapy will prolong life, reduce recurrence rates and, on occasions, result in cure.

Two rare ovarian tumours:

- dysgerminoma
- granulosa cell tumours

are very radiosensitive and postoperative supervoltage therapy should always be given to such patients except in the earliest and most localized lesions.

Cytotoxic drugs may be used in ovarian carcinoma and may produce worthwhile remissions in a proportion of patients.

Conclusion

Radiotherapy has an important place in the management of endocrine disorders.

The approach should be multidisciplinary with close team work between the endocrinologist, neurosurgeon or surgeon and radiotherapist.

REPLACEMENT HORMONES

The principles of hormone replacement therapy are to:

- establish the need for therapy
- select the appropriate replacement therapy
- monitor the response to therapy
- avoid adverse drug reactions
- ensure adequate compliance of the patient with the regime

Establishing the need for therapy

Hormone deficiency leads to recognizable clinical syndromes. Once the diagnosis is suspected on clinical grounds, the deficiency can be confirmed by laboratory tests. These may confirm deficiencies by demonstrating:

- decreased plasma levels of the hormone itself
  e.g. low plasma T4 in hypothyroidism
- increased stimulating hormones due to the failure of feedback inhibition
  e.g. if the hypothyroidism is primary, plasma TSH levels will be high
- alterations in plasma or urinary concentrations of the products of endocrine tissues affected by the disorder
  e.g. the rise in urinary levels of precursors in congenital adrenal hyperplasia
- other biochemical abnormalities associated with deficiency of the hormone
  e.g. the hyperglycaemia of diabetes mellitus, the hyponatraemia and hyperkalaemia associated with Addison's disease

Selection of appropriate replacement therapy

The type of therapy chosen may depend on a number of factors including:

- route of administration
- metabolism of the hormone preparation
- required plasma level
- availability of non-hormonal alternatives

Route of administration

In a number of cases, the hormone may only be available for administration by one particular route but, where possible, oral therapy should be given. A good example of this is the use of cortisol and thyroxine to replace ACTH and TSH deficiency in hypopituitarism. Likewise, fludrocortisone, which is orally effective, is given in cases of hypoaldosteronism rather than aldosterone itself which is effective only by parenteral administration.

Drug metabolism

An example of this is the use of cortisone rather than cortisol in hypopituitarism or hypoadrenalism. Cortisone, after absorption, has to be converted in the liver to the active metabolite cortisol. This conversion is variable and can lead to variation in the plasma levels of cortisol produced. The rate of conversion may also be reduced by concomitant androgen therapy.

Plasma level requirements

These vary with the deficiency being treated. It may be that the aim of therapy is to provide a constant plasma level within a certain range. An example of this is the use of thyroxine in hypothyroidism. Thyroxine has a long half-life, and consequently can be given once daily to produce a fairly constant plasma level. On the other hand, the intention may be to mimic normal physiological variation in plasma levels. Thus cortisol is given as two thirds of the total daily dose in the morning and one third at night to mimic normal diurnal variation.

Non-hormonal alternatives

It is worthwhile to remember while considering replacement hormones that non-hormonal therapy may be available. Oral hypoglycaemic drugs may be effective in 'maturity onset' diabetes, chlorpropamide and carbamazepine may be effective in diabetes insipidus, and vitamin D may be used in hypoparathyroidism.

Monitoring the response to therapy

This can be accomplished by:

- assessment of clinical response
- measurement of plasma hormone levels
- measurement of feedback inhibition
- measurement of other biochemical parameters

Assessment of clinical response

Symptomatic improvement may be elicited, such as loss of cold intolerance in treated hypothyroidism. Clinical signs of hormone deficiency, such as coarsening of skin and hair and myxoedema in hypothyroidism, may improve.

Measurement of plasma hormone levels

Where these estimations are available, they can be used as a ready guide both to efficacy of therapy and patient compliance (see below). Thyroxine replacement therapy can be monitored by estimation of serum thyroxine levels, the ideal being to produce a value within the upper half of the normal range.

Measurement of feedback inhibition

For example, TSH levels fall to within the normal range when primary hypothyroidism is adequately treated with thyroxine.

Measurement of other biochemical parameters

Examples of this are the reversions to normal of electrolytes when Addison's disease and hypoaldosteronism are treated with appropriate replacement therapy.

Avoidance of adverse drug reactions

Initiation of therapy

Adverse reactions may arise because of inappropriate dosage, or because of the patient's physical state.

Inappropriate dosage may be suspected by the clinical effects of under- or overdosage, and confirmed by the methods described above.

The patient's physical state can be a factor of great importance in determining initial dosage schedules. As an example, if a patient has ischaemic heart disease and hypothyroidism, thyroxine replacement therapy must be introduced in a minimal dose (as little as 0.025 mg/day) and increased very gradually thereafter over a long period, to avoid an exacerbation of the symptoms or sequelae of ischaemic heart disease.

The initiation of therapy for insulin dependent diabetes has special features worthy of comment.

After the initial presentation, which may range from a chance finding at routine examination to ketoacidosis secondary to infection, insulin dependent diabetics enter the 'honeymoon phase', when control is easy, and insulin requirements are low. Gradually, however, requirements increase, and twice-daily injection regimes are now considered mandatory for good control (see p. 206). There is evidence that initiation of therapy with 'monocomponent' insulins (see below) may significantly lengthen the duration of the 'honeymoon period'.

Continuing long term therapy

Again, inappropriate dosage – usually overdosage – may produce adverse effects. A good example of side-effects in hormone replacement therapy is when too much cortisol is given and the pharmacological effects of steroids, such as hypertension, osteoporosis, diabetes, truncal obesity, and striae formation become apparent.

Variation of physical state is again important in long term replacement therapy, the most important example of this being the need to increase cortisol dosage in response to the stress of operation or infection.

Drug interactions with hormone replacement therapy are rare. One example already quoted is the alteration of metabolism in the liver of cortisone to cortisol by the concomitant administration of androgen therapy.

It should be remembered that hormones used NOT for true replacement purposes may lead in the long term to unexpected and severe side-effects, as was shown by the increased incidence of vaginal and cervical carcinoma in the children of women who took diethinyl stilboestrol in the first trimester of pregnancy.

Ensuring adequate compliance

One important factor which can affect the type of replacement therapy selected, the interpretation of results of monitoring of therapy, and the incidence of adverse reactions is the degree of compliance of the patient with the regime.

Studies in general therapeutics reveal two groups of patients who do not comply with their regime.

The first group do so because they cannot accurately describe their own regime. Perhaps surprisingly, age, sex, social class, education, social isolation and the patient's own satisfaction with information received do not significantly influence this group. Total number of drugs, and frequency of dosage, do, however, affect compliance in this group.

The second group can accurately describe their regime, but still do not comply with it. Again, the only factor significantly influencing this group is the total number of drugs in the regime, and especially remarkable is the fact that difficulty or unpleasantness of administration of therapy is not a contributory feature to their non-compliance.

It is possible that overall non-compliance rates for drugs in general may reach up to 50%. Studies in the endocrine field are more favourable. In one study of patients on thyroxine replacement therapy, approximately 10 per cent had low serum thyroxine levels which were later found to be due to lack of compliance.

A particular problem related to compliance with hormone replacement therapy is that the patient, or even the doctor, may not consider the treatment necessary.

Patients may consider this on the grounds that they feel entirely well, and may have been receiving therapy so long, or indeed may have started therapy before the effects of deficiency were apparent (see discussion of replacement therapy post hypophysectomy p. 117), that they have forgotten, or were never aware of, the contribution of their therapy to their well-being.

Equally, doctors may encounter patients, especially on long term thyroxine therapy, who are apparently very well. They may not be acquainted with the original grounds for diagnosis, and may wonder whether their treatment is actually necessary. They may be tempted to initiate a trial off therapy, which may lead to unfortunate results.

ADRENOCORTICAL STEROIDS

Corticosteroids are used therapeutically in two separate ways:

- Potent synthetic corticosteroids such as prednisolone are given in greater than physiological doses to achieve a pharmacological response, particularly an anti-inflammatory effect.

- Less potent steroids are used as replacement therapy to supply a physiological need in varying degrees of adrenocortical insufficiency.

The former is only of interest to endocrinologists from the point of view of resultant suppression of the hypothalamic–pituitary–adrenal axis and we will deal here with replacement therapy only.

Chronic adrenal insufficiency

Hydrocortisone (cortisol) is the drug usually used in replacement regimes and is given in divided doses two to three times daily. Others suggest that dosage schedules should attempt to mimic the circadian rhythm in average doses of 20 mg in the morning and 10 mg in the evening. It is questionable whether this method of therapy offers any true advantage to patients. Cortisone acetate is still used in North America, but it has the disadvantage that it has to be converted to cortisol before it is active – a conversion that may be incomplete.

Side-effects are inevitable when steroids are used in large doses to achieve a therapeutic response. However, side-effects occurring in patients on replacement doses of steroid indicate incorrect dosage.

The most important aspect of management of patients receiving replacement therapy is to explain in detail the implications of therapy. The stressing of lifelong treatment, and that doses of steroid must never be omitted, is of utmost importance. Relatives and patients must be aware that well-being depends entirely on obsessional attention to regular tablet taking.

Patients should be advised to increase their dosage at times of illness and, if relatives are reliable, it may be feasible to provide syringes and hydrocortisone for injection in the event of gastrointestinal disturbances resulting in vomiting of medications. All patients should carry a steroid card giving details of their therapy, and it is wise to wear a Medic Alert bracelet or chain.

In patients with primary hypoadrenalism, in addition to glucocorticoid deficiency there may be inadequate mineralocorticoid production. The mineralocorticoid of choice for replacement therapy is, at present, fludrocortisone in a daily dose varying between 0.05 mg and 0.2 mg/day. Such patients require regular review initially to check for signs of mineralocorticoid overdosage such as hypertension, peripheral oedema or headaches.

Acute adrenal insufficiency

Addisonian crisis or acute adrenal insufficiency is a medical emergency requiring prompt intravenous injection of 100 mg hydrocortisone hemisuccinate and rapid rehydration with normal saline and 5% dextrose. Thereafter, 100 mg hydrocortisone is given 8-hourly until the patient's condition shows improve-ment. Replacement therapy can be re-established when any precipitating illnesses have subsided or have been treated.

Major surgery

During major surgery all patients receiving replacement steroids require adequate 'cover'. Hydrocortisone hemisuccinate should be administered in a dose of 100 mg 6-hourly preoperatively, throughout surgery, and for several days of the postoperative recovery phase. Following bilateral adrenalectomy for Cushing's disease, hydrocortisone may require to be administered in high doses for considerably longer periods of time prior to establishing replacement oral therapy.

Withdrawal of steroids

The feasibility of withdrawing steroids when given for therapeutic purposes depends largely on the activity of the disease process for which they were given. The ease with which such medication can be withdrawn, however, depends on the dose and duration of therapy and, therefore, the degree of hypothalamic–pituitary–adrenal axis suppression. Suppression is more a function of dosage than duration and usually patients receiving small doses can have their drugs withdrawn without difficulty. However, large doses of steroids given for long periods may result in varying degrees of suppression of ACTH release and consequent atrophy of the adrenal cortex. In the circumstances, withdrawal must be carefully monitored and prolonged. Tests of hypothalamic–pituitary–adrenal axis function must be carried out to confirm whether patients will respond to stress after steroid withdrawal. A normal plasma cortisol under resting conditions may delude the physician that full restoration of hypothalamic–pituitary–adrenal function has taken place, with catastrophic effects in the face of new illness.

SEX HORMONES

Oestrogens

It is not possible to define an oestrogen in chemical

terms. Although the 20 or 30 natural oestrogens which have been isolated from human sources have a recognized common structure, synthetic oestrogens are greatly different and can be classified as oestrogens only in terms of a biological definition, i.e. they cause uterine enlargement.

The primary hormone secreted by the ovary is oestradiol-17β. It is readily interconvertible with oestrone from which it differs only by one hydrogen atom. Whichever steroid is administered or secreted, the result in the circulation is an equilibrium mixture of the two.

Oestriol is an irreversible metabolic product of oestradiol which is synthesized in large amounts during pregnancy by a route not involving oestradiol.

Natural oestrogens are rapidly inactivated by being metabolized to sulphates and glucuronides and are excreted in urine. Synthetic oestrogens such as stilboestrol or mestranol owe their potency to the fact that they are metabolized and excreted much more slowly and therefore remain in the circulation in an active form much longer. A common pharmacological device is to attach a substituent to the oestrogen molecule as in ethinyloestradiol. Such substituents hinder the fit of the enzymes which metabolize oestrogens so that inactivation is impeded.

Guidelines for use

Although the widespread use of oestrogens has decreased recently as their dangers have become more evident, they are still a very common form of medication. Some controversy attends their use on healthy women as oral contraceptives or in hormone replacement therapy for the menopause. There is no doubt that the:

- hot flushes
- senile vaginitis
- insomnia
- restlessness

of the menopausal syndrome are often controlled by oestrogens. They easily provoke postmenopausal bleeding and are associated with venous thrombosis and endometrial cancer. Their use outside of disease states is a matter of judgement and should be confined to situations where medical supervision is possible.

There is room for the exercise of skill in selecting the dosage and type of oestrogen. Susceptibility to thrombosis is dose related. Menopausal symptoms can be controlled by small amounts of oestrogen and it should be possible to remain in the safety zone. The evidence directly relating oestrogens to endometrial cancer may not apply to easily metabolized steroids natural to humans. For the menopausal woman:

- oestriol
- oestradiol valerate
- oestrone sulphate

are the compounds of choice.

Indications

Oestrogens are useful in many forms of menstrual irregularity.

Dysfunctional uterine haemorrhage

Given in high dosage they are one of the simplest and surest means of stopping a prolonged bleeding phase in dysfunctional uterine haemorrhage. The disadvantage is that further bleeding is often precipitated by their withdrawal. Such withdrawal bleeding can be tempered by adding a progestogen to the oestrogen. When menorrhagia is treated with oestrogens, it is wise to use such a mixture as it found in the oral contraceptive rather than oestrogen alone.

Amenorrhoea

Here bleeding can be provoked by a short course of oestrogen. There is little cause to believe that such oestrogen withdrawal bleeding often sets the menstrual cycle going again and this form of treatment is best reserved for women much in need of the psychological reassurance of having a period or as a means of testing the responsiveness of the endometrium.

Suppression of menstruation

A continuous high dosage of oestrogen will suppress menstruation. When given in combination with nor-testosterone derivatives or other synthetic progestogens, ectopic endometrial tissue eventually atrophies. This form of therapy has proved most successful in the treatment of metriosis.

Prostatic carcinoma and hypertrophy

Because of their feminizing actions oestrogens can be given to males only in special circumstances such as carcinoma of the prostate and prostatic hypertrophy, where they are useful in controlling the disease.

Lactation

Certain indications for the use of oestrogens have now

virtually disappeared. Prominent among these is the suppression of lactation where the high dosage needed leads to an unacceptable risk of thromboembolism. Suppression of lactation is now most commonly achieved by the use of prolactin inhibitors such as bromocriptine.

Oral contraceptives

Oral contraceptive preparations contain one of two synthetic oestrogens:

- ethinyloestradiol or
- mestranol – the 3-methyl ether of ethinyloestradiol

together with a progestogen.

The progestogens are derivatives of nortestosterone such as:

- norethylnodrel
- norethisterone
- lynoestrenol
- ethynodiol diacetate
- norgestrel

or are derivatives of 17-α-hydroxyprogesterone such as:

- chlormadinone
- megestrol acetate

Given in combination these compounds:

- suppress ovulation
- alter cervical mucus so that it becomes impenetrable to sperm
- render the endometrium hostile to nidation

The most important contraceptive effect arises from the suppression of ovulation. This is largely due to the oestrogen component.

In one form of oral contraceptive, a progestogen only is used. It has a high failure rate and, not surprisingly, has not proved popular. According to market statistics more than a million women in Britain are taking oral contraceptives. This is mass medication on an unprecedented scale and the use of these compounds should be looked at in a different light from ordinary medication.

One consequence is that even rare side-effects will affect substantial numbers of women. Although the true biological effect of these contraceptives may be disguised by fanciful allusions to the resemblance

between the state brought about by oral contraceptives and pregnancy, the truth is that women receiving oral contraceptives are physiologically castrate, all ovarian activity being substituted by the ingredients of the pill. The only justification for their use lies in their capacity to prevent pregnancy.

Side-effects

Oral contraceptives may have untoward effects on:

- blood clotting mechanisms
- mood
- metabolism

They may therefore cause venous thrombosis with a danger of embolism, or headaches, dizziness and hypertension or depression, tiredness and decreased libido. Some women put on weight.

Although a great many women can take oral contraceptives without untoward effects, such effects are sufficiently common to have led to increasing restriction in the types of women for whom it is wise to prescribe them. Thus, hypertension or a history of previous thrombosis are clear contraindications. The side-effects are due almost entirely to the oestrogen component and are age and dose related.

Combined oral contraceptives should not be prescribed for women over 35 years old (especially if associated with smoking). The oestrogen component should not exceed 50 μg per day.

Anti-oestrogens

The first step in the biological activity of an oestrogen is the binding of the steroid by a receptor protein in the cytoplasm of the cells.

Some non-steroidal compounds have the peculiar property of also binding to such receptor proteins but without setting in train the cascade of intracellular events which results from oestrogen binding.

When the binding protein is occupied by such an anti-oestrogen, true oestrogens cannot bind and exert their effect on the target tissue.

Several such anti-oestrogens have been synthesized. Clomiphene citrate and tamoxifen are two in common use.

Clomiphene is used largely for its effect on oestrogen receptors in the hypothalamus.

Tamoxifen particularly is also used as an anti-oestrogen in oestrogen dependent breast cancer.

The use of clomiphene or of tamoxifen results in

the secretion of gonadotrophin releasing hormone with consequent FSH and LH release from the pituitary. In some anovular or amenorrhoeic women, this results in ovulation.

Anti-oestrogens may also have a considerable therapeutic role in protecting other oestrogen sensitive tissues from untoward oestrogen effects. Although they have been very little tried in hyperoestrogenic metrorrhagia, this would seem a good indication for their use.

Androgens

Although androgens were formerly commonly used in gynaecological treatment, they are apt to cause unwanted masculinizing effects such as beard growth. They have a protein anabolic effect and improve libido as well as imparting a sense of well-being.

They are still regarded in many clinics as a valuable therapeutic adjunct in the treatment of menopausal women.

Androgen deficiency states in the male, where they may safely be used in heavy doses, are rare. Testosterone, the androgen secreted by the testes, is easily synthesized and readily available. Like oestradiol, it suffers from the therapeutic disadvantage of being rapidly metabolized and inactivated. As with the oestrogens, the pharmacological trick of adding substituents to the molecule which slow down metabolism is commonly used. Two testosterone derivatives:

- methyl testosterone
- testosterone propionate

serve this purpose well.

The supposed rejuvenating and aphrodisiac effects of androgens are mythological. Sexual appetite in older men resides in the cerebral cortex, not the testes.

TREATMENT OF DIABETES MELLITUS

Aims of treatment

First Aim

The first aim of treatment is to maintain a normal or near normal blood glucose level at all times. It follows that a pattern of intermittent large meals, so common

in the obese, is unsuitable in diabetics, since they cannot produce the extra insulin that such meals require.

The food must be provided in fairly regularly spaced meals which are planned to counterbalance the effect of any hypoglycaemic treatment with insulin or tablets. Refined carbohydrate should be avoided as it is absorbed too quickly and the inclusion of foods containing non-absorbable residue or fibre is valuable in prolonging the absorption time of the dietary carbohydrate.

Second aim

The second aim of treatment is to achieve a healthy weight.

Third aim

The third aim is to postpone or prevent the complications of diabetes. This is difficult and the methods of achieving it are controversial. It appears that the microangiopathic complications of diabetes are related mainly to poor control of the blood glucose but the macroangiopathic ones may well be associated with other risk factors that affect non-diabetics such as raised levels of cholesterol and triglyceride in the plasma.

DIETARY TREATMENT

Diet is the only treatment common to all diabetics and, in the mildest cases, is all that is needed to restore carbohydrate tolerance to normal.

Planning a diet

The energy content must first be decided on the basis of the physical activity of the patient. This will normally determine the dietary carbohydrate content at a rate of about 10 g for each 100 kcal (0.42 MJ) required. Each 1 g of carbohydrate contributes about 4 kcal. An energetic young man might reasonably be given a dietary carbohydrate allowance of up to 300 g/day.

Normally at least 60 g of protein is included with more for high energy diets and again each 1 g contributes about 4 kcal. The calorific deficiency is made up by fat (1 g = 9 kcal).

In the past, specialists in diabetes were satisfied to control only the carbohydrate content of diabetic diets. Traditional diabetic diets contained 40% carbohydrate content. Recently however, in an attempt to reduce heart disease, diets with a larger (55–60%) proportion of carbohydrate derived energy have been favoured, with a corresponding reduction in fat consumption to 30% of energy intake. It is essential, however, that the major part of carbohydrate consumed should not be in the form of refined sugar or fibre free food if significant postprandial hyperglycaemic peaks are to be avoided.

Carbohydrate content is regulated by listing the amounts of foods such as:

- bread
- vegetables
- cereals
- milk

that contain 10 g of carbohydrate. Domestic measures are usually employed, although a scale weighing in grams is much more accurate. At first, an insulin dependent diabetic will need to weigh the correct proportions of the various exchange foods in order to memorize how they look. Thereafter, most assess their proportions visually, although it is wise to have periodic checks by weighing the main exchanges again.

Patients on sulphonylurea tablet treatment need to phase their meals to prevent undue swings in blood glucose, and those on the long-acting chlorpropramide should have a bedtime snack to reduce the risk of nocturnal hypoglycaemia. This same principle is even more important for those taking long-acting insulins such as Ultralente or Ultratard.

Dietary fibre

There has been much interest recently in the use of dietary bran and high fibre residue, and its association with improvement in glucose tolerance in all types of diabetes. Carbohydrate from foods rich in fibre tend to be absorbed more slowly from the gut and help to reduce to postprandial glucose peak. Diabetics should be encouraged to take carbohydrate known to be rich in fibre, e.g. wholemeal bread and biscuits. The use of fibre supplements such as guar and bran, whilst of proven value, is perhaps of less practical importance.

Children

Considerable conflicts arise regarding the current dietary approach for young children with diabetes. It has been suggested that strict dietary discipline inflicts lasting psychological damage. Some flexibility is required and, as the child grows older, he or she should be encouraged to share in dietary planning.

Special arrangements should be made for school lunches to be taken with readily assessed exchanges such as fruit or biscuits and cheese instead of sweet puddings, and 'sugarless' pastilles can be substituted for sugary confectionery. Teenage revolt may include rejection of dietary discipline with resultant diabetic emergencies but many diabetic children grow through this phase avoiding such a crisis.

Alcohol

Although alcohol can be metabolized without increasing the requirement for insulin, it can lead to hyperglycaemia, both because of the sugar so often associated with it (e.g. wines and beers) and also because the energy it provides obviates the need to metabolize the normal amount of glucose. It can also lead to hypoglycaemia on 'the morning after', for the effect of alcohol is to interfere with gluconeogenesis on which the fasting level of glucose normally largely depends.

A potential risk for an insulin dependent diabetic in hypoglycaemia, with a smell of alcohol on his breath, is that his behaviour, which may closely mimic that of a drunken man, may be attributed solely to alcohol and be dealt with accordingly by the police.

Artificial sweeteners

Some diabetics have been very partial to sugar before the diabetes was diagnosed, and such people wish to know what alternatives are available to them.

Artifical sweeteners are of two main types. One contains negligible energy, such as saccharin or saxin. The other is just another form of carbohydrate, such as the polyhydric alcohol, sorbitol, which is so gradually absorbed and metabolized through fructose in the liver that it produces no quick rise in blood glucose. It is expensive and not suitable as a sweetener for those on reducing diets. In amounts more than about 30 g/day, it tends to produce an osmotic type of diarrhoea. Recently an amino acid derivative – aspartame – has been introduced.

Fats and blood lipids

The levels of both cholesterol and triglyceride are raised in the plasma of poorly-controlled diabetics but return to normal when the diabetes is well cared for.

It is known that polyunsaturated fats are about half as effective in lowering serum cholesterol levels as saturated fats are in raising them, so that a ratio of at least 2:1 of unsaturated:saturated fat should be recommended. This means substituting a polyunsaturated margarine for butter and avoiding cream and fat on meat as well as fatty cheeses. Cottage cheese, made from skimmed milk, is a substitute, and frying should be done with oil rather than lard.

Another step which favours a low serum cholesterol level is to increase the residue in the diet with a high content of vegetable fibre and bran. The non-absorbed fibre binds the bile salts, derived from cholesterol, and partially prevent their reabsorption.

Weight reducing diets

The obese diabetic stands to benefit by dietary weight reduction, not only by improving his carbohydrate tolerance, but also by reducing the risk of the many serious complications of obesity. Nevertheless, marked reductions in weight are difficult to attain and require a strong and long-sustained motivation by the patient. Limitation of carbohydrate intake with little or no loss of weight often suffices to improve the diabetes, although a greater weight loss is needed to get a worthwhile improvement in carbohydrate tolerance.

There is a considerable individual variation in efficiency in conservation of the energy content of food so that some obese people need very low diets to achieve weight loss.

The principle of dietary weight loss is to omit, or greatly curtail, concentrated sources of energy such as:

- refined carbohydrates
- fat
- alcohol

and to substitute foods low in energy and containing adequate:

- protein
- minerals
- vitamins

Foods to be favoured include clear soups and most vegetables other than those which form a store of energy for new growth, e.g. potatoes, peas, sweet corn, beans (as opposed to their pods), beetroot and many fruits.

The actual diet prescribed should depend on the history of the individual patient and be altered only in the essential respects since too radical a change will invite its rejection. Most such diets contain about 1000 kcals (4.2 MJ), but in very inactive individuals, a smaller diet may well be needed.

Once adequate weight loss has been achieved and if the diabetes is still not controlled, the patient will require either oral hypoglycaemic agents or insulin.

ORAL HYPOGLYCAEMIC AGENTS

The aim in treatment of maturity onset diabetes or NIDD is to achieve normal or near normal levels of blood sugar while reducing other factors which increase the risk of large and small vessel disease. Among these, obesity is a key factor and should be controlled by dietary modification. When such measures fail to control the blood sugar, the oral hypoglycaemic compounds are appropriate.

There are two main types of oral hypoglycaemic drugs:

- sulphonylureas
- biguanides

They share the property of being unable to control the diabetes if ketosis is present. Their mode of action is strikingly different.

Sulphonylureas

The hypoglycaemic action of such compounds was found by chance in the search for better sulphonamides to control infection. They were later modified to retain or increase their hypoglycaemic potency but to lose their antibacterial action (LD-59).

Mode of action

At an early stage the sulphonylureas were found to depend on the presence of a pancreas which could still release insulin. Later they were shown to have extra pancreatic effects including effects on insulin in receptor numbers and sensitivity accounting for more of their chronic hypoglycaemic action. They sensitize the beta cells of the pancreatic islets to the insulin-releasing effect of hyperglycaemia.

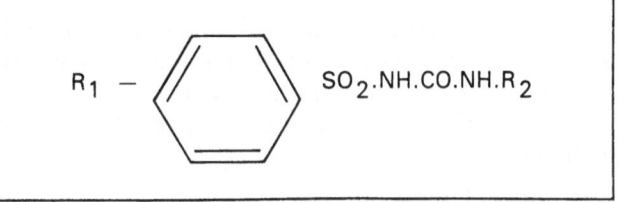

LD-59. *General formula of sulphonylureas.*

Clinical experience has shown that patients who initially responded to sulphonylurea treatment may later become secondary failures. In general, the sulphonylureas produce similar effects to those of insulin in lowering the plasma levels of lipids and amino acids.

Clinical application

The clinical use of the sulphonylureas has become controversial since the publication of the reports of the University Group Diabetes Program (UGDP) in the USA in 1970.

This controversial study suggested that oral hypoglycaemic agents increased the mortality from cardiovascular events. The study has led, in North America, to a marked decrease in the use of such agents. In other parts of the world, the effect has not been quite so dramatic, the results of the study being interpreted with some scepticism.

Nevertheless, a salutary result of the UGDP study has been to stress the value of purely dietary treatment for such mild maturity onset diabetics.

The choice between the sulphonylureas and the biguanides should depend mainly on the patient's weight. The sulphonylureas tend to produce weight gain and the biguanides in suitable doses encourage weight loss. Even in those who will clearly need some supplement to dietary treatment, it may well be best to start treatment with diet alone as this emphasizes to the patient its beneficial effect and key importance.

Patients with postprandial ketonuria under basal conditions are candidates for treatment with insulin from the beginning.

There is a wide choice of sulphonylureas and the main variables are:

- potency
- duration of action
- cost

There is little to choose between them in respect of side-effects. The first generation includes:

- tolbutamide
- chlorpropamide
- acetohexamide

In recent years a second generation of sulphonylurea agents has been developed. These are believed to have fewer side-effects and are claimed to be more potent than their firstgeneration relatives. This group includes:

- glibenclamide
- glipizide

- glibornuride
- gliclazide
- gliquidone

It is reasonable to aim to achieve normal pre-meal blood sugar levels in sulphonylurea-treated patients, and the smallest dose which achieves this result should be used.

Chlorpropamide stands out from all the other sulphonylureas because of its much longer duration of action. 50–60% is excreted in the urine unchanged and, for this reason, it is unsuitable for use by patients with significant renal impairment, e.g. blood urea above about 10 mmol/l (60 mg/dl).

Fasting blood sugar levels are more likely to become normal with a long-acting drug, like chlorpropamide, than with the short-acting members of the group. For the same reason, chlorpropamide needs to be given only once a day, while medium-to-large doses of the shorter-acting drugs should be given two or three times a day shortly before meals.

Sulphonylurea failure

Failure of the sulphonylurea in adequate dose to control the hyperglycaemia within a month of starting treatment is termed primary failure. Subsequent deterioration of control is called secondary failure. Some secondary failures can be overcome by adding a biguanide but, if there is definite ketonuria, insulin should be used.

In general it is probably better to accept insulin therapy early rather than allowing chronic hyperglycaemia. Clearly, other factors – particularly the age of the patient – need consideration, but the feeling of well-being associated with insulin therapy and the relative ease of insulin administration by community nurses has resulted in acceptance of insulin injections even in aged patients who are inadequately controlled by use of a sulphonylurea drug.

Pregnancy and sulphonylureas

Although there is a natural reluctance to use drugs of any type in pregnancy, many women of child-bearing age continue to be treated with sulphonylureas often before they realize they are pregnant.

There is no evidence that the sulphonylureas in ordinary dose are teratogenic in humans but, since they cross the placenta, they are potentially dangerous and can produce severe hypoglycaemia in the neonate.

Stress

In the presence of intercurrent illness such as infections or surgery, it is best to substitute a quick-acting insulin for a sulphonylurea. In the case of minor operations, which do not interfere with feeding by mouth for more than a day, it is enough merely to discontinue treatment on the day of the operation.

Side-effects

These are not frequent and are seldom serious. Some of the toxic reactions are similar to those of the sulphonamides from which the sulphonylureas are derived. These include dyspeptic symptoms and skin reactions which usually appear in the first 2 months of treatment.

Flushing of the face shortly after taking even modest amounts of alcohol may occur in up to a third of sulphonylurea-treated subjects. Chlorpropamide is most likely and the second generation sulphonylureas least likely to produce this disulfiram-like effect.

Toxic hepatitis and depression of the bone marrow occur rarely and chlorpropamide in particular may occasionally cause reversible cholestatic jaundice.

Clinically unsuspected microgranulomas have been found in several organs of patients who received tolbutamide and have been attributed to a sensitivity reaction. Their significance may be their occurrence in the heart, in view of the reports of increased mortality from myocardial infarction in sulphonylurea-treated diabetics.

The action of sulphonylureas to increase the contractility and automaticity of the heart beat may also be harmful after myocardial infarction.

Chlorpropamide may also be associated with hyponatraemia during long term treatment, the picture being one of inappropriate secretion of antidiuretic hormone (SIADH).

Hypoglycaemia

This is an overeffect rather than a side-effect of the sulphonylureas. It most typically appears with a long-acting agent such as chlorpropamide in the presence of previously unsuspected renal impairment. Another contributory cause is the concurrent administration of drugs which interfere with the excretion of or displace the sulphonylureas from protein-binding sites.

These include:

- sulphaphenazole
- phenylbutazone
- oxyphenbutazone
- clofibrate
- chloramphenicol
- dicoumarol
- large doses of salicylate

Monoamine oxidase inhibitors may produce hypoglycaemia in association with sulphonylureas by a pharmacodynamic effect and similarly β-adreno-receptor blocking drugs may enhance the hypoglycaemic effect of sulphonylureas.

In contrast, other drugs such as:

- glucocorticoids
- oestrogens
- thiazides
- diuretics

antagonize the antidiabetic effect of the sulphonylureas and tend to raise the blood sugar.

It is important to remember that hypoglycaemia due to chlorpropamide tends to persist and to recur due to the long duration of action of this compound so that a single administration of intravenous glucose is usually not enough. Either an intravenous infusion of glucose is needed or the patient, after consciousness has been restored, will need to have between-meal and nocturnal snacks for about 24 hours.

Biguanides

The hypoglycaemic action of the biguanides, like that of the sulphonylureas, was discovered fortuitously.

Mode of action

Unlike other hypoglycaemic agents, these do not of themselves bring the blood sugar level significantly below normal. Furthermore, non-diabetic animals and humans are resistant to their action on the blood sugar. Diabetics responding to biguanides have residual insulin-producing capacity. Indeed, obese patients may be hyperinsulinaemic (LD-60).

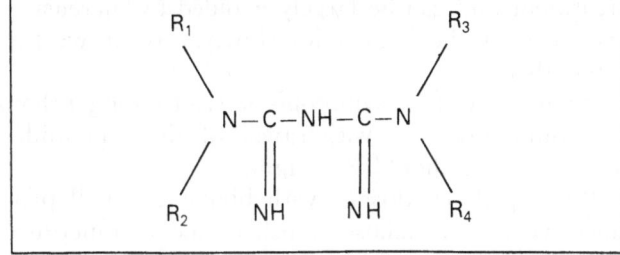

LD-60. *General formula of biguanides.*

The biguanides tend to favour weight loss when given in appropriate dose. The main mode of action involves reduction of glucose absorption from the gut. Another important mechanism of their hypoglycaemic action is a reduction of gluconeogenesis by the liver and increasing glucose uptake into peripheral tissues by facilitating the action of insulin.

Clinical application

Only two biguanides, metformin and phenformin, are available in the UK but buformin is available as well in some countries.

The dose required to achieve optimal control of hyperglycaemia in many maturity onset diabetics is often uncomfortably close to the dose which starts to induce dyspeptic side-effects. Thus, when the biguanides are used near their limit of tolerance, patients tend to feel less well than when on sulphonylureas or when changed over to insulin treatment.

The most common indication for biguanide treatment is for those obese, maturity onset diabetics who either remain hyperglycaemic in spite of dietary restriction or simply refuse to diet effectively. Carefully and slowly adjusted increments of biguanide treatment will tend to lower not only the blood sugar but also the weight. Metformin rather than phenformin should be used because of the major risk of lactic acidosis with phenformin.

Side-effects

The biguanides are associated with more frequent side-effects than the sulphonylureas but they are often transient and dose-related. Such effects are:

- a metallic taste in the mouth
- anorexia
- nausea
- vomiting
- diarrhoea

The effects are usually seen early in the course of treatment and can be largely avoided by increasing the dose of the biguanides slowly, say at weekly intervals.

If gastrointestinal symptoms persist for longer than 12 hours after the withdrawal of the biguanide, another cause should be sought.

Prolonged overdosage with biguanides will produce weakness, malaise and drowsiness, sometimes associated with headaches and dizziness.

Rashes are rarely attributable to biguanides.

Lactic acidosis has largely been associated with phenformin, probably due to its concentration in the liver which is the main site of removal of lactate in the body.

It tends to occur within the first few weeks of therapy and is by far the most serious side-effect of biguanide treatment for, dependent on associated circumstances, such as shock, the mortality may be up to 50%. The use of phenformin has been abandoned in many countries but metformin, which is seldom associated with serious lactic acidosis is still widely used.

INSULIN TREATMENT

Insulin is required not only for Type I (insulin dependent) diabetes (IDD) but also for many diabetics with postprandial hyperglycaemia which is not controlled by diet and oral hypoglycaemic tablets. If not only the control of the symptoms of diabetes, but also a delay or prevention of its complications is to be achieved, then the aim should be to keep the 2 hour postprandial plasma glucose levels as near normal values as possible during the 24 hour period. In elderly patients, these standards can be somewhat relaxed.

Since it has been possible to prevent diabetics dying from ketoacidosis by adequate treatment with insulin and fluids, the full picture of the formidable problem of late complications has become manifest. The prevention and treatment of these complications is now the main challenge in the treatment of diabetes.

Classification

At present, insulins can be classified according to their antigenicity in man and according to their duration of action. The main sources of insulin used for diabetics are:

- cattle
- pigs

Pig insulin differs from human insulin by only one amino acid and is less antigenic than ox insulin which differs from the human type by three amino acids. Thus, insulins in the least immunogenic group are mainly of porcine origin.

Synthetic human insulin is now available for experimental use and within a few years will replace that from animals.

The removal of impurities of larger molecular size than insulin itself has produced the so-called proinsulin-free or 'single peak' category of insulins.

The advantages of the highly purified insulins are:

- the rare incidence of fat atrophy
- the absence of local allergy at the site of the injection
- smaller doses are often required.

Twenty-four varieties of insulin are at present marketed in the United Kingdom.

For clinical purposes insulins may be divided by their duration of action (LD-61) into:

- short acting
- medium acting
- long acting

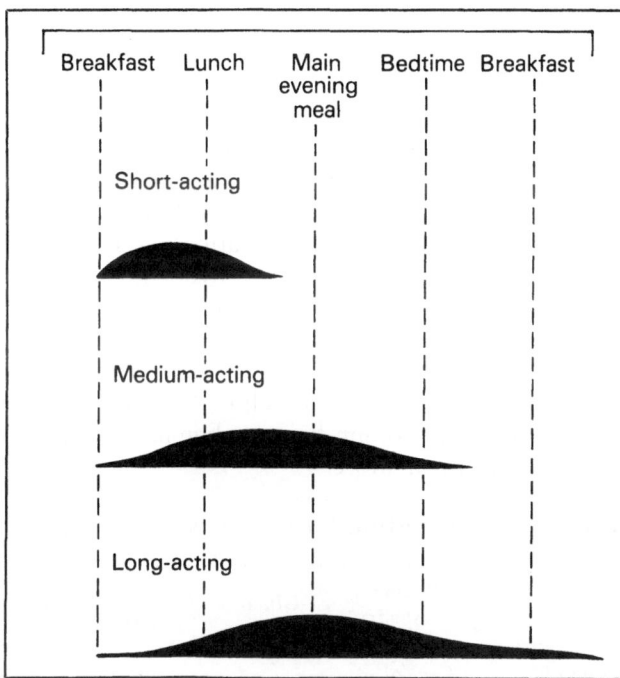

LD-61. *Diagrammatic representation of peaks and duration of action of types of insulin.*

Aim of insulin therapy

In the non-diabetic the level of plasma glucose is maintained approximately within the range of about 4.5–8.0 mmol/l (80–140 mg/dl) even after feasting or fasting, thus achieving a constancy far surpassing that of other circulatory fuels or metabolites. The evolutionary value of such stability derives from the brain's dependence on glucose, by far its most important source of energy.

Insulin preparations

These are summarized in Table 58 and provide a variety and scope to suit older diabetics who may be adequately controlled on a single dose a day, and shorter-acting preparations, mainly for the Type I diabetics who may have little endogenous insulin secretion.

When patients receiving large doses of conventional insulin are changed to a highly purified (H.P.) porcine insulin, the insulin requirement will often be reduced by as much as 20%. This reduction may be quick, due to failure to combine with antibody, or occur slowly over several weeks, presumably due to less antibody being formed. Any changeover from large doses of conventional insulin to a highly purified insulin should be done under close medical supervision, if potentially dangerous hypoglycaemia is to be avoided.

All insulin preparations marketed in the UK are available in the:

- 40 u
- 80 u/ml strengths

Soluble insulin is available also in very much higher strengths.

In the United States of America and Canada, there

Table 58. Duration of action of insulin preparations

Duration of action	'Conventional' insulin type	Highly purified insulin type (Conventional insulin equivalent)
Short acting	Soluble insulin Neutral (Nuso) insulin	Actrapid MC (Neutral, Nuso) Velosulin (Neutral, Nuso)
Medium acting	Insulin zinc suspension (semilente) Isophane (NPH) insulin Globin zinc insulin	Semitard MC (semilente) Insulatard (isophane, NPH) Rapitard MC (biphasic) Insulin Leo Mixtard (biphasic)
Long acting	Protamine zinc insulin Insulin zinc suspension (lente) Insulin zinc suspension (ultralente)	Monotard MC (lente-like) Lentard MC (lente) Ultratard MC (ultralente)

has been a changeover to the 100 u/ml strength and this change will take place in the United Kingdom in the near future.

Starting insulin therapy

Data exist suggesting that the 'honeymoon period' of reduced insulin requirement following diagnosis may be significantly prolonged with H.P. insulins. Further, many workers believe that the incidence of long term complications may be diminished by the use of these insulins. It is therefore recommended that all new, young, insulin dependent diabetics (diabetics facing many years of insulin therapy), should be treated from the start with H.P. insulins.

In young, newly diagnosed, insulin dependent diabetics the object of therapy is to achieve the best possible metabolic control throughout the day. In the first few months of diabetes when endogenous insulin secretion is often temporarily re-established (the honeymoon phase) it is frequently possible to achieve adequate control by a single daily injection of almost any insulin. However, in the long term the great majority of patients will progress to a requirement of twice-daily injections of short and medium acting insulins as the endogenous insulin secretion fails. As habits of treatment are best established at the outset it is best to start patients on a twice-daily regime.

Suitable combinations for twice-daily injection are:

- Velosulin/Insulatard
- Actrapid MC/Monotard MC
- Actrapid MC/Semitard MC

A comparable twice-daily regime using conventional insulins is a soluble/isophane combination (LD-62).

Note that the duration of action for each of the longer acting preparations in the above combination is not the same. Further, the exact duration of effect will depend on individual patient variation, dosage and the nature of the injection site.

In elderly diabetics and those with coincidental chronic medical conditions a lesser degree of diabetic control may be acceptable, e.g. single daily injections of a longer acting 'conventional' insulin preparation.

Patients with a temporary insulin requirement, e.g. during pregnancy, should be treated with H.P. insulins.

Self-testing

Although it is the plasma glucose which needs to be maintained at a satisfactory level, and although self-monitoring techniques are now available, for plasma glucose, in the majority of patients, reliance is placed on indirect measurements based on urine testing for glucose or reducing substances (Table 59). Unfortunately, there is considerable variation in the so-called renal threshold for glucose above which plasma glucose spills over into the urine. The threshold tends to be low in pregnancy and to rise with increasing age and the reduction in glomerular filtration rate. Thus a sugar-free urine implies a much more strict standard of control in a pregnant diabetic than in a diabetic of 70 years of age or more.

Table 59. Urine testing for diabetics.

Name of test for	Reducing sugar	Glucose	Ketones (acetoacetate)
Clinitest	⊗		
Clinistix		×	
Diastix		⊗	
Testape		⊗	
Acetest			⊗
Ketostix			⊗
Ketodiastix		⊗	⊗

× = qualitative
⊗ = semiquantitive

Diabetics with renal glucose thresholds at an accepted norm of about 10 mmol/l (180 mg/dl) should show little or no glycosuria in recently formed urine passed before meals. The best time of the day to test depends on the insulin regime in use. In general, it is best to base the insulin dose on the result of a sugar test done on urine formed when it is expected that the insulin will be acting maximally since, if much glucose is shown at that time, it should be safe to

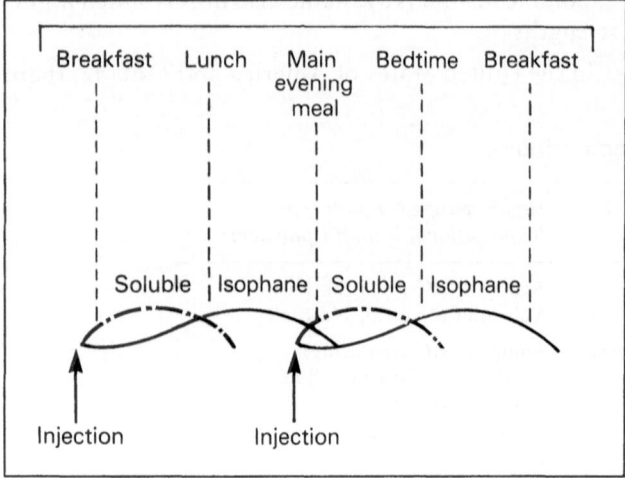

LD-62. *Peaks of insulin action on a twice-daily insulin injection regime.*

increase that insulin dose without risking too low a plasma glucose level at other times.

If the sugar test of the main index time is negative, and the tests at other times show much sugar, it is an indication to change the distribution of carbohydrate or the type of insulin regime in order to obtain more even control. With a scale of insulin doses, the more intelligent patients will be able to alter the dose of the appropriate insulin up or down as the occasion demands, e.g. with:

- infection
- stress
- reduced food intake

The majority of new patients should be taught to self-adjust their insulin dosage from the time of diagnosis.

Urine testing before all the main meals will provide information on the adequacy of dosage of the insulins in use. For instance, the pre-lunch specimen will reflect the action of the morning dose of short acting insulin. The pre-dinner specimen provides information about the corresponding medium acting insulin. Similar information can be obtained during the second half of the day in relation to the evening insulin dosage. **Example:**

	Before breakfast	Before lunch	Before tea	Before bed
Tuesday	0.5 %	1%	0%	0.25%
Wednesday	0.25%	2%	0%	0%
Thursday	0%	2%	0.5%	0%

Consistently 2% urine tests at the pre-lunch specimen would indicate either:

- that an increase in dosage of the morning short acting insulin is required, or
- if hypoglycaemic episodes are occurring mid-morning, that carbohydrate is being taken to counteract the associated features, i.e. a reduction in dosage is required

Home blood glucose monitoring

Blood glucose monitoring allows much more accurate assessment of diabetic control and consequently 'fine tuning' of the insulin regime. Blood is applied to a reagent strip for 60 seconds before washing or wiping off. Then by the use of several alternative reflectance meters or by comparison with a visual colour code it is possible to obtain accurate glucose assessment. The portable reflectance meters have a built-in 1 minute timer and are a convenient pocket size for carrying. Patients are encouraged to maintain preprandial blood glucose values in the 4–7 mmol/l range.

The use of home blood glucose monitoring has allowed much tighter diabetic control in insulin dependent diabetic patients and is proving of particular value in diabetic control during pregnancy and in patients with abnormalities of their renal threshold for glucose.

The role of exercise in diabetic control

Exercise lowers the plasma glucose in general, although the catecholamines released at the start may initially raise it slightly.

Vigorous exercise thus has the effect of reducing the requirement for insulin and, as a result, many manual workers need a larger dose of insulin in the evening than before breakfast. Nevertheless, extra exertion will lead to weight loss unless compensated by an increase in the food intake. It should be noted that extra carbohydrates are needed before, during and usually after prolonged exercise to prevent hypoglycaemia.

Stable control

All insulin dependent diabetics must know that they must never stop their insulin even during times of decreased food intake.

'Brittle' is a term applied to some diabetic patients who have notoriously unstable plasma glucose levels which may swing for no obvious reason between manifest hypoglycaemia and gross hyperglycaemia. It is easy to assume that such instability is due to the fickle behaviour of the patient but, in fact, there are few such insulin dependent diabetics who cannot be helped to achieve at least symptomatically satisfactory control of their diabetes.

An important and often unrecognized mechanism for wide swings in plasma glucose is the reactive hyperglycaemia which regularly follows even mild hypoglycaemia, such as when the plasma glucose sinks below about 2.8 mmol/l (50 mg/dl). This brings into action counterregulatory hormones such as:

- catecholamines
- growth hormone
- glucose
- cortisol

which raise the level of plasma glucose for many hours (Somogyi effect). Thus, if the patient is seen by the doctor in this state of heavy glycosuria, and a raised plasma glucose level, the tendency is to raise the dose of insulin. This then precipitates worse hypoglycaemia and greater reactive hyperglycaemia follows.

Local reactions to insulin

Local reactions to insulin include:

- acute allergic reactions
- painful lumps
- insulin tumour
- fat atrophy or hypertrophy

Allergic reactions

These are rare, but may occur when insulin treatment has been stopped and then restarted after a few weeks or months. Allergy has not been reported with the highly purified insulins.

Painful lumps

Local reactions may consist of painful lumps which appear at the injection sites, but not with the highly purified insulins.

Insulin tumour

Another type of local reaction is an insulin tumour which may occur when conventional insulins are repeatedly injected into the same area which becomes insensitive.

Fat atrophy or hypertrophy

Fat atrophy is the commonest local reaction and is seen more in women than men, but also commonly in children. It appears in the form of shallow excavations in the subcutaneous fat at sites of repeated injections of conventional insulin (CP-5). Insulin is poorly absorbed from areas of fat hypertrophy or atrophy.

Intradermal injections produce first induration and then necrosis, ulceration and scars.

Surgery in diabetic patients

The risk of surgery on diabetic patients can be much reduced if the diabetes can be well controlled over this period. Two main points need consideration before deciding on treatment of the diabetes at this time:

- is the patient insulin dependent?
- how long will the patient be prevented from taking food by mouth?

The surgeon should put any diabetic patient first on his operating list unless there is some specific contraindication, such as gross infection. It is then possible to phase the medical and surgical treatments with each other.

Shortly after major operations, there is a period of low renal loss of water and salt, so it is better to use a 10% rather than a 5% glucose infusion run in at approximately 500 ml in 4 hours. This should avoid fluid overload in elderly patients and the 12.5 g of glucose an hour should prevent ketosis if enough insulin is available. Usually no normal saline is needed in the first 24 hours postoperatively, but thereafter some normal saline and potassium should be included – always in association with at least 5% glucose to prevent hypoglycaemia or ketosis while the patient is not eating.

If the patient's diabetes has been controlled by diet alone, no special treatment should be required, but the urine should be tested regularly for sugar and, if this is present, for ketones too. If in doubt and no urine is available, the plasma glucose should be estimated. If the value is above 11 mmol/l (198 mg/dl) with no more than a slow rate of infused glucose, soluble insulin about 8 u eight-hourly should be given.

Oral hypoglycaemic agents

If the patient's diabetes has been controlled by diet and a biguanide or sulphonylurea, or both, the medication should be omitted on the day of operation when only a minor procedure is contemplated. Chlorpropamide can continue to lower the blood glucose for a day or two after stopping such treatment, and it is probably advisable to change to a shorter acting sulphonylurea preparation several days prior to operation. Urine should be tested regularly for sugar and ketones. If there is more than 1% glycosuria on two or more consecutive tests, the patient should be started on a small dose of soluble insulin 8 hourly. If there is definitive ketonuria, at least 25% more insulin should be given. After a few days, it should be possible to resume treatment with oral hypoglycaemic tablets. In the case of major operations it is probably advisable to treat as for insulin dependent diabetes.

Insulin dependent diabetics

These patients should be admitted to hospital for assessment and stabilization of diabetes at least 48–72 hours prior to surgery.

In recent years there was a vogue for reducing the insulin dose on the day of operation. However, despite no major rise in blood glucose levels it has been shown that alarming rises of ketones may occur during such regimes and most doctors now prefer a system of continuous glucose infusion with intermittent or continuous insulin therapy.

The most satisfactory and easily manageable regime is that attributed to Alberti recommending:

- Careful stabilization using subcutaneous short acting insulin only on the day prior to surgery.
- No subcutaneous insulin is given on the day of operation.
- An infusion of 10% glucose 500 ml containing 10 u short acting insulin is set up early morning and run through over 4 h. One gram of KCl can be added to the infusion and adjusted according to electrolyte levels obtained.
- The rate of insulin infusion can be adjusted according to blood glucose values checked with the aid of a reflectance meter. The aim should be to keep the blood glucose level between 5 and 10 mmol/l.

After the operation, the regime of 4 hourly bottles can be continued until oral feeding is reintroduced and a three times daily insulin subcutaneous regime is reinstituted.

Diabetes and pregnancy

Before insulin was available, insulin dependent diabetics had a high maternal mortality, seldom carried their babies to term and even less often produced normal babies. The maternal mortality is now virtually nil, and in the teaching centres the perinatal mortality has been reduced below 5%, but is still more than twice that of the normal population. These advances have depended on many factors but most of all on the recognition of the importance of really strict control of the maternal blood glucose throughout pregnancy and especially in the last trimester.

Fertility in well-controlled diabetics is probably normal and, if young diabetics seek genetic advice, it is true to say that, even when both parents are diabetic, the chances of their progeny getting diabetes early in life are very small. Nevertheless, there are a few types of diabetics, such as the relatively benign maturity onset diabetes of the young (MODY) which seem to be inherited as a dominant character.

In normal pregnancy there is a reduction of glucose tolerance. The fasting plasma glucose falls and the level at 2–2½ hours after an oral load rises. Serum insulin values rise, its biological half-life is shortened and resistance to it is increased. These changes quickly revert to normal after delivery.

Classification

Potential diabetics

These are non-diabetics at known greater risk of developing diabetes during pregnancy. They include those:

- with glycosuria in a second fasting urine specimen
- with a family history of diabetes in a parent, sibling or child
- who have previously borne a heavy-for-dates baby
- who have had unexplained fetal loss or neonatal deaths in the past

Such patients should have regular tests for fasting glycosuria. If present, a blood glucose should be measured 2 hours postprandially to exclude definite diabetes (e.g. greater than 8.5 mmol/l). If this is normal, a glucose tolerance test should be undertaken. If this also is normal, the test should be repeated early in the last trimester.

Impaired glucose tolerance diagnosed in pregnancy

These patients may have a normal fasting blood glucose level but an abnormal oral or intravenous glucose tolerance test. If the test returns to normal post partum, as it usually does, some centres call these patients 'gestational diabetics' but others use that term independently of any subsequent glucose tolerance test result.

The importance of these early stages of diabetes is that they are associated with a significant increase in perinatal mortality and there is evidence that this can be largely prevented by suitable diabetic and obstetric care.

The effect of pregnancy on diabetes

Renal threshold

This falls from usually between the 3rd and 5th months of pregnancy. This means that, when the insulin dosage is based on semiquantitative urine tests for sugar, the blood glucose will be more tightly controlled than in the past. If attempts to keep the urine sugar-free are associated with frequent hypoglycaemia, then it is best to adjust the Clinitest by taking 3 instead of 5 drops of urine and 12 instead of 10 drops of water.

Insulin requirements

The insulin requirement rises from the 4th or 5th

month in most pregnant insulin dependent diabetics and may double the prepregnant dose. This is probably related to insulin's more rapid destruction, especially in the placenta, and the higher concentration of hormonal antagonists in pregnancy.

Ketonuria and hypoglycaemia

These can be prevented by taking small volumes of glucose drinks, flavoured with lemon, to replace breakfast.

The effect of diabetes in pregnancy

The fetus of a diabetic or even a prediabetic mother tends to grow faster than normal from about the 28th week of gestation. This may lead to a difficult delivery of a baby that is not as mature as its size might suggest (CP-81).

The explanation for fetal macrosomia, which is due largely to excess fat, is probably maternal hyperglycaemia. This leads to hyperglycaemia and hyperinsulinism in the fetus with beta-cell hyperplasia. The hyperinsulinism is a powerful stimulus to growth and fat formation.

Hydramnios and pre-eclamptic toxaemia occur with increased frequency in diabetic pregnancies and there is an increased chance of intrauterine death in the last few weeks of pregnancy. All these features of diabetic pregnancy can be largely prevented by accurate control of the maternal blood glucose.

Effect of diabetes on the fetus

In diabetic pregnancies there is about three times the normal prevalence of congenital anomalies (which also tend to be more severe) and also two or three times the normal perinatal mortality. Good control of the maternal diabetes can reduce the perinatal mortality and there is evidence to suggest that good control from the time of conception can reduce the fetal abnormality rate.

Congenital defects are now the commonest causes of the increased perinatal mortality in diabetic pregnancies.

Hyaline membrane disease was a common cause of neonatal deaths from 'idiopathic respiratory distress syndrome' until amniocentesis and measurement of lecithin and sphingomyelin ratios in the amniotic fluid (as an index of surfactant in the fetal lungs) became available. These are now used as a guide as to when to induce labour in the pregnant diabetic.

Management of diabetic pregnancy

The best results are obtained when there is a high degree of co-operation between:

- physician
- obstetrician
- paediatrician

who are all specially interested in the subject.

The role of the physician is to normalize the blood glucose of the mother, and this involves training her well and motivating her strongly. The chief role of the obstetrician is to judge the maturity of the pregnancy accurately and to decide with the physician the optimal time for delivery.

Impaired glucose tolerance diagnosed in pregnancy is treated by diet. The energy value of the diet should be about 5 MJ (1200 kcals). If the patient is not overweight, insulin is normally used, although chlorpropamide in a dose not exceeding 100 mg a day has been shown to be effective and safe and often to be associated with a reversal of the impaired glucose tolerance later in pregnancy. A repeat glucose tolerance test is done at least a week after withdrawal of the chlorpropamide. The sulphuronylureas can cross the placenta to the fetus and larger doses can induce severe neonatal hypoglycaemia and should therefore be avoided.

Patients receiving treatment with a sulphonylurea before conception should be retested off the drug in the first trimester. If any degree of diabetes is found, they should be changed over to a single daily dose of a quick, plus a slow acting, insulin, given before breakfast. Initially, no more than 20 u/day is likely to be required.

If the patient was previously receiving a single daily dose of insulin, this should be changed to a twice-daily dose, preferably of a mixed quick and slow insulin with about two thirds of the total dose before breakfast and the remaining one third before the evening meal.

Normally, the insulin dose is altered within predetermined limits by the patient on the basis of semiquantitative urine sugar tests done about four times a day, but self-monitoring of the blood glucose is now being done to an ever-increasing extent. Using this technique up to six times a day, in patients with a poor prognosis for the pregnancy, a mean plasma glucose can be kept within strictly normal limits with obvious benefit to the progress of the pregnancy and the quality of the baby.

The diet needs to be followed carefully and increased about 0.8 MJ (200 kcals) a day from about the 4th month. If at any time during pregnancy significant ketonuria or marked hyperglycaemia occurs, the patient should be brought into hospital urgently for stabilization. Some physicians do this routinely at about 32 weeks gestation for careful supervision before delivery. However home blood glucose monitoring has allowed well-controlled patients to be maintained out of hospital until nearer term.

Urinary infections are treated actively and a close watch is kept on the blood pressure. The better the control of the diabetes, the safer it is to allow the pregnancy to continue to term, but tests for fetal distress are important. These include:

- maternal urinary or plasma oestrol
- abnormal changes in fetal heart rate

A sudden drop in insulin requirement is an ominous sign.

Fetal maturity is judged clinically or by ultrasound measurements of the fetal skull at what is judged to be 36 weeks gestation.

Before a decision is made to deliver earlier than at 38 weeks, a check is performed on the surfactant level in the amniotic fluid as a measure of fetal lung maturity. The placenta is first localized by ultrasound and amniocentesis is then performed. The lecithin:sphingomyelin ratio is measured. Values above 2.0 give some assurance that severe respiratory distress will not occur.

Diabetes in childhood and adolescence

In the management of diabetic children, the same basic principles of dietary control and insulin usage apply as for adult diabetics. Meals must be regular and the carbohydrate content of the diet measured in 10 g exchanges should be evenly distributed throughout the day. It is essential to provide a normal calorie intake for the age and sex in order to ensure normal growth and sexual development, as inadequate nutrition leads to short stature and delayed puberty. A restricted calorie intake should be prescribed only for obese children.

Insulin requirements

Insulin is used optimally to ensure good utilization of the food eaten and to minimize hyperglycaemia and glycosuria. However, it must be remembered that children indulge in sudden bursts of exercise throughout the day; some leeway must therefore be allowed to avoid hypoglycaemic attacks which, if severe, may take the form of fits.

In young children, in whom a degree of pancreatic insulin production remains intact, it may be quite satisfactory to control the diabetes on one dose of insulin a day taken before breakfast. This is usually in the form of a mixture of a soluble and a delayed action insulin. Older children and certainly pubertal adolescents require insulin injections twice daily to achieve good control. The evening injection is given before the main meal and again a mixture of a soluble insulin and a delayed action insulin is usually employed. Most children learn to give their own injections some time after the eighth birthday. The injection sites should be checked from time to time as sometimes the children choose one area and inject it over and over again, resulting in irregular absorption of insulin from the site.

The children should be taught that taking exercise has a similar effect to insulin and tends to lower the blood sugar. It is essential, therefore, that they carry sugar with them at all times in their pocket ready to take when the first symptoms of hypoglycaemia appear. Fortunately, many children recognize the early warning symptoms. In view of the danger of hypoglycaemic reactions, children should always wear an identity disc or other means of recognition. A supply of glucagon at home is occasionally useful. The parents give it by injection if the child is unable to take sugar orally.

The importance of good control is emphasized and every child has a hard-covered book in which urine tests are recorded every day before breakfast and before the evening meal. The bladder is emptied about half an hour before the urine is collected for these tests so that the result reflects the blood sugar level at the time. In addition, the morning sample is tested for acetone. The results should be brought to the clinic, added up for each time in terms of no sugar, a trace, 1% or 2% of sugar and this information is useful in determining the optimal insulin dose. Further information may be obtained from time to time by estimating capillary blood glucose sugar levels taken at home, using Boehringer blood test strips or reflectance meters. Measurement of glycosylated haemoglobin (HbACI) has proved a useful guide to longer term control. The height and weight and stage of pubertal development when relevant are recorded at each clinic visit. These reflect the adequacy of the child's nutrition. Finally, it is essential to discourage the teenagers from starting to smoke and to insist that their parents and siblings set them a good example.

MANAGEMENT OF THE ACUTE COMPLICATIONS OF DIABETES

The acute complications of diabetes include:

- ketoacidosis
- hyperosmolar (non-ketotic or aketotic) coma
- lactic acidosis
- hypoglycaemia

Diabetic ketoacidosis

Insulin lack leads to:

- hyperglycaemia
- dehydration
- electrolyte and mineral loss

The essentials of treatment are therefore to:

- replace fluid and electrolytes
- replace insulin

Fluids and electrolytes

It is unwise to rely on giving fluids by mouth if there has been any history of vomiting, since they are likely to cause further vomiting which may well be aspirated by a drowsy patient. A tube should be passed into the stomach which may be distended.

The first litre of intravenous fluid should be given in 30–45 minutes and the second in about 1½ hours but, thereafter, the rate moderated to 1 litre every 2 or 3 hours until the dehydration is corrected.

A slow infusion should then be maintained until it is apparent that small amounts of oral fluid are being well tolerated.

The first fluid to be administered should be isotonic saline. Even though a hypotonic fluid might seem to be indicated, it is unwise to reduce the osmolality of the patient too rapidly.

If the patient is markedly acidotic, e.g. plasma pH less than 7.1 (or plasma bicarbonate remaining below 3 mmol/l), sodium bicarbonate can be given cautiously, such as 250–400 ml or 3% solution in 30–60 minutes. Potassium should be given at the same time, e.g. 1–2 g KCl (13–26 mmol/l) hourly, since sodium bicarbonate will increase urinary losses of potassium and favour its uptake by cells.

When the blood glucose falls to approximately 14 mmol/l (250 mg/dl), glucose should be adminis-tered intravenously at a 5% concentration. By this time, much of the extracellular fluid deficit will have been made up, and it is wise to provide the electrolytes for the process of cellular repair while continuing to give some saline.

Serum electrolytes should be measured about 3 hourly until at least two successive levels are obtained with serum potassium between 4 and 5 mmol/l. Average requirements for intravenous potassium are about 50 mmol (4 g KCl) in the first 6 hours and 90 mmol (7 g KCl) in the first 12 hours. If the patient develops oliguria, potassium should be given very cautiously and under close laboratory control to avoid dangerous hyperkalaemia.

Although no syndromes due to acute deficiency of magnesium and phosphate have been described during the treatment of severe ketoacidosis, their serum level falls markedly. Phosphate is needed for the first stage of intracellular glucose metabolism when glucose is made into glucose-6-phosphate and magnesium is an important cofactor for several stages of carbohydrate metabolism.

Insulin

All patients should be treated in hospital and, if a long preliminary journey is involved, the general practitioner should give about 20 u soluble insulin intra-muscularly and document the patient accordingly.

The insulin level in the plasma needs to be kept at an appropriate level for the degree of hyperglycaemia. This will be approximately 50–200 u/ml. It is better to get a smooth decline in the raised plasma glucose than an intermittent one such as is produced by large doses of intramuscular insulin given every few hours. The latter will also produce fluxes of potentially toxic metabolites such as lactic acid.

Insulin is best administered in isotonic saline by continuous infusion so as to deliver 3–6 u per hour or, in the case of children, 0.1 u/kg per hour.

The infusion pump should be attached by a four-way tap to the infusion line. This allows the insulin to be given at a constant rate while that of the infusion fluid can be varied independently. Some of the insulin may be adsorbed to the plastic syringe and tubing but this is unlikely to be significant at the dose levels used. Such an insulin infusion usually lowers the plasma glucose by about 6 mmol/l (approximately 108 mg/dl) per hour although, rarely, insulin-resistant patients with acute infections have been found to need con-siderably larger doses.

As an alternative to the intravenous route, soluble insulin can be given by hourly intramuscular injec-tion. The dosage is similar but should be preceded by a loading dose of approximately 20 u. In either event, the blood glucose and electrolyte levels should be measured at 2 or 3 hourly intervals until the blood glucose is down to approximately 14 mmol/l (250 mg/dl). At this stage, the insulin infusion may be stopped and the patient given subcutaneous soluble insulin 4 hourly according to urine or blood tests.

After some 12 hours, soluble insulin can be given 8 hourly and, when the patient is having regular meals, this is best given at first three times a day in relation to food.

In newly diagnosed diabetics, the total daily dose will probably not exceed about 40 u and this can be given in equally divided doses and control assessed retrospectively in relation to urine tests or blood glucose measurements.

In an established diabetic, the initial daily dosage will approximately equal the preketosis total daily dosage, but extra insulin may be required if the patient is infected or taking other medications such as steroids.

Hyperosmolar (non-ketotic or aketotic coma)

Patients with this condition tend to be more sensitive to insulin than those with ketoacidosis and the intravenous infusion of insulin is best given at a rate of approximately 3–4 u per hour intramuscularly with an initial loading dose of 10 u. Treatment should be planned to produce a slow fall in osmolality as cerebral oedema may occur if the drop is too rapid. Potassium supplements will be needed and should be given as indicated by the results of glucose and potassium values at intervals of approximately 3 hours.

In view of the considerable risk of thrombosis, some clinicians use a low dose heparin regime until rehydration and clinical recovery are complete.

Lactic acidosis

Since lactic acidosis in diabetic patients has been found largely in association with the use of phenformin, many physicians and some countries have now stopped using phenformin and have substituted metformin which is much safer in this respect. In any case, significant:

- renal failure
- hepatic failure
- heart failure

are strong contraindications to the use of phenformin and to a lesser extent of metformin. Patients on biguanide treatment should be warned also about the dangers of even a moderate intake of alcohol precipitating lactic acidosis.

The treatment of lactic acidosis consists of:

- reversal of hypoxia and circulatory collapse
- correction of the acidosis
- attempts to increase removal of lactate

Efforts to correct the acidosis are usually made by infusing isotonic sodium bicarbonate in amounts of 150–1000 mmol or more.

The mortality of lactic acidosis remains about 50% and even higher when there is associated shock or dehydration.

Hypoglycaemia

If the patient is too stuporose to take glucose orally, this method should not be attempted since persistence may lead to aspiration of hypertonic glucose into the lungs with marked pulmonary oedema. The alternatives are to give:

- 30–50% glucose slowly into a vein
- glucagon intramuscularly or subcutaneously

When glucose is infused, the recovery is often dramatic, the patient regaining consciousness while the glucose is still being injected, although recovery may be delayed when the patient has been long in coma.

Glucagon is in many ways preferable to intravenous glucose, since it can be given single-handed even to an unco-operative patient. Within 15–20 minutes it will raise the plasma glucose sufficiently for the patient to be able to take a further 20 g or so of carbohydrate by mouth.

TREATMENT OF DIABETIC EYE DISEASE
Diabetic retinopathy

There is some evidence that certain HLA tissue types are more associated with retinopathy than others.

Good diabetic control may delay the onset and slow the progress of retinopathy.

Various drugs have been tried and most of them later rejected. Clofibrate has been used in an attempt to prevent the development of waxy exudates. Its use has been largely abandoned however.

Photocoagulation

This has been used over the past 15 years. It is essentially a destructive intervention designed largely to prevent haemorrhages into and in front of the retina but has offered more hope than any treatment previously tried.

The treatment depends on focusing a beam of intense light on the part of the retina which is coagulated by absorption of the light energy. Usually, white light from a xenon arc is absorbed mostly by the pigment layer which is where most damage is produced. Small, and even fairly large, blood vessels can be occluded, although not if they grow forward into the vitreous. It is not safe to use the xenon arc near the macula or the optic disc.

More recently, the argon laser has been preferred because its green light is absorbed well by the haemoglobin the moving columns of blood in the vessels. It is safe to use an argon laser near the optic disc or macula or to coagulate vessels growing forward from the retina.

It is used for three distinct reasons:

- direct destruction of new vessels which have bled or are likely to bleed
  Coagulation of feeder and smaller vessels reduces oedema in the retina, most importantly near the macula.
- destruction of new vessels
  To prevent progressive, irreversible retinitis proliferans and to prevent retinal detachment by forming chorioretinal adhesions
- obliteration of vessels to ischaemic areas of retina
  It is assumed that some vasoproliferative factor is formed there which stimulates the formation of new vessels elsewhere in the retina.

On this basis, large areas of retina have been photocoagulated in some centres. The evidence is good that such treatment at the periphery of the retina improves the circulation of the macular region. It may, however, produce night blindness.

Controlled studies have shown significant benefit from photocoagulation, especially for maculopathy.

Pituitary ablation

This treatment was used mainly for rapidly advancing haemorrhagic or proliferative retinopathy so long as the macular vision of at least one eye remained good and was not threatened by retinitis proliferans. It has now been largely replaced by photocoagulation. Pituitary ablation did not improve renal microangiopathy and rendered the patient dependent on permanent endocrine replacement therapy.

Treatment of glaucoma

Diabetic secondary glaucoma is due to retinitis proliferans with new vessels in the angle of the anterior chamber and vitreous haemorrhage, and it may destroy the sight of the eye completely. This type of glaucoma occurs usually in younger diabetics and differs from the thrombotic glaucoma which follows occlusion of the central retinal vein in middle-aged or elderly patients who, if diabetic, are usually of the maturity onset type. The pain of glaucoma often responds to acetazolamide (Diamox) and local hydrocortisone, but leaves the patient with a blind, painful eye.

Any penetrating operation on the eye is likely to produce more intraocular haemorrhage but, if persistent pain leads to surgery, cyclodiathermy may be preferable to removal of the eye.

Palsy of the external ocular muscles

Although uncommon, diabetes is associated with about 20% of all isolated third nerve palsies. These are nearly always unilateral. The sixth nerve is affected as often as the third and the fourth nerve only rarely.

Such palsies may be presenting symptoms of the diabetes and are usually seen in patients over 50 years old. The prognosis is good, recovery usually occurring within 1–3 months. In most diabetics with a third nerve palsy only part of the nerve is affected since the pupillary reactions and lid-closing reflex may remain normal.

GLOSSARY
Special Tests of Endocrine Function

CLOMIPHENE STIMULATION TEST

Clomiphene is a drug which will block oestradiol receptors in the hypothalamus leading to increased gonadotrophin secretion and a resultant increase in LH and to a less extent FSH release from the pituitary. It can therefore be administered and used to assess hypothalamic gonadotrophin reserve.

Procedure

Clomiphene is given a dose of 3 mg/kg body weight up to a maximum of 200 mg daily for 7–10 days. Blood is sampled for LH and FSH on days 4, 7 and 10.

Interpretation

A normal response is indicated by a rise in LH and FSH outside the normal range for the laboratory.

Lack of response in a patient who has an otherwise normal response to exogenously administered LHRH is suggestive of a diminished hypothalamic gonadotrophin reserve.

The test will be abnormal in hypothalamic disease and often also in anorexia nervosa.

DEXAMETHASONE SUPPRESSION TEST

This is used both as a screening test and in the differential diagnosis of Cushing's syndrome.

Some centres use betamethasone instead of dexamethasone.

Procedure

- Single dose test
  2 mg dexamethasone is administered orally between 11 pm and midnight. A blood sample is obtained at 9 am the following morning for plasma cortisol estimation.
- High dose test
  0.5 mg dexamethasone 6 hourly is administered orally for 2 days. 2 mg dexamethasone 6 hourly is administered orally on days 3 and 4.

Blood samples are obtained daily at 9 am for plasma cortisol estimations for 5 days.

Interpretation

The single dose test is a useful screening test. A 9 am plasma cotrisol of less than 190 mmol/l during this test is normal.

During the high dose test normal subjects will suppress their cortisol levels during the first 2 days in response to the low dose of dexamethasone. Patients with adrenal hyperplasia due to excess pituitary ACTH will not suppress during the first 2 days but may achieve suppression of cortisol levels in response to the high dose of dexamethasone on days 3 and 4.

Patients with ectopic ACTH syndrome or a pituitary adrenal lesion characteristically do not suppress their cortisol levels during the test.

GLUCOSE TOLERANCE TEST (ORAL) FOR DIAGNOSIS OF ACROMEGALY

Since GH is released during stress a single raised level is insufficient in the diagnosis of acromegaly. In normal subjects GH levels should suppress in response to a glucose level.

Procedure

Following a basal blood sample for glucose and GH estimate 75 g of glucose in water is given orally.

Blood samples are obtained at 30 minute intervals until 3 hours for glucose and GH estimation.

Interpretation

Acromegalic patients may show a degree of impaired glucose tolerance. The GH level should be suppressed below 2 mμ/l 90 minutes to 3 hours after the glucose load in normal subjects. A high basal GH level which does not suppress is indicative of acromegaly.

HCG TEST

Used in the diagnosis of primary testicular failure.

HCG is capable of stimulating interstitial cells of the testis to secrete testosterone.

Procedure

A blood sample is obtained for testosterone estimation. 5000 units of HCG is administered intramuscularly twice weekly for two weeks. A second blood sample is obtained 4 hours after the last dose.

Interpretation

In adults with primary testicular disease the testosterone response is reduced or absent. In hypogonadotrophic hypogonadism due to pituitary or hypothalamic dysfunction the response may be normal.

INSULIN HYPOGLYCAEMIA TEST

Stress due to hypoglycaemia occurs when the blood glucose level falls below 2 mmol/l.

Growth hormone and ACTH will be released in response to stress in normal subjects. The test is of considerable value in the diagnosis of hypopituitarism. It is also used in the assessment of GH deficiency in children.

Procedure

(Adult patients should be generally fit, with no history of ischaemic heart disease or epilepsy and have a normal fasting glucose level and normal synacthen test.) Following an overnight fast adequate hypoglycaemia is achieved by initially administering soluble insulin in a dose of 0.1 μ/kg body weight intravenously. A dose as high as 0.3 μ/kg may be necessary in obese patients. A repeat dose can be administered at 45 minutes if adequate hypoglycaemia has not been achieved. Blood samples are obtained at baseline and subsequently at 15 minutes, 30 minutes and every 30 minutes thereafter for 2 hours for GH and cortisol estimations.

Interpretation

With adequate hypoglycaemia stress normal subjects will achieve

- a cortisol level in excess of 4.50 mmol/l
- a GH level in excess of 20 mμ/l

Peak levels normally occur 60 minutes after the insulin infusion.

A low basal cortisol level which does not rise is indicative of hypopituitarism in a patient with a normal synacthen test.

Children achieving a rise to 7 mμ/l–15 mμ/l are felt to have 'partial' growth hormone deficiency. Failure to rise above 7 mμ/l during the test is indicative of growth hormone deficiency.

LHRH STIMULATION TEST

Used to assess the anterior pituitary reserve to secrete LH and FSH.

Procedure

Following a basal blood sample a dose of $100\,\mu g$ of LHRH is administered intravenously. Further blood samples are obtained at 20 minutes and 60 minutes for LH and FSH estimations.

Interpretation

An inadequate rise in LH and FSH values is indicative of anterior pituitary dysfunction in post pubertal patients. It is important to note however that hypothalmic dysfunction may lead to secondary unresponsiveness of the anterior pituitary.

METYRAPONE TEST

Metyrapone inhibits 11β-hydroxylase preventing the final step in the synthesis of cortisol. There is a consequent rise in 11-deoxycortisol and other cortisol precursors. The reduced plasma cortisol results in a rise in plasma ACTH resulting in further stimulation of the adrenal cortex to produce cortisol precursors. These are excreted in the urine as 17-oxogenic steroids. The test is of limited value in the differential diagnosis of Cushing's syndrome.

Procedure

Metyrapone is administered in a dose of 750 mg four hourly for 24 hours.

Urine is collected for 17-oxogenic steroids for two days prior to metyrapone, on the day of administration and for 24 hours subsequently.

Interpretation

Patients with adrenal adenomas show no rise in 17-oxogenic steroid output because ACTH secretion is suppressed. In patients with bilateral adrenal hyperplasia due to pituitary ACTH there is an exaggerated 17-oxogenic steroid output.

STANDARDISED EXERCISE TEST FOR GROWTH HORMONE SECRETION

This test is used to assess GH deficiency in children of short stature.

Procedure

The child is exercised on a bicycle ergometer against a known resistance for 10 minutes. The exercise is varied to suit the age and stamina.

The blood sample for growth hormone should be taken 25 minutes after commencement of exercise.

Interpretation

A rise of growth hormone to $15\,m\mu/l$ or more is a normal response.

SYNACTHEN TEST (PROLONGED)

In patients suspected of primary adrenocortical hypofunction (Addison's disease) in whom there is an equivocal response.

Procedure

A basal blood sample is taken at 9 am for plasma cortisol estimation. 1 mg of depot tetracosactrin (synacthen) is administered intramuscularly and further samples obtained at 1, 4, 8 and 24 hours.

Interpretation

In primary adrenocortical insufficiency plasma cortisol levels remain low throughout the test.

In adrenocorticol insufficiency secondary to pituitary dysfunction, the cortisol levels remain low in the initial part of the test, but may show a gradual rise in the later stages.

SYNACTHEN TEST (SHORT)

This test is usually used in the diagnosis of primary adrenocortical insufficiency (Addison's disease).

It is occasionally of value in the diagnosis of adrenocortical hyperfunction (Cushing's syndrome).

Procedure

A blood sample is taken for plasma cortisol at 9 am. 250 μg of tetracosactrin (synacthen) is given intramuscularly and further blood samples are taken at 30 minutes, 60 minutes and 90 minutes.

Interpretation

In normal subjects
- (a) the basal cortisol should exceed 170 mmol/l.
- (b) the increment rise should be at least 190 mmol/l.
- (c) a cortisol level of 580 mmol/l should be achieved during the test.

In primary adrenocortical failure cortisol levels will remain low throughout the test.

The test is occasionally of value in the diagnosis of adrenocortical hyperfunction due to bilateral adrenal hyperplasia. The synacthen test will show an exaggerated and prolonged rise in cortisol response.

TRH STIMULATION TEST

This test may be used in the assessment of pituitary function or as an aid in the diagnosis of primary thyroid disease.

Procedure

Following a basal blood sample 200 μg of TRH is given intravenously. Further blood samples are obtained at 20 minutes and 60 minutes for TSH estimation.

Interpretation

Patients with primary hypothyroidism have an elevated basal TSH level with an exaggerated and prolonged rise in TSH response. In patients with borderline hypothyroidism the initial TSH level may be in the normal range but an exaggerated rise in TSH in response to TRH provides evidence of some degree of thyroid failure.

The TSH response to TRH is suppressed in patients with high circulating thyroid hormone levels, i.e. in thyrotoxicosis and patients on thyroxine replacement.

Other causes of an impaired TSH response include
autonomous single or multiple thyroid nodules
opthalmic Graves disease
hypopituitarism

N.B. A variety of drugs will alter the TSH response to TRH. These include
corticosteroids
propranolol
L-dopa
estrogens
antithyroid drugs

TRIIODOTHYRONINE SUPPRESSION TEST

When circulating thyroid hormone levels rise BH secretion is suppressed and in normal subjects there should be a consequent fall in thyroid uptake of iodine. This test is of value in conditions where the thyroid-pituitary relationship is abnormal:

Thyrotoxicosis
Ophthalmic Graves' disease
Autonomous thyroid nodule
Multi-nodular goitre

Procedure

Thyroid uptake of ^{99m}Tc is measured at 20 minutes. Triiodothyronine is given in a daily divided dose of 60 μg for 7 days. A second ^{99m}Tc uptake is performed. N.B. T3 should not be administered to the elderly or patients with heart disease.

Interpretation

Normally the second uptake should fall to less than 50% of the first or the uptake fall to less than 10% of the administered dose.

TSH STIMULATION TEST

This test is occasionally of value in the diagnosis of primary hypothyroidism. It is particularly useful to assess thyroid reserve in patients on thyroxine replacement therapy.

Procedure

Following an initial 24 hour ^{131}I uptake TSH is administered by intramuscular injection in a dose of 10 u on 3 consecutive days.

A further tracer dose of ^{131}I is given on the third injection day and a second uptake performed.

Interpretation

A doubling of the control value or an increase in uptake of 15% should be regarded as normal.

WATER DEPRIVATION TEST

This is used in the diagnosis of diabetes insipidus.

Procedure

This must be done under careful supervision – patients with severe deficiency of ADH may become dangerously dehydrated – 'compulsive water drinkers' may 'steal' water thus invalidating the test.

The patient should not smoke during the test.

Starting at 8 am no fluid is administered for eight hours.

Dry food is allowed.

The patient should be weighed before the test and after 4, 6 and 8 hours – if more than 3% body weight is lost **STOP**.

Urine is passed hourly – measure volume and osmolality.

Plasma is obtained at the mid point of each hour for osmolality.

Interpretation

Normally the urine osmolality should reach 600 mOsm/kg or more and the plasma osmolality should not rise above 300 mOsm/kg.

In diabetes insipidus

- plasma osmolality exceeds 300 mOsm/kg
- urine remains dilute at less than 270 mOsm/kg

Correction of this abnormality after administration of DDAVP confirms the diagnosis of neurogenic diabetes insipidus. Failure to concentrate after DDAVP is suggestive of nephrogenic diabetes insipidus.

COLOUR PICTURES

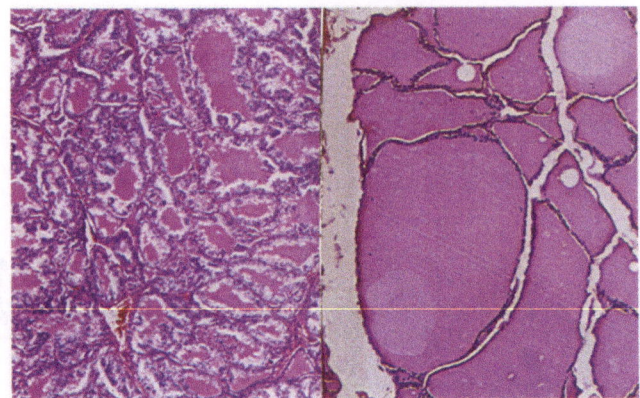

CP-1. Normal thyroid histology showing comparison with that of thyrotoxicosis.

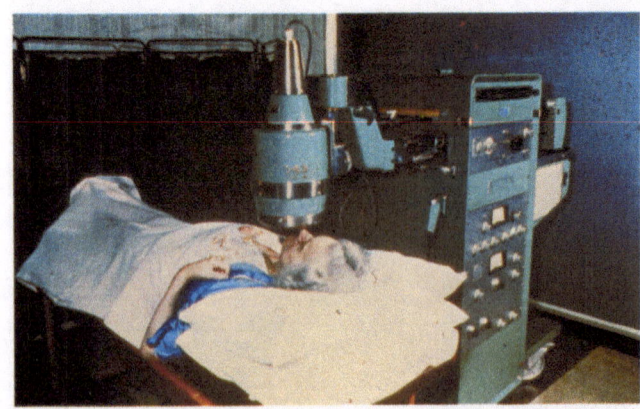

CP-2. A rectilinear scanner.

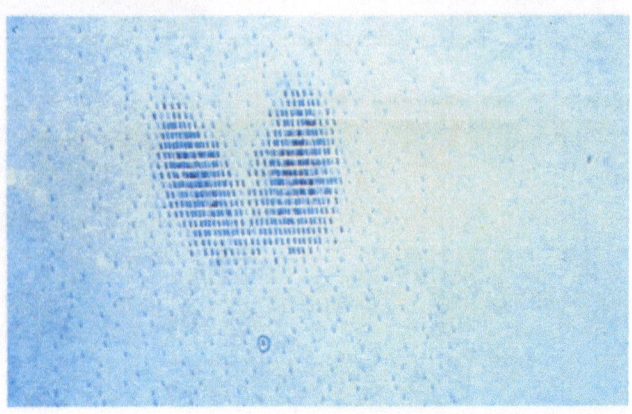

CP-3. Thyroid radio iodine ^{131}I scintiscan showing diffuse uptake in a normal gland.

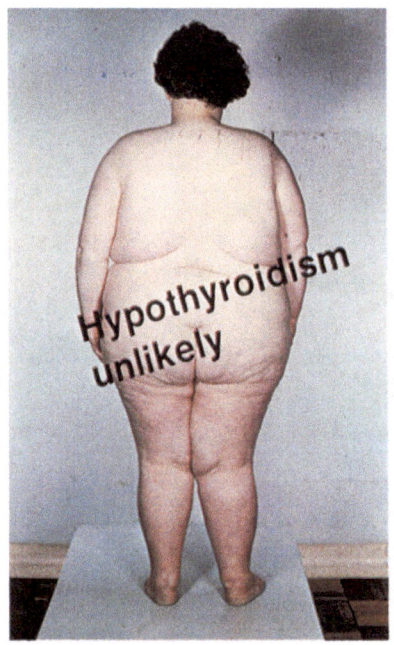

CP-4. *Patient with simple obesity showing relatively even distribution of adipose tissue.*

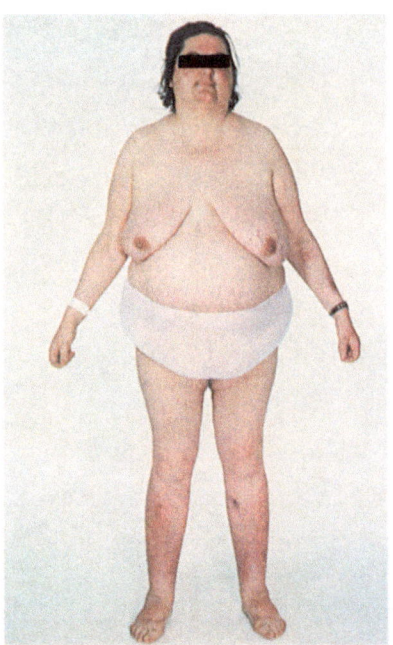

CP-5. *This patient has Cushing's syndrome due to a pituitary basophil adenoma – note the centripetal pattern of adipose tissue.*

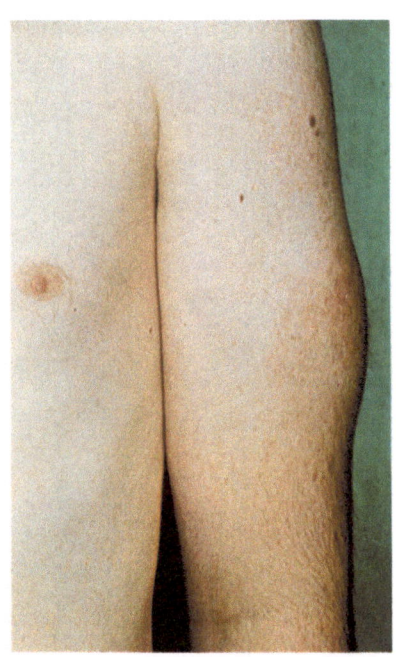

CP-6.1. *Hypertrophic lipodystrophy.*

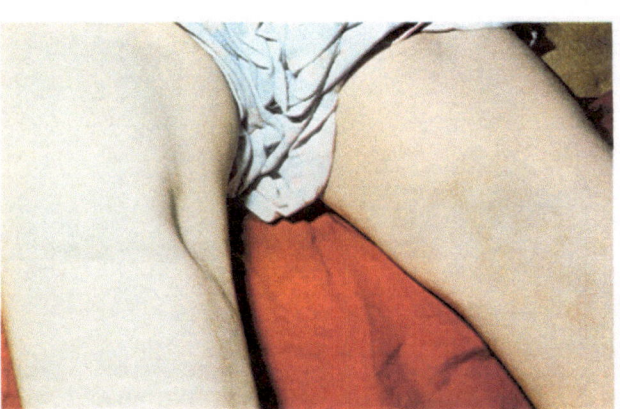

CP-6. *This girl had an area of fat atrophy at the insulin injection site – the use of highly purified insulin material resulted in considerable regression of this feature.*

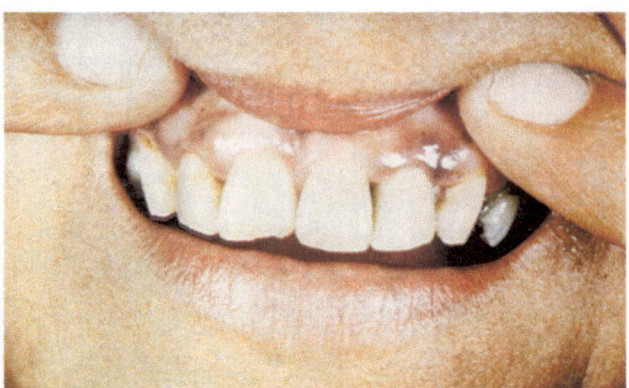

CP-7, 8. *This woman had Addison's disease. Apart from skin changes she had pigmentation of her buccal mucosa affecting (**CP-7**, above) gums and (**CP-8**, next page) inside of her cheek.*

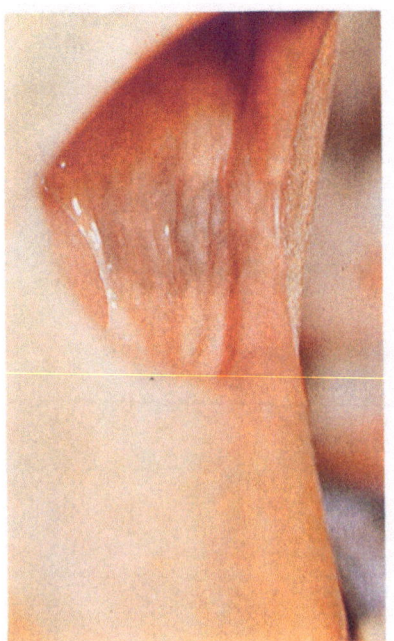

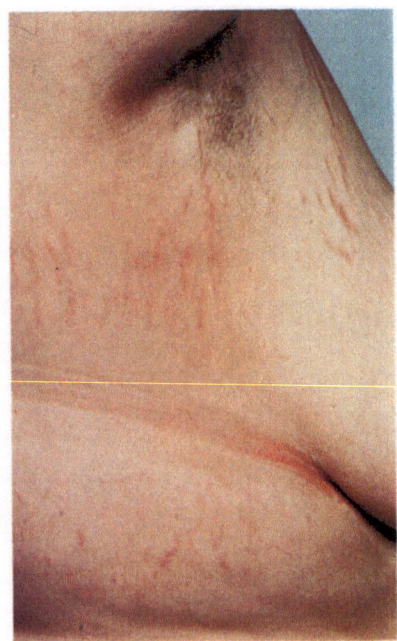

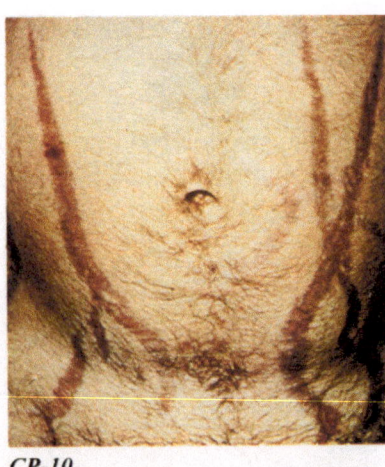

CP-10

CP-8. See legend to CP-7.

CP-9, 10. Both of these patients have Cushing's syndrome. In *CP-9* (above left) the striae are less extensive and are more representative of what clinicians may expect to see in the earlier stages of disease.

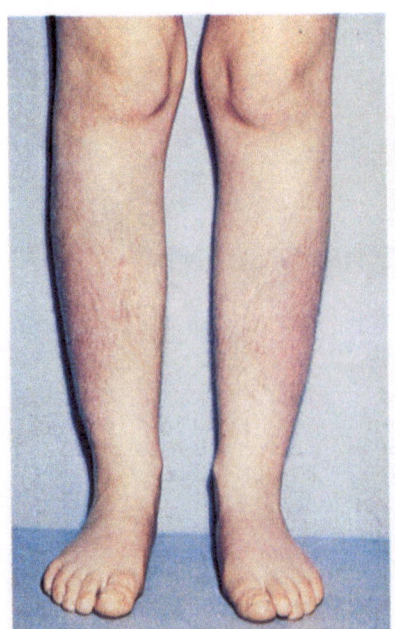

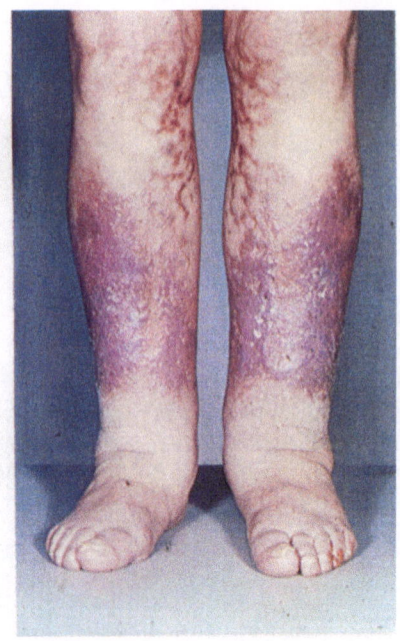

CP-11. This patient had Graves's disease with mild pretibial myxoedema.

CP-12. This patient had a much more extensive pretibial myxoedema. This severity of lesion is less frequently seen.

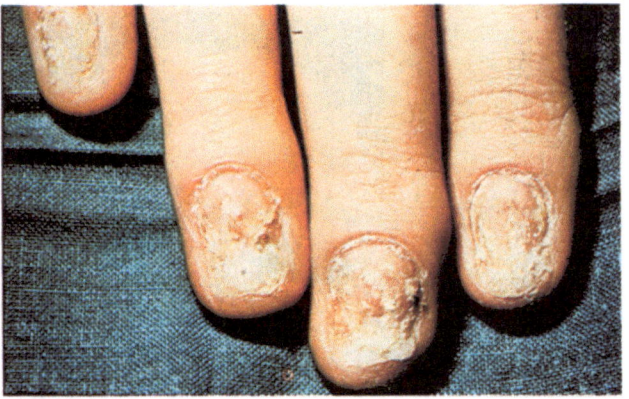

CP-13. *Typical fungal infection of the nails occuring in a patient with hypoparathyroidism.*

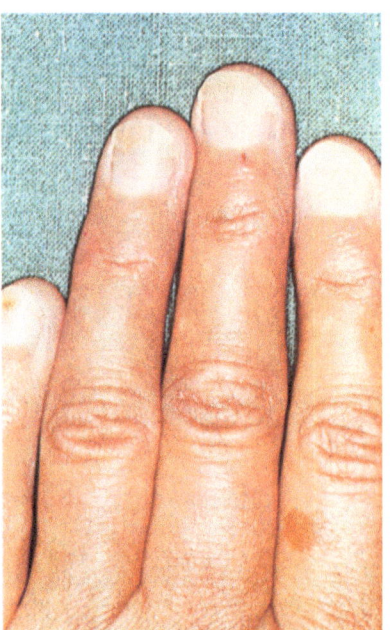

CP-14. *This patient was hyperthyroid and had characteristic onycholysis.*

CP-14.1. *Thyroid achropachy – clubbing of the fingers.*

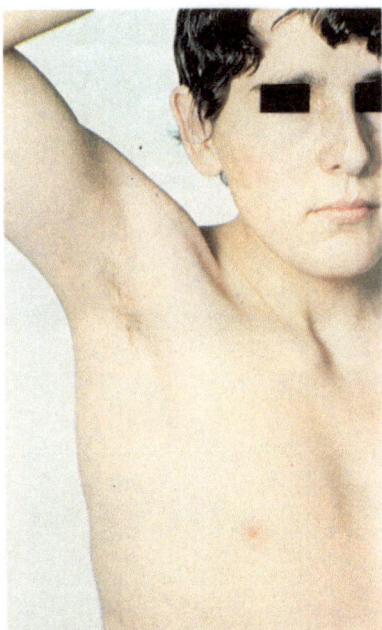
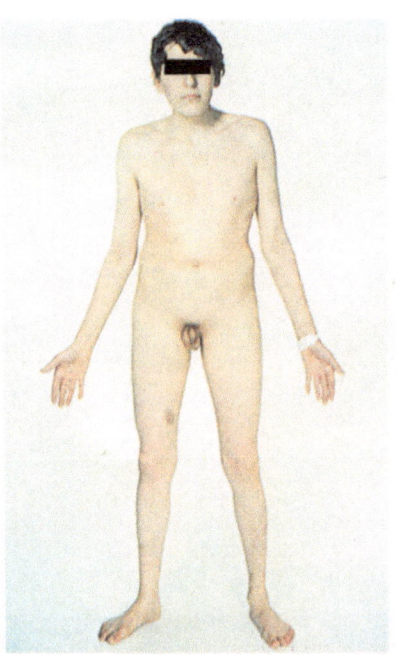

CP-15, 16. *This man, aged 27 years, had sparsity of pubic hair and underdeveloped genitalia due to primary hypogonadotrophic hypogonadism – a specific hypothalamic deficiency of LHRH. NB. eunuchoid stature.*

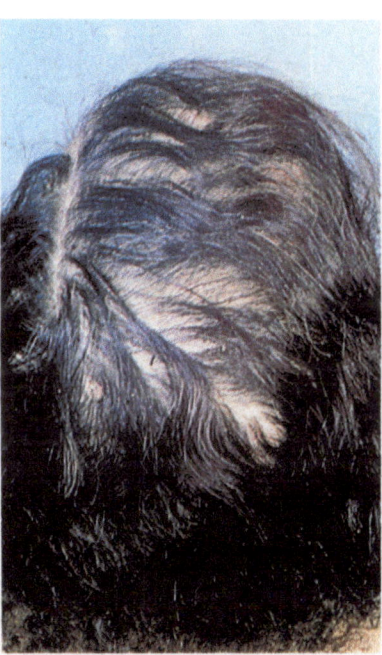

CP-17. *This female patient had thinning and loss of hair due to primary hypothyroidism.*

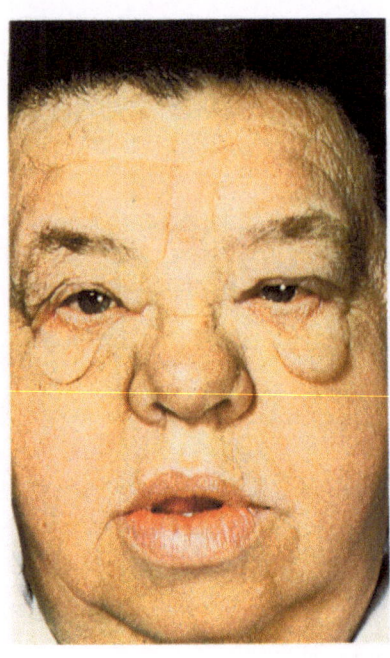

CP-18. This patient had severe hypothyroidism and had characteristic myxoedematous facies with periorbital puffiness.

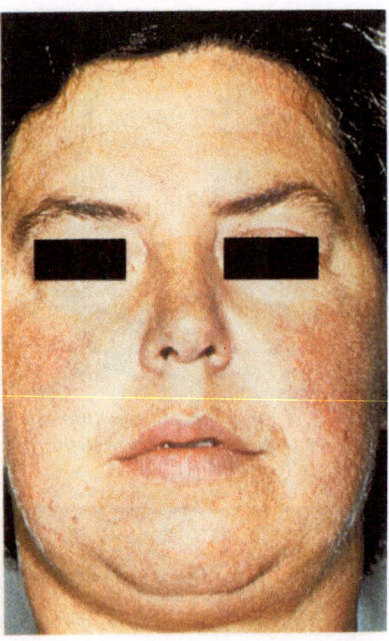

CP-19. The characteristic mooning of the face, plethora and malar telangectasia in this patient is typical of Cushing's syndrome.

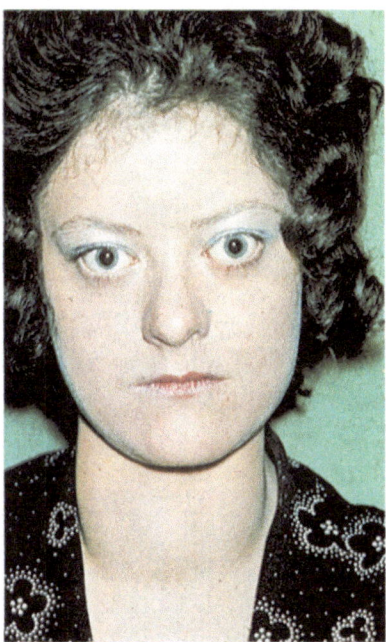

CP-25. *This hyperthyroid girl had lid retraction with a characteristic staring expression rather than true exophthalmos.*

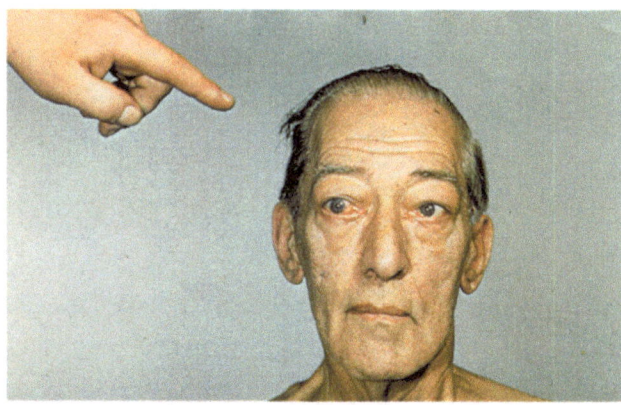

CP-26. *This patient with ophthalmic Graves's disease had lid retraction and limitation of upward and outward movement of the eyes (ophthalmoplegia).*

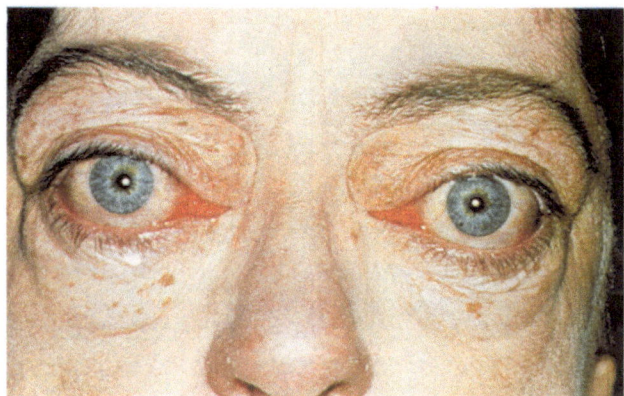

CP-27. *This patient with early exophthalmos has chemosis with conjunctival injection. In addition there is bulging and oedema of the eyelids.*

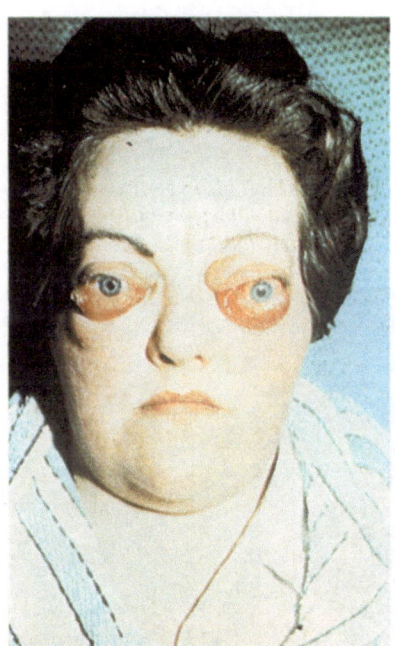

CP-28. *Malignant exophthalmos with inadequate lid closure, oedema and chemosis. This severe form of exophthalmos with visual deterioration requires urgent therapy, usually in the form of orbital decompression.*

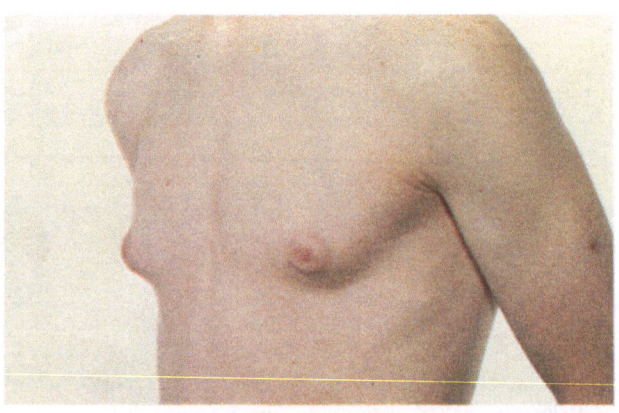

CP-20. *This 14-year-old boy had significant gynaecomastia. It was assumed to be the physiological effect of sex hormone production at puberty and this feature gradually subsided.*

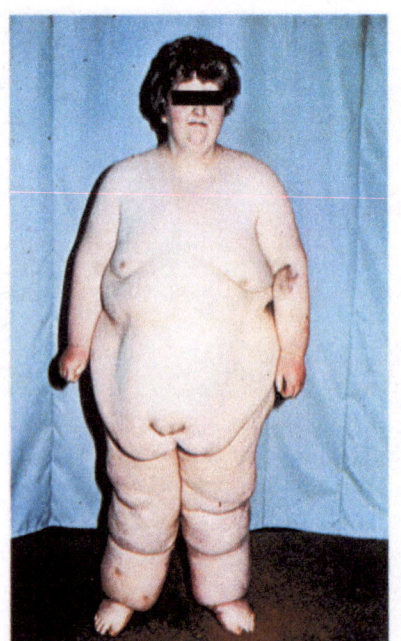

CP-21. *This female patient with gross obesity had no underlying endocrinological cause. The morbidity from this type of weight excess is considerable.*

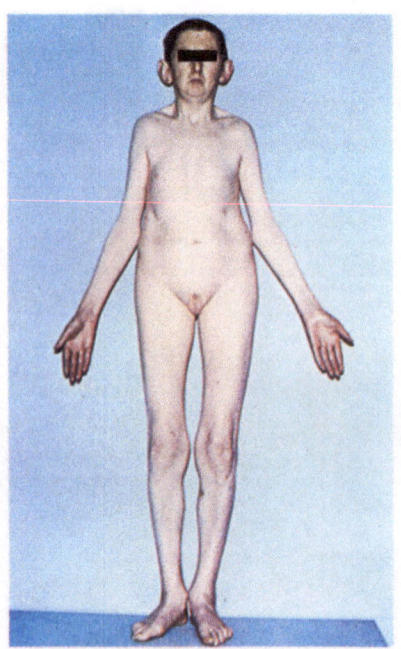

CP-22. *This male patient with hypogonadism shows very poor muscular development and has a characteristic body habitus.*

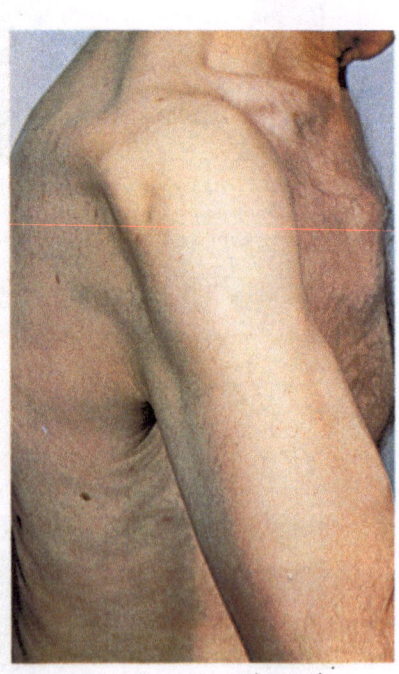

CP-23. *This hyperthyroid man has considerable weight loss with loss of muscle bulk.*

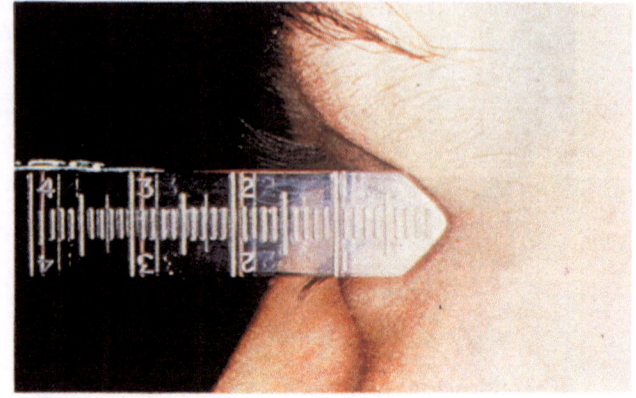

CP-24. *One type of exophthalmometer in use in a normal subject.*

CP-29. Aged 16 years.

CP-30. Aged 18 years.

CP-31. Aged 22 years.

CP-32. Aged 25 years.

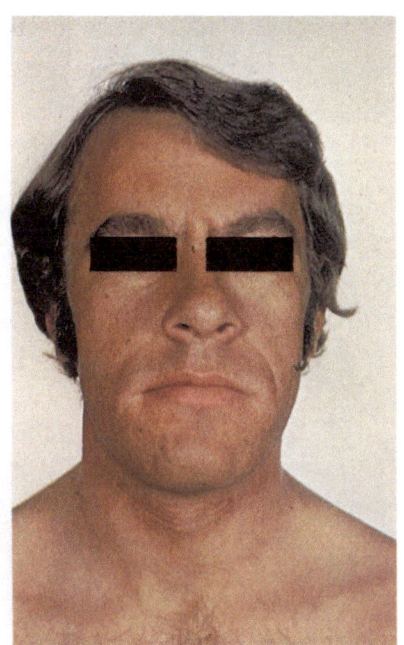

CP-33. Aged 29 years.

CP-29–33. These illustrations together with X-7 and X-8 shows the gradual development of acromegaly in a 29-year-old man. During a 13-year period, he developed coarsening of his features, prognathism and enlargement of his hands and feet. After biochemical and radiological confirmation, transphenoidal hypophysectomy resulted in complete suppression of growth hormone secretion.

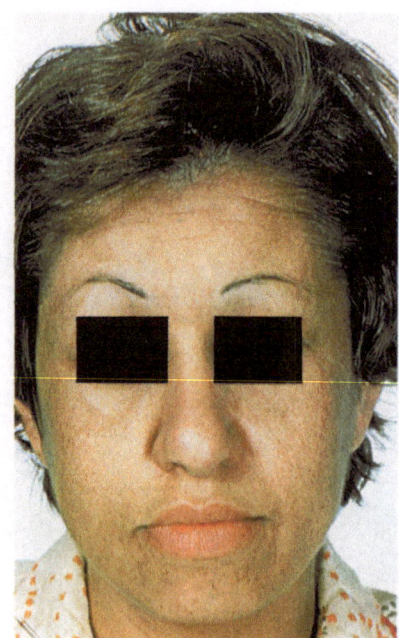
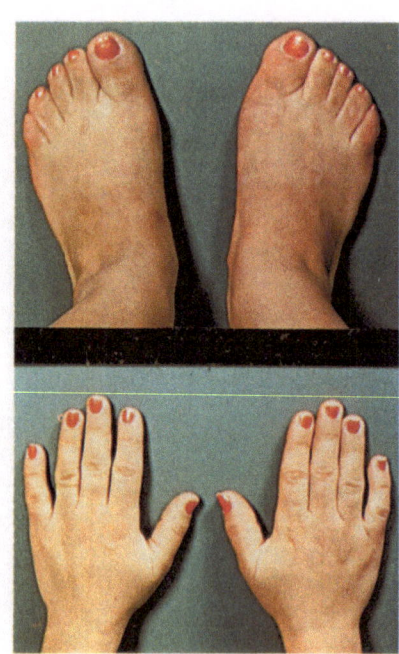

CP-34–36. *These illustrations demonstrate the development of acromegalic features in this female patient. Hypophysectomy followed by radiation of the pituitary fossa resulted in adequate growth hormone suppression.*

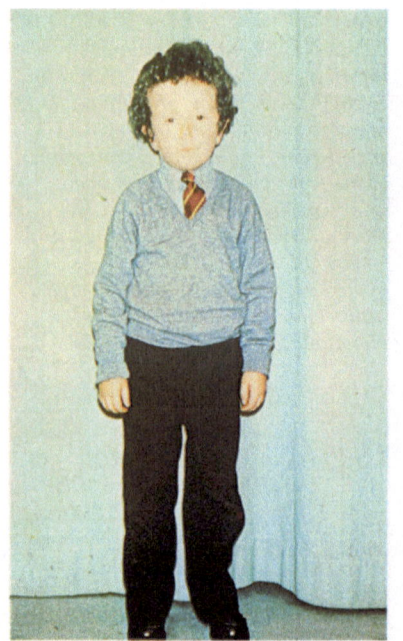
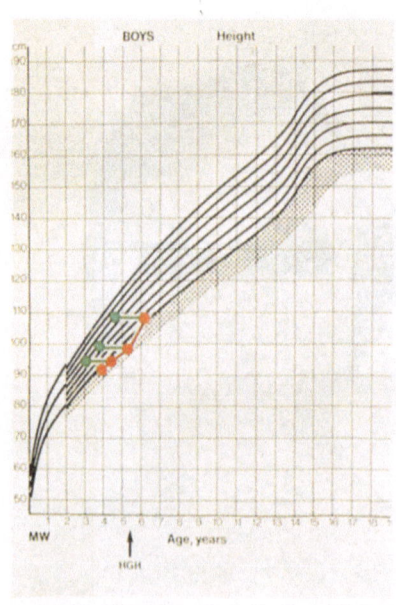
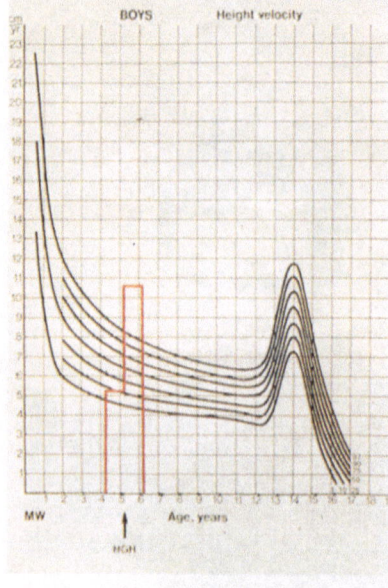

CP-37.1. *Standing.*

CP-37.2. *Height chart.*

CP-37.3. *Height velocity chart.*

CP-37.1–3. *Growth hormone and short stature. Boy, aged 5 years 4 months, with deficiency of growth hormone, which presented with hypoglycaemic convulsions. The hypoglycaemia responded to growth hormone therapy and its effect on height velocity is shown.*

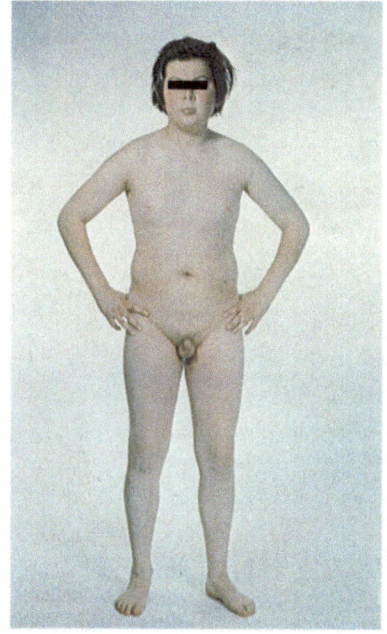

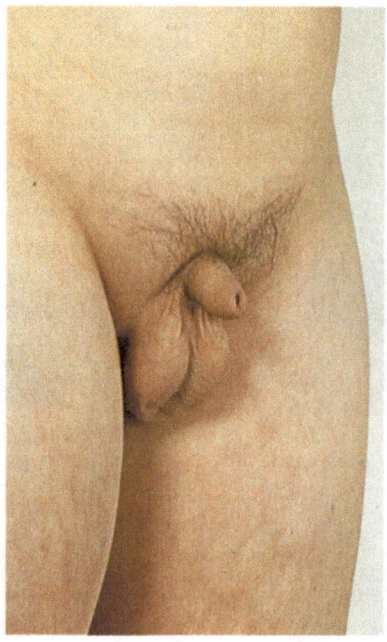

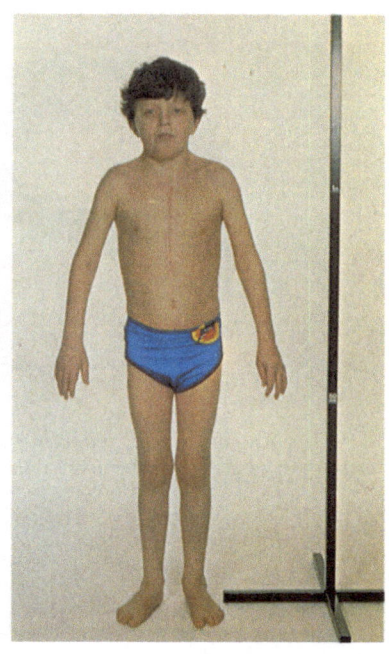

CP-38, 39. This 19-year-old boy was of short stature and had typical hypogonadal features. His hypopituitary state related to a large prolactin-producing chromophobe adenoma.

CP-39.1, 2. Delayed puberty. Boy aged 16 years 6 months, with short stature and delayed puberty due to intrauterine infection with rubella virus. *CP-39.1* (above) standing; *CP-39.2* (below) genitalia.

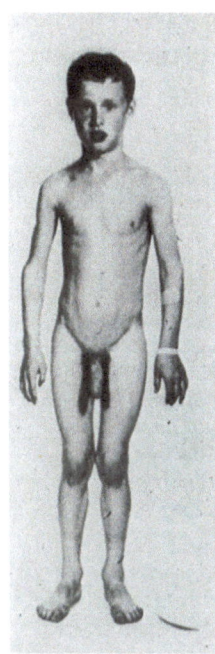

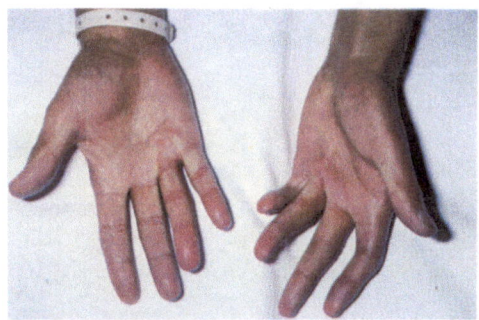

CP-41.1 Hyperpigmentation in a patient with Addison's disease – note in particular the increased pigmentation of the palmar creases.

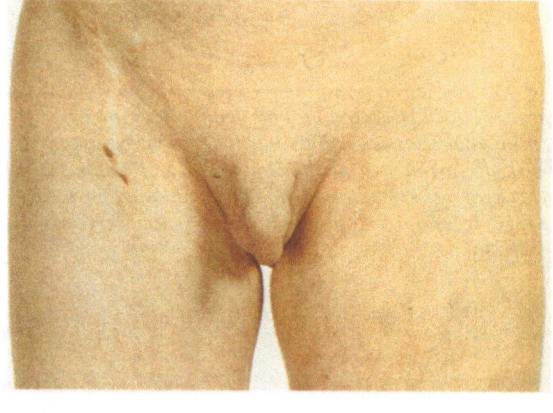

CP-42. These photographs show the gradual development of Cushingoid features over several years in a young woman.

CP-40. Precocious puberty. Boy, aged exactly 8 years, with precocious puberty due to an astrocytoma involving the hypothalamus. Standing.

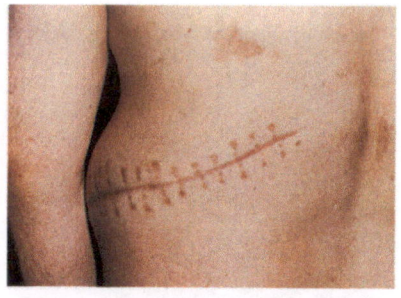

CP-41.2 A patient with Addison's disease showing pigmentation in a scar.

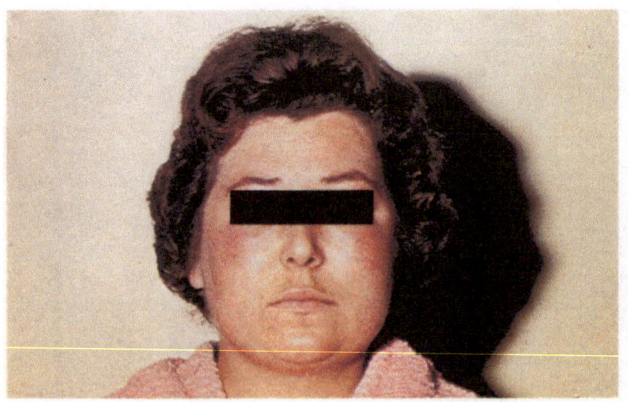

CP-43. *This woman had characteristic Cushingoid features as a result of a glucocorticoid-secreting adrenal adenoma.*

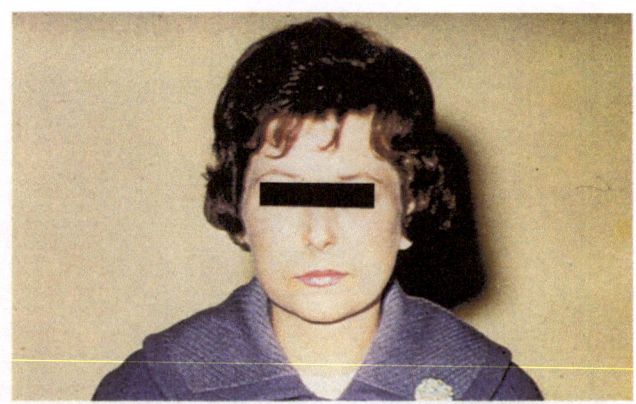

CP-44. *Surgical removal of the adenoma resulted in complete regression of her clinical features.*

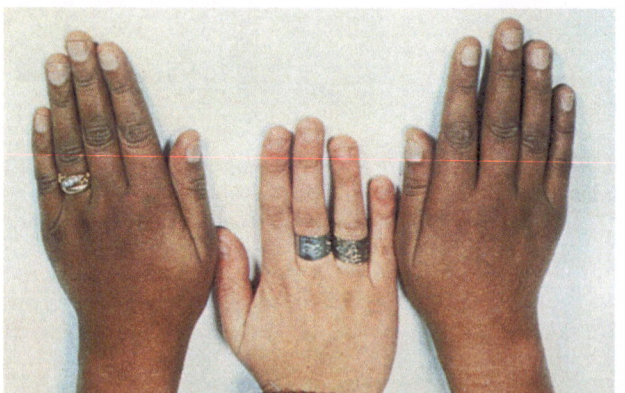

CP-45. *The initial abnormality here was due to increased pituitary ACTH secretion. Following operation, the Cushingoid features regress but the patient became markedly pigmented as the consequence of enhanced release of ACTH and melanocyte stimulating hormone (MSH) from a developing pituitary adenoma (Nelson's syndrome).*

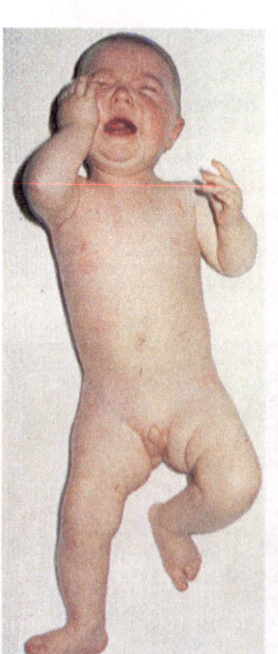
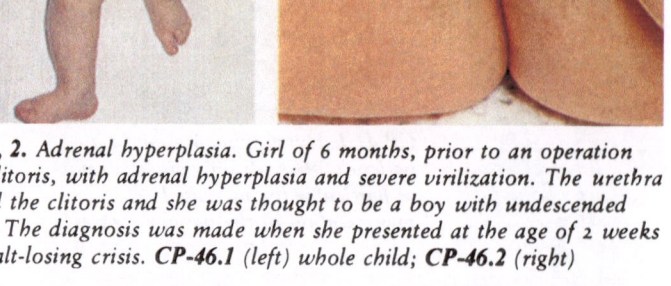

CP-46.1, 2. *Adrenal hyperplasia. Girl of 6 months, prior to an operation on the clitoris, with adrenal hyperplasia and severe virilization. The urethra traversed the clitoris and she was thought to be a boy with undescended testicles. The diagnosis was made when she presented at the age of 2 weeks with a salt-losing crisis.* **CP-46.1** *(left) whole child;* **CP-46.2** *(right) genitalia.*

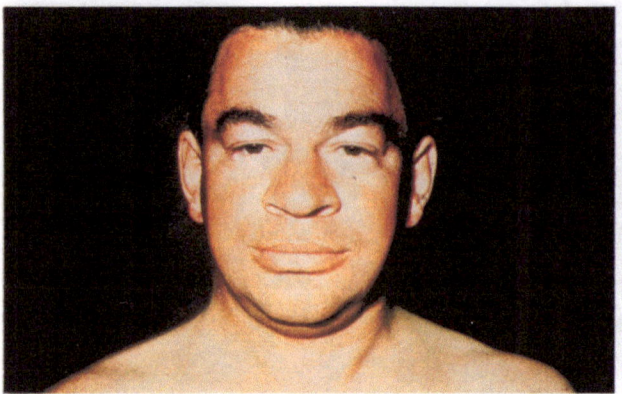

CP-47. *This man had a typical hypothyroid facies with periorbital puffiness, broadening of the nose and thickening of the lips.*

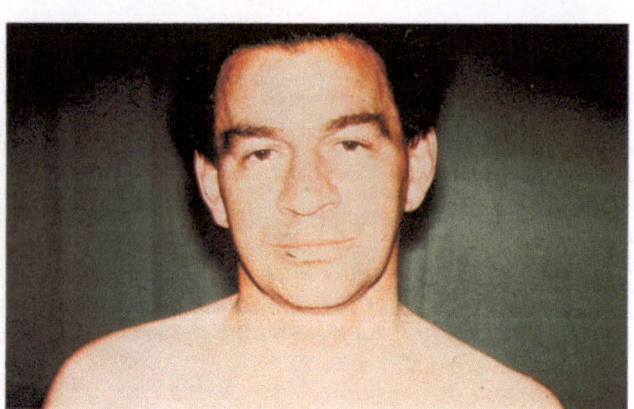

CP-48. *On thyroxine therapy these features have regressed considerably.*

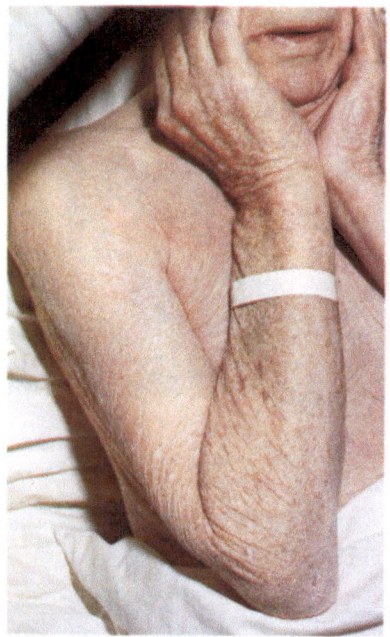

CP-49. Note the dry waxy appearance of the skin in this hypothyroid lady.

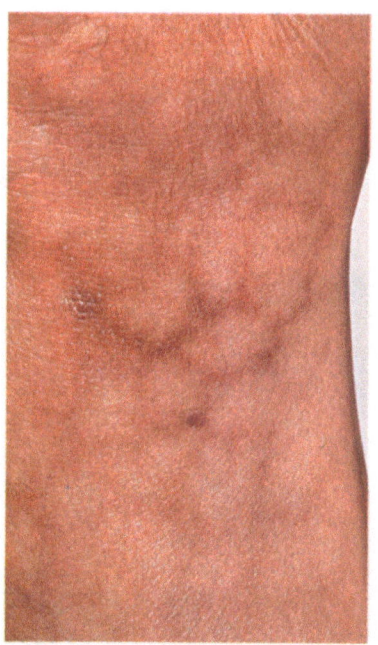

CP-50. Erythema ab igne due to sitting close to a fire in a patient with hypothyroidism.

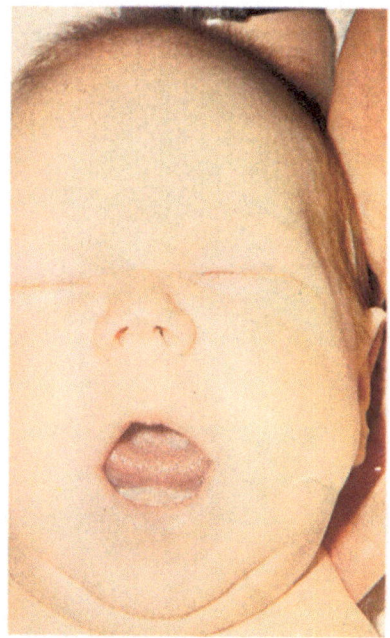

CP-51.3, 4. Hypothyroidism in childhood. Boy with congenital hypothyroidism when first seen at the age of 3 months (**CP-51.3**, above) and, following thyroxine therapy, at the age of exactly 2 years (**CP-51.4**, below). In spite of the late diagnosis, his intelligence was within normal limits.

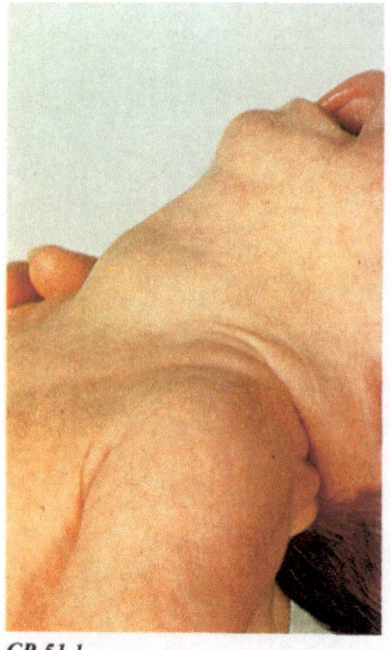

CP-51.1

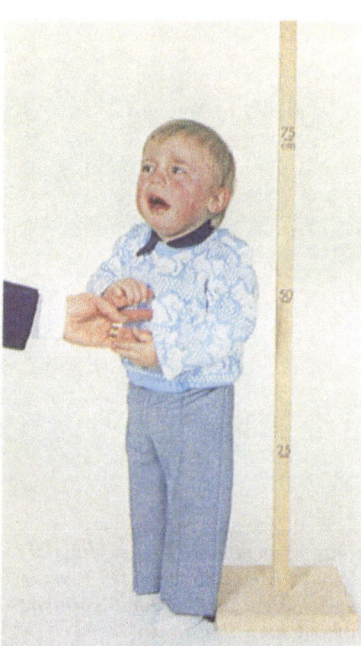

CP-51.2

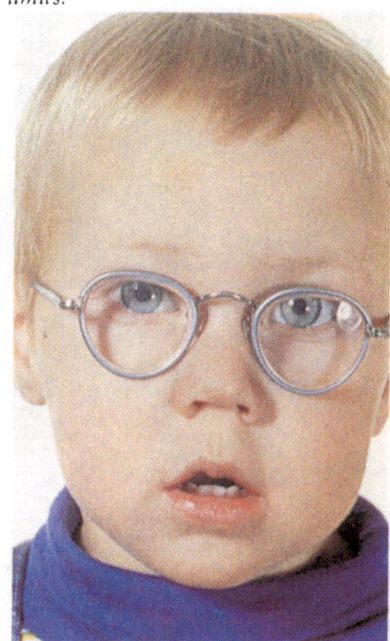

CP-51.1, 2. Hypothyroidism in childhood. Boy with congenital hypothyroidism and goitre due to dyshormonogenesis at the age of 4 days (**CP-51.1**) and at exactly 2 years of age (**CP-51.2**) following treatment with thyroxine. He was of normal intelligence.

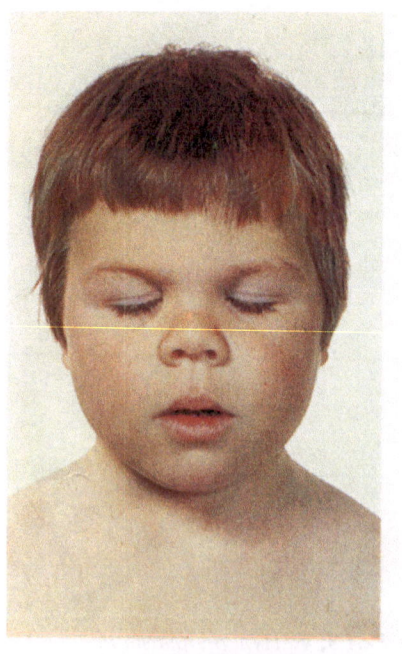

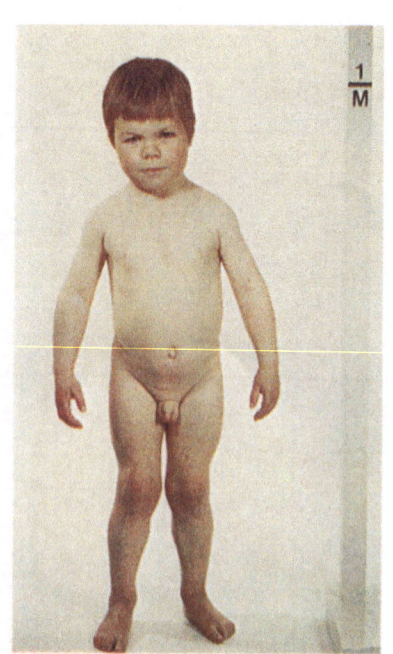

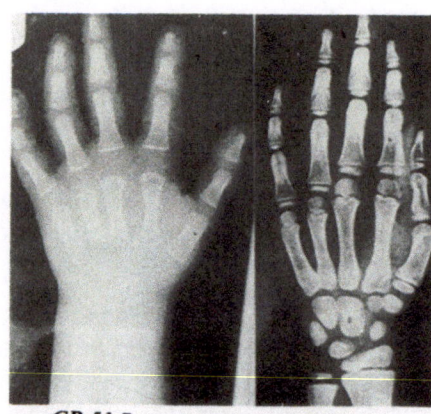

CP-51.7

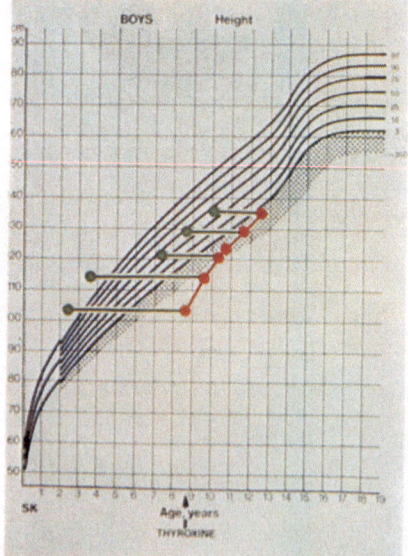

CP-51.10

CP-51.11

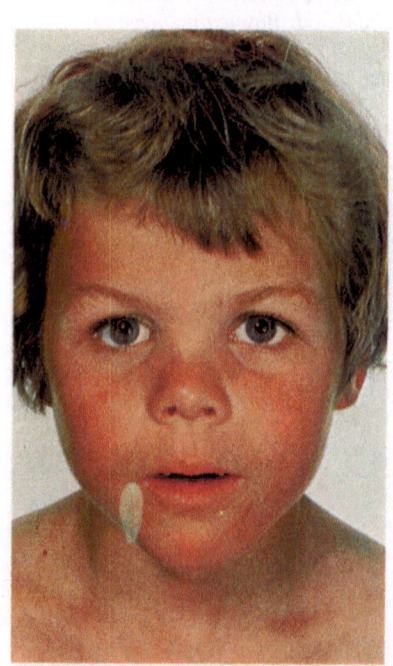

CP-51.8

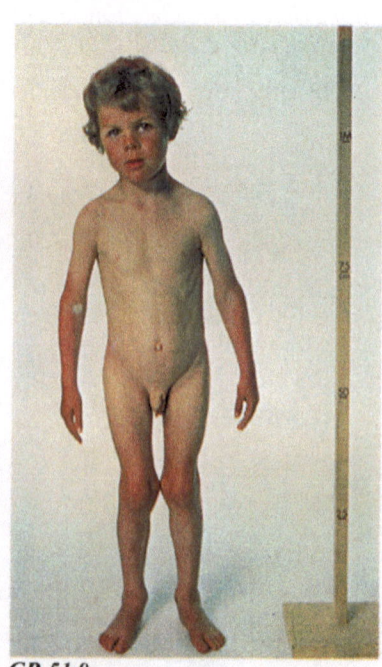

CP-51.9

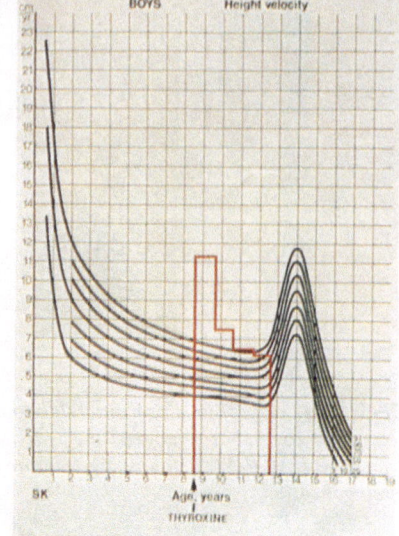

CP-51.5–11. *Boy with deficiency of thyroid hormones associated with a sublingual thyroid gland, before and after treatment. **CP-51.5** and **51.6** show him aged 8 years 7 months, before treatment. **CP-51.7** is an X-ray of his left wrist and hand showing his bone age of 2 years 8 months, on the left, compared with that of a normal control on the right. **CP-51.8** and **51.9** show the boy aged 9 years 7 months after treatment with thyroxine. The effect of thyroxine therapy on height (**CP-51.10**) and height velocity (**CP-51.11**) is seen on the charts for a period of 4 years.*

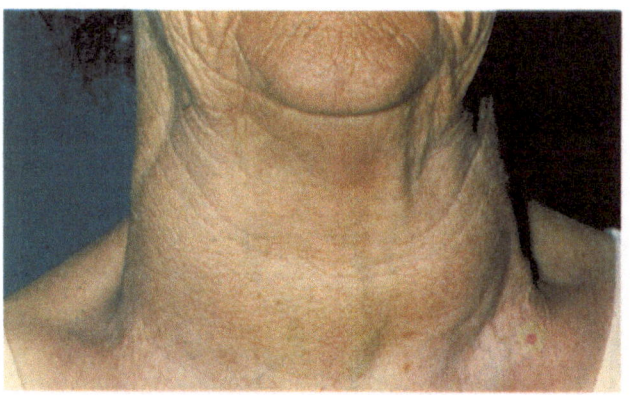

CP-52. *This patient had a firm diffuse enlargement of her thyroid due to autoimmune thyroiditis (Hashimoto's).*

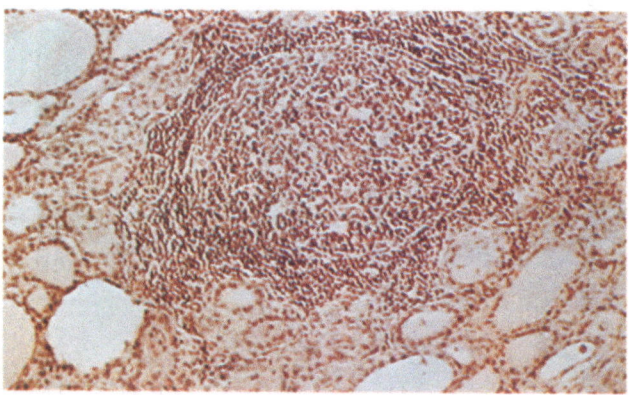

CP-53. *Hashimoto's thyroiditis. Note the lymphocyte and plasma cell infiltration in this lymphoid follicle. There is a marked increase in lymphoid tissue.*

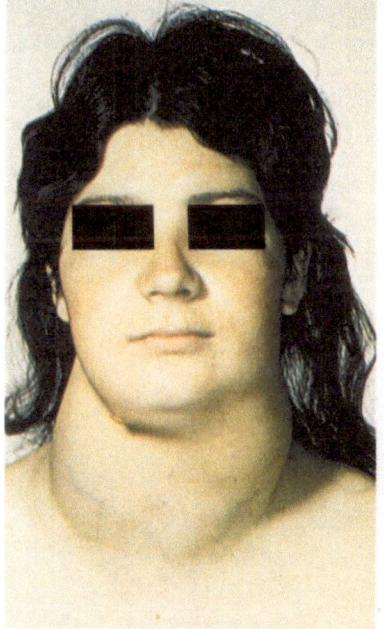

CP-54. *This girl has a large dishormonogenic goitre. She is one of several affected members of her family.*

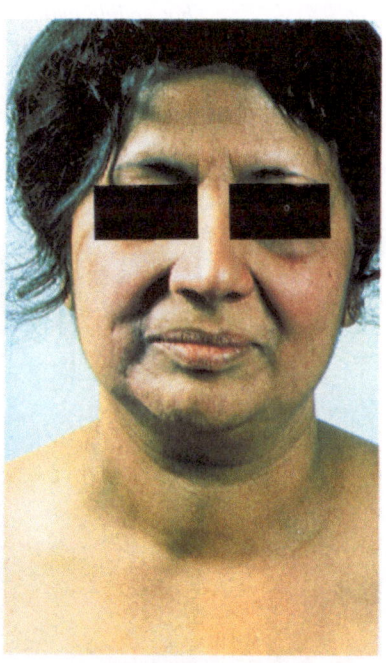

CP-55. *Subacute thyroiditis. This patient had a flu-like illness associated with pain in her neck. She developed diffuse symmetrical enlargement of her thyroid over the ensuing 2 weeks which was confirmed as subacute thyroiditis.*

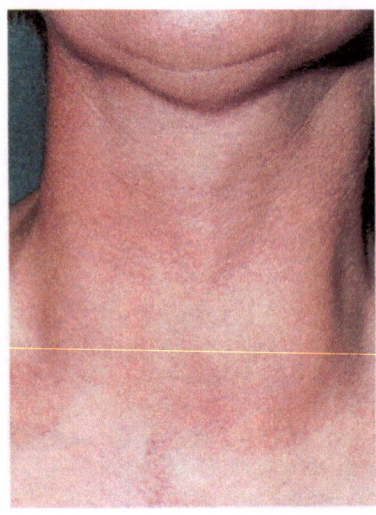

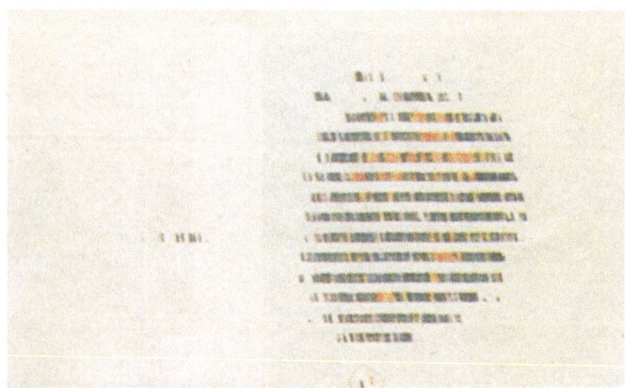

CP-56, 57. *This female patient had a single toxic adenoma. Surgical removal of the lesion resulted in complete reversal of her hyperthyroid features. The thyroid remnant subsequently functioned normally. The radioiodine scan (*[131]*I) shows a single hot nodule with suppression of iodine uptake in the rest of the thyroid gland prior to removal of the adenoma.*

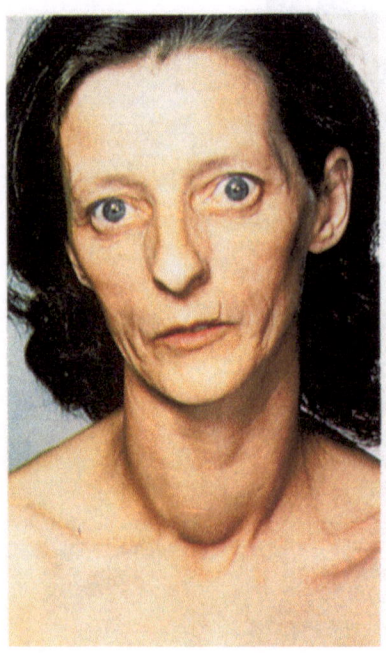

CP-58. *This patient has Graves's disease with a diffusely enlarged thyroid gland.*

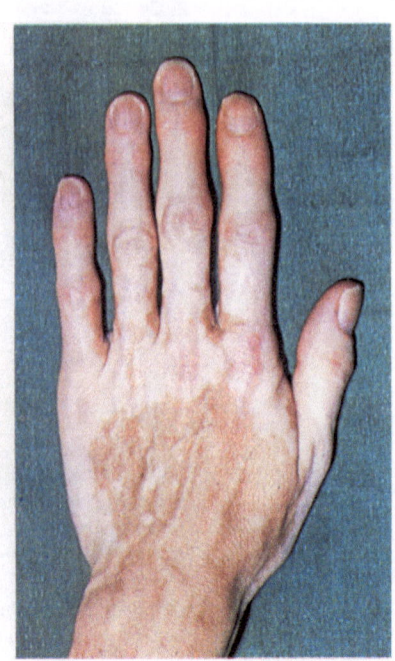

CP-59. *Vitiligo in a patient with thyrotoxicosis.*

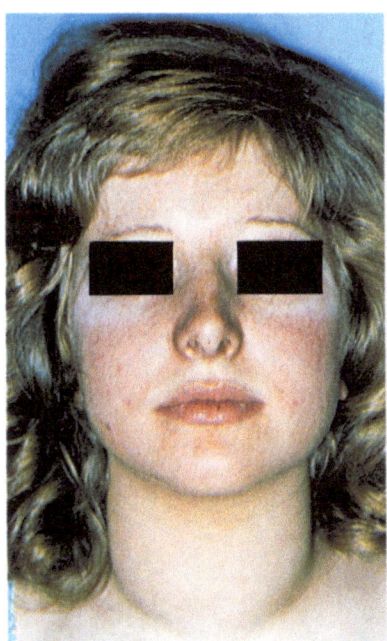

CP-60. This thyrotoxic girl had a moderate sized goitre. Partial thyroidectomy was undertaken with a good clinical and cosmetic result.

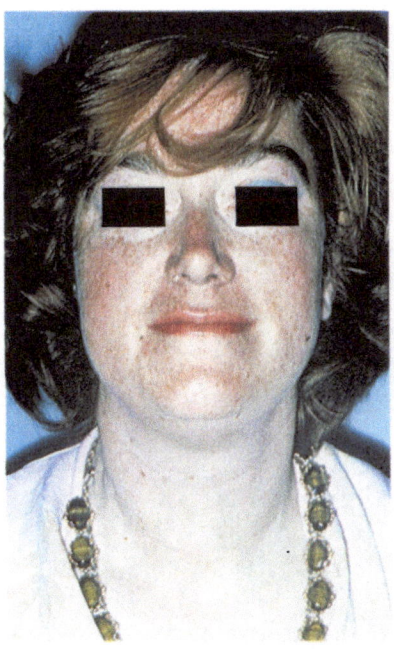

CP-61. This thyrotoxic patient has had a thyroidectomy with a good clinical and cosmetic result. The scar is only just visible.

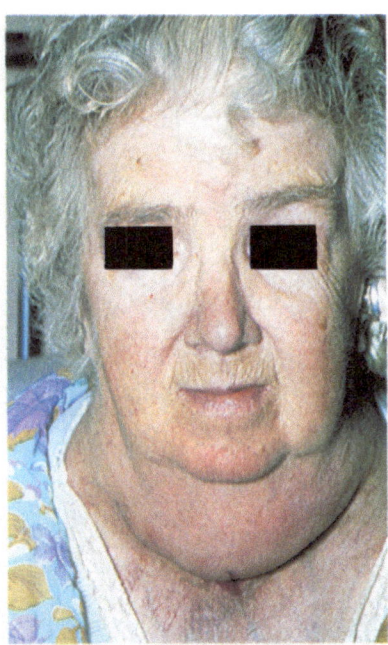

CP-62. This lady had an anaplastic carcinoma of the thyroid.

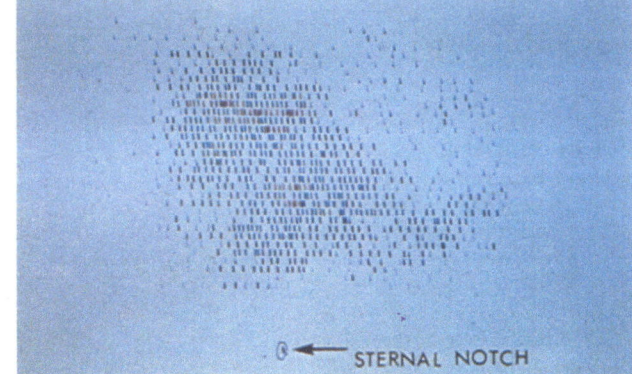

CP-63. Radiodine (^{131} I) scanning of the patient in CP-62 revealed patchy iodine uptake in the region of the enlarged thyroid gland.

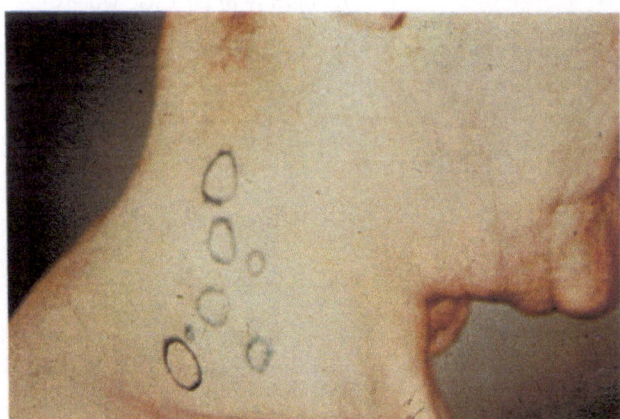

CP-64. Cervical lymphadenopathy due to a spread of a papillary carcinoma.

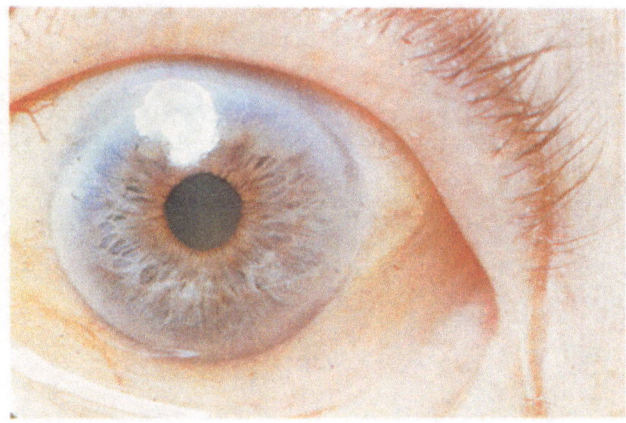

CP-65. Corneal calcification in a patient with long standing primary hyperparathyroidism. The white area above the pupil is an artefact from the lighting.

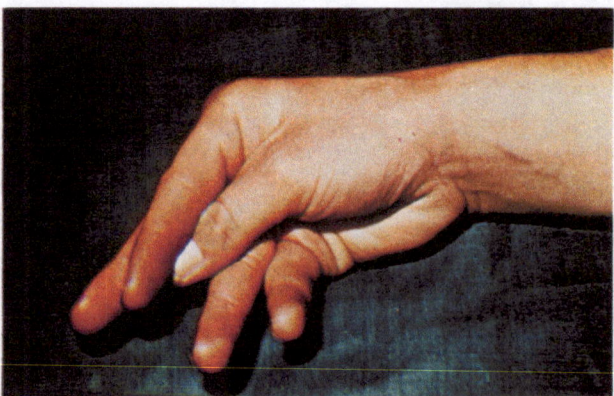

CP-66. Trousseau's sign.

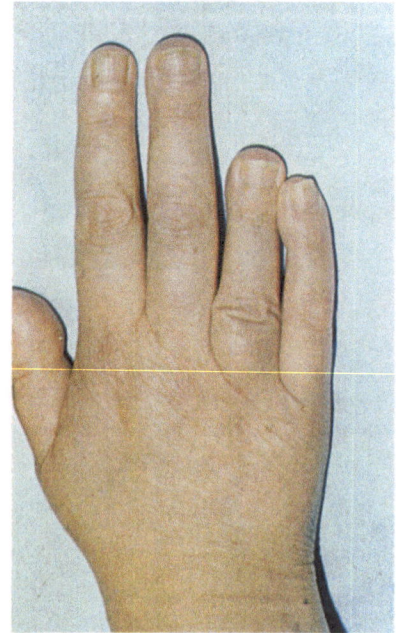

CP-67. Right hand of a 40-year-old lady with pseudohypoparathyroidism.

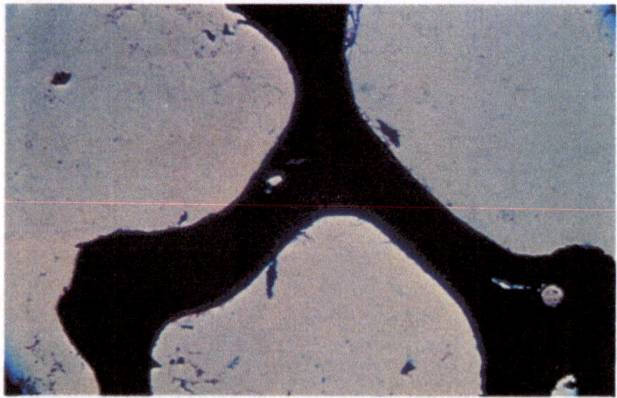

CP-68. Normal trabecular bone stained with von Kossa technique. This shows the calcified bone as black. (Courtesy of Dr C.G. Woods.)

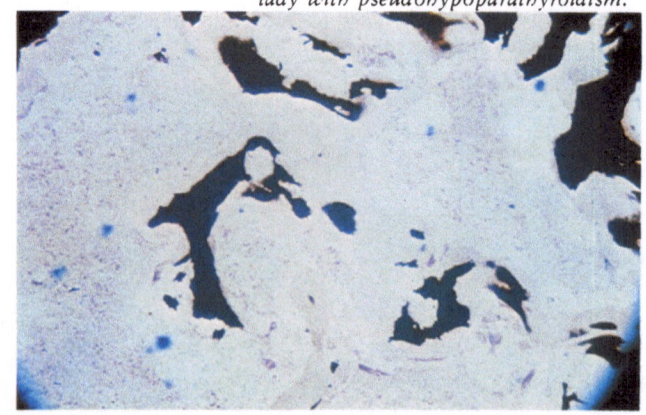

CP-70. Deformity of the knees in a 15-year-old Asian patient with nutritional rickets who presented with pain in the knees.

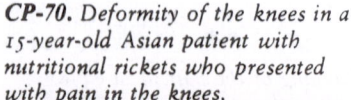

CP-69. Severe osteomalcia in a 50-year-old woman with postgastrectomy osteomalacia. The black is the calcified bone, the pink is the uncalcified bone matrix. (Courtesy of Dr C.G. Woods.)

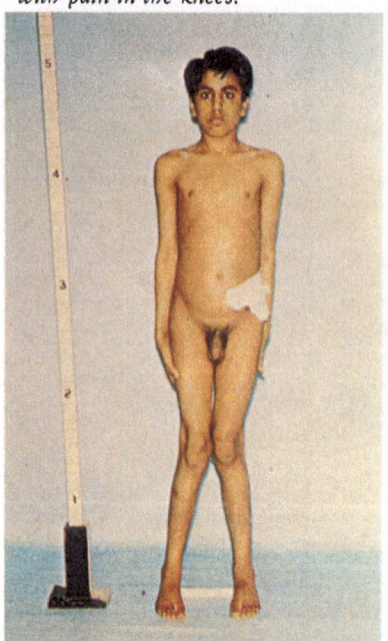
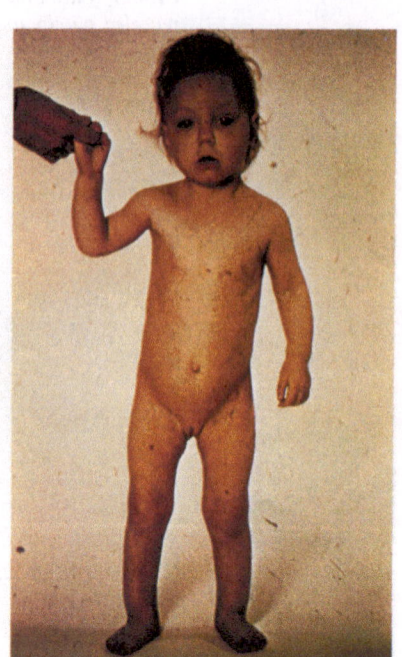

CP-71. Severe deformity of the ankles, enlargement of the wrists and 'rickety rosary' in a 2-year-old girl with nutritional rickets. (Courtesy of Dr S.G.F. Wilson.)

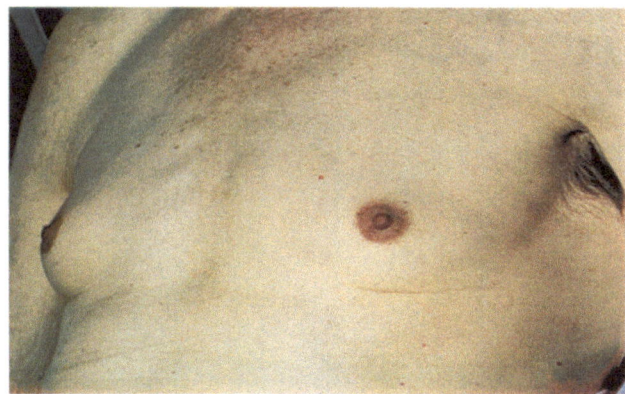

CP-72. Bilateral gynaecomastia in an elderly patient on therapy for cardiac failure. He was receiving digoxin and spironolactone, amongst other drugs.

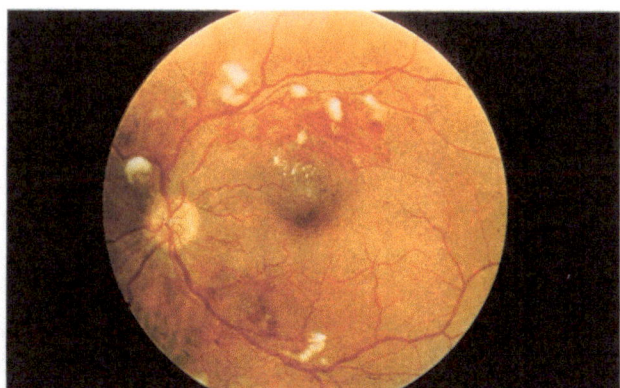

CP-73. Diabetic retinopathy showing hard and cottonwool exudates, blotch haemorrhages and proliferative vascular changes.

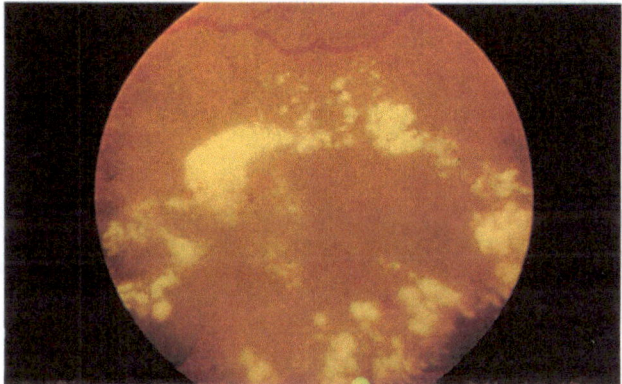

CP-73.1. Hard exudates occurring as a result of lipid deposition in sites where leakage of plasma into the retina has been caused by increased vascular permeability.

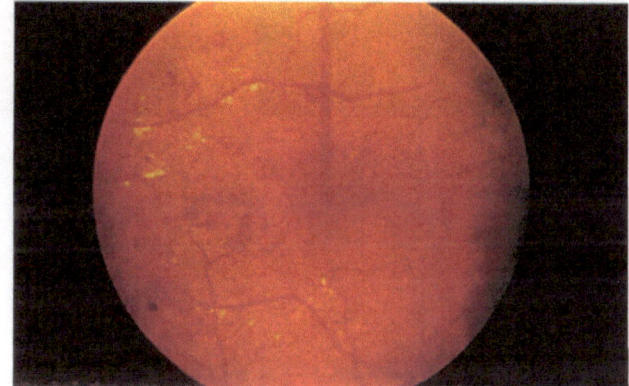

CP-73.2. Fundus photograph showing (a) microaneurysms, (b) haemorrhages, (c) hard exudates and (d) cottonwool spots.

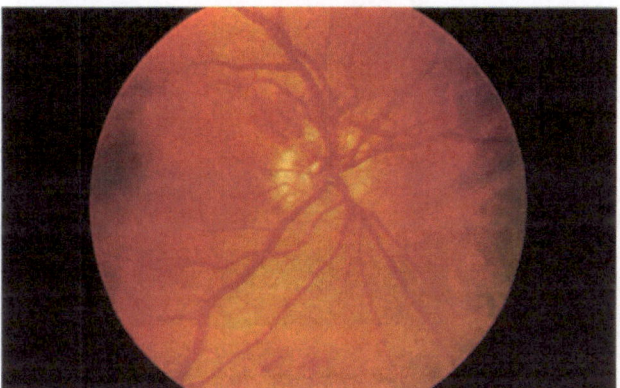

CP-73.3. Proliferative retinopathy (neovascularization). A massive network of new vessels is seen arising from the optic disc. About half such patients left untreated will be blind after 5 years.

CP-73.4. An example of fibrous proliferation is illustrated here. Bleeding new vessels become organized by glial tissue which, on contraction, may produce traction on the retina with visual loss. These changes represent advanced diabetic eye disease.

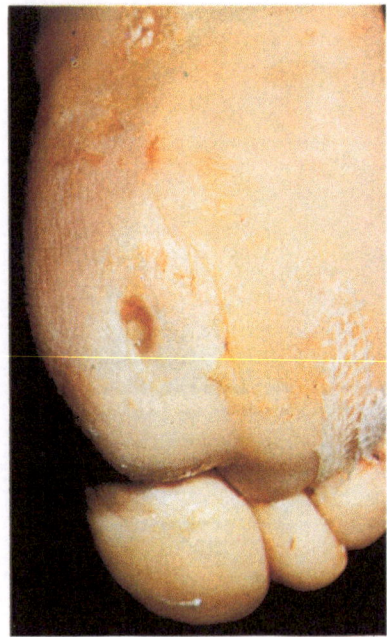

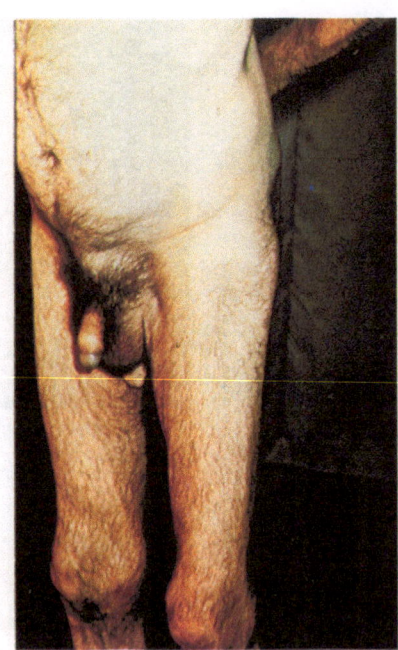

CP-75. *Diabetic amyotrophy in a middle-aged patient. This man presented with considerable weakness and wasting affecting the quadriceps muscles of both legs. Control of his diabetes over the following 3 months resulted in complete regression of these features.*

CP-74. *Neuropathic ulcer in a diabetic patient. The combination of claw toe deformity due to intrinsic muscle weakness and loss of pain sensation frequently leads to ulceration below the metatarsal heads.*

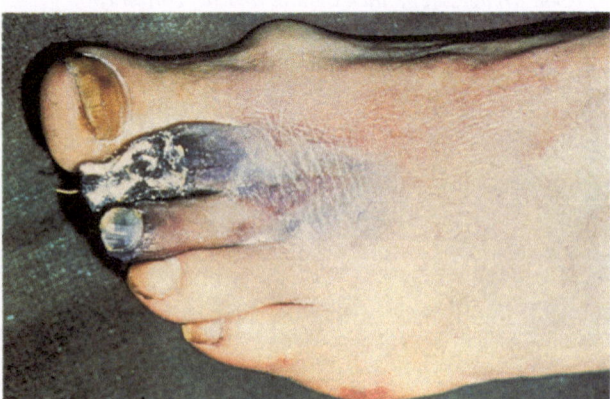

CP-76. *Dry gangrene occurring in a diabetic foot.*

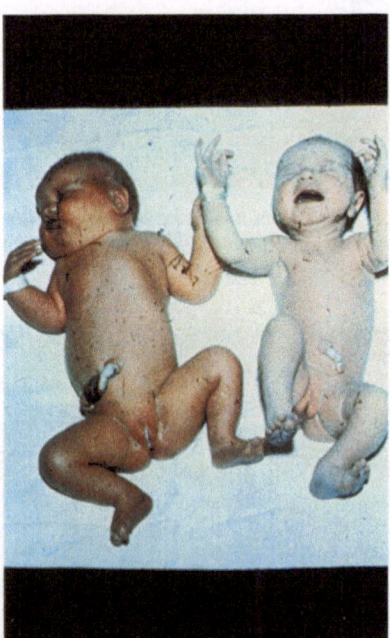

CP-77. *The large healthy-looking baby of a diabetic mother is seen with a normal newborn baby. The increased adipose tissue covering is deceptive as such neonates have a high incidence of complications following delivery.*

 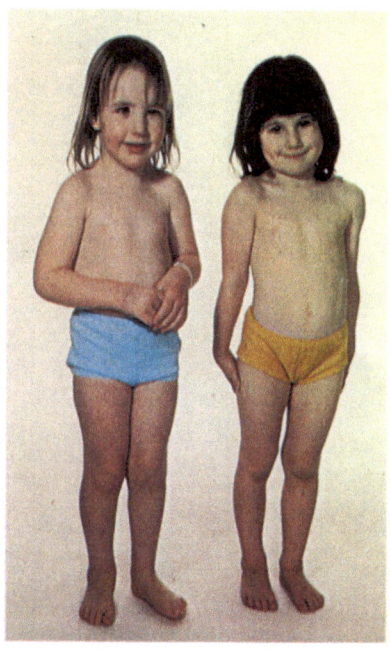

CP-78. *Girl on left with short stature due to Crohn's disease, aged exactly 13 years, with her brother, aged 8 years, 6 months, and her school friend, on the right, aged 12 years, 11 months.*

CP-79. *Girl with cerebral gigantism, on the left, aged 3 years and 10 months, with her normal older sister on the right, aged 4 years and 10 months.*

X–RAYS

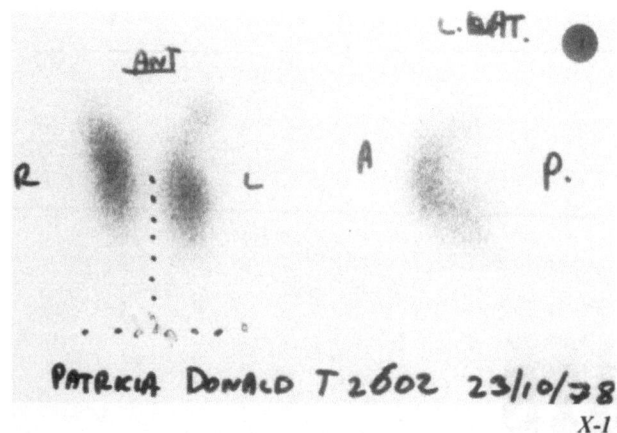

X-1

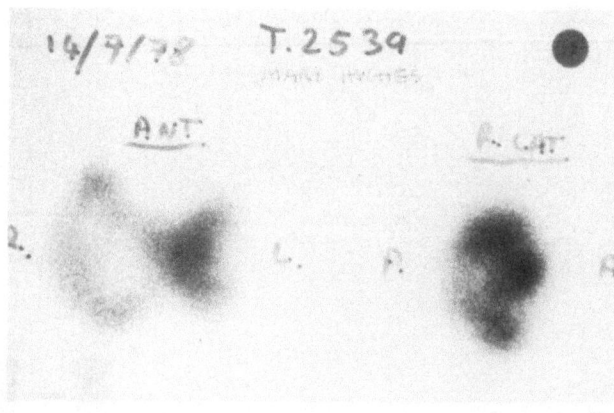

X-2

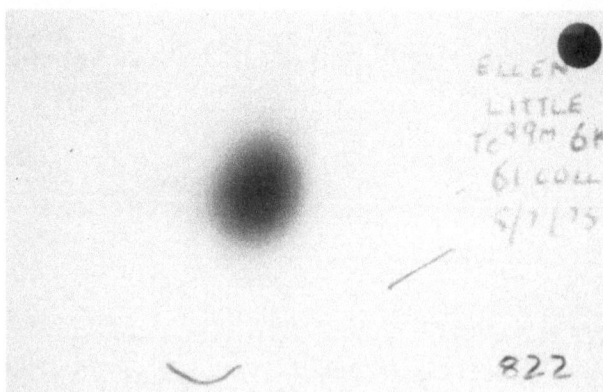

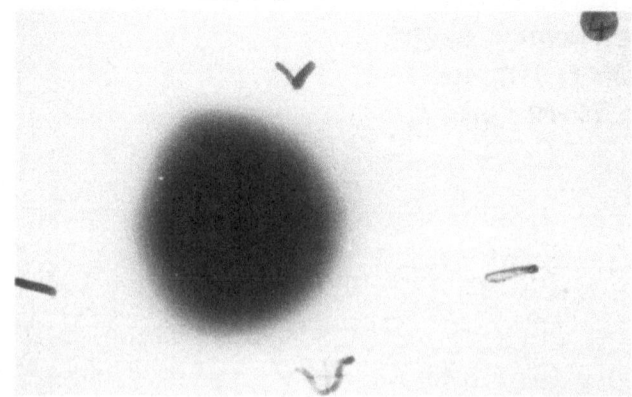

X-1–4. Technetium pertechnetate (⁹⁹ᵐTc)
scans of the thyroid showing:

X-1. A 'cold' area indenting the left lobe – this
patient had a thyroid cyst (left half only).

X-2. A 'cold' area occupying the central area of
the right lobe – a lateral scan shows two separate cold
areas found subsequently to be thyroid cysts.

X-3. Technetium pertechnetate (⁹⁹ᵐTc) of thyroid
showing an area of increased radioisotope uptake in the
left lobe. Uptake in the rest of the thyroid gland is
suppressed, implying autonomy of the thyroid nodule ('hot
nodule'). This patient was euthyroid, and presented with a
painless lump in the left side of her neck.

X-4. Thyroid (¹³¹I) scan showing a large functioning
nodule. Uptake in the rest of the gland is suppressed. This
patient presented with features of thyrotoxicosis and was
found to have a thyroid nodule. Removal of the toxic
nodule resulted in complete reversal of hyperthyroidism
and a subsequent thyroid scan showed the remnant of the
thyroid functioning normally.

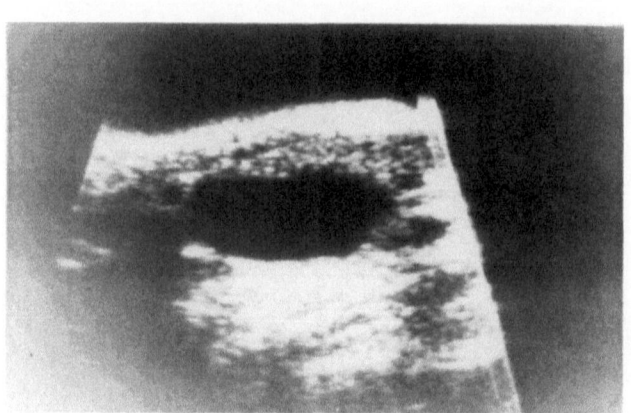

X-5. Longitudinal ultrasound scan of thyroid lobe, showing
an echo-free simple cyst. Note the marked increased through-
transmission deep to the cyst.

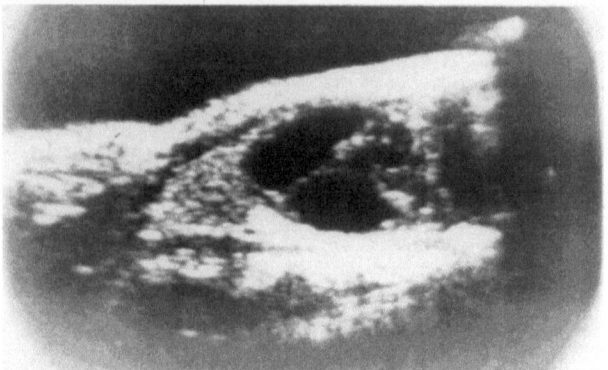

X-6. Longitudinal ultrasound scan, showing multiple simple
cysts in a case of multinodular colloid goitre.

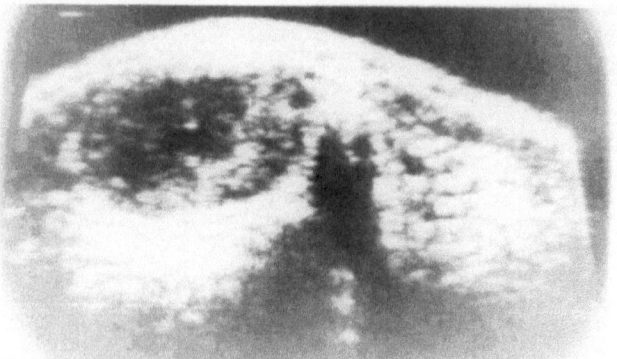

X-6a. Transverse ultrasound scan, showing a solid nodule in
the right lobe of thyroid. Note the areas of decreased
echogenicity due to tissue necrosis.

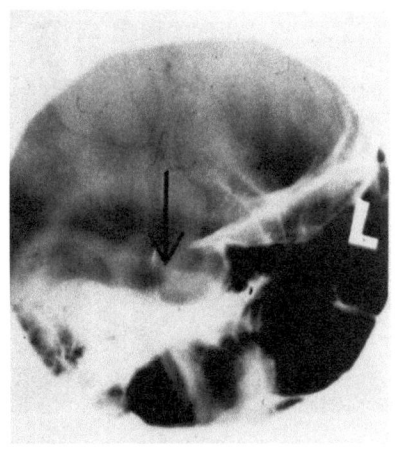

X-7. *Lateral skull X-ray of patient in X-7a. This shows the expanded pituitary fossa.*

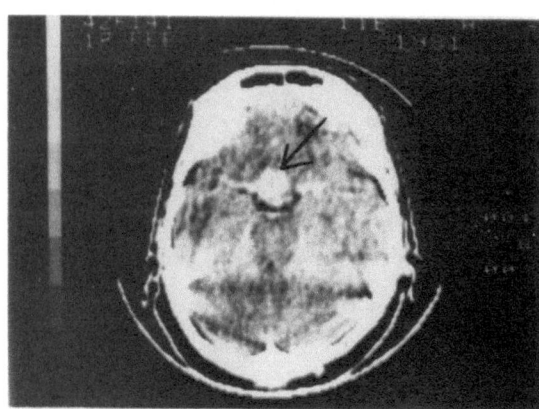

X-7a. *CAT scan with enhancement showing suprasellar extension of pituitary tumour.*

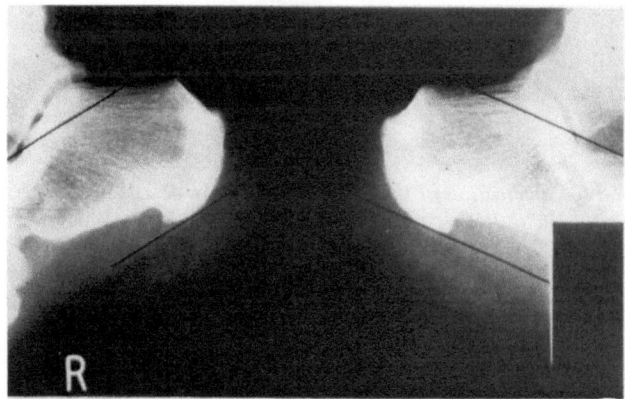

X-8. *An X-ray in acromegaly showing increased heel pad.*

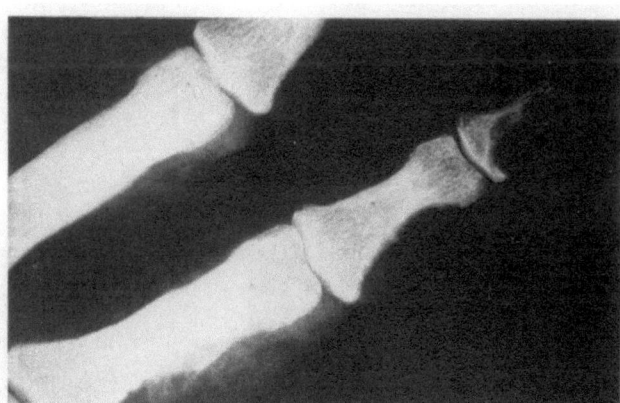

X-10. *Close-up view of a patient with subperiosteal erosions in a patient with severe primary hyperparathyroidism. (Courtesy of Dr O.L.M. Bijvoet.)*

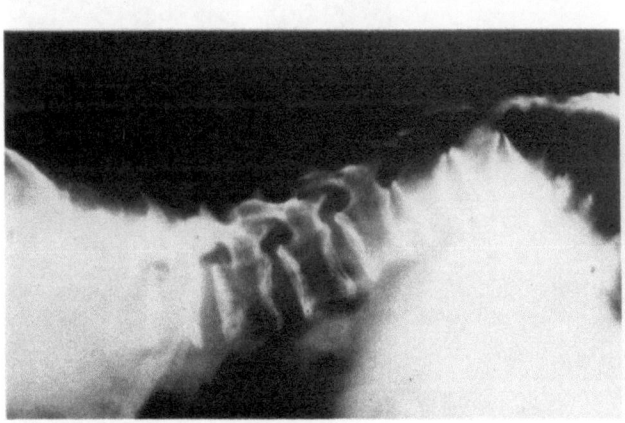

X-9. *Osteoporosis with vertebral collapse from Cushing's syndrome.*

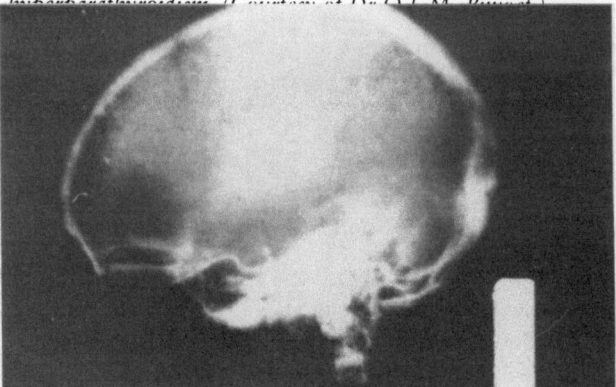

X-11. *A 'pepper pot' skull in a woman with severe hyperparathyroidsim. Her plasma calcium pre-operatively was 4.6 mmol/l (18.2 mg/dl) and her plasma alkalinephosphatase was 80 KAU. (Courtesy of Dr T.C.K. Marr.)*

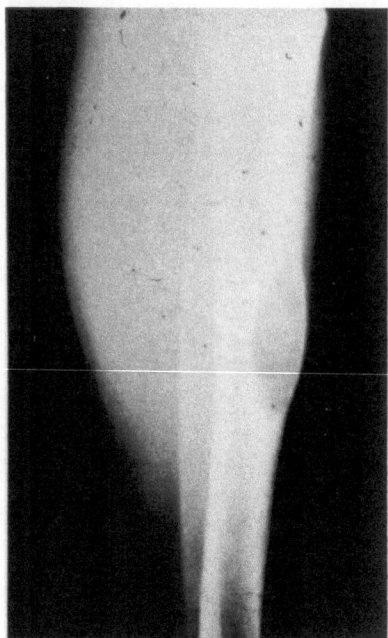

X-12. Bone cyst (osteoclastoma) in a patient with hyperparathyroidsim.

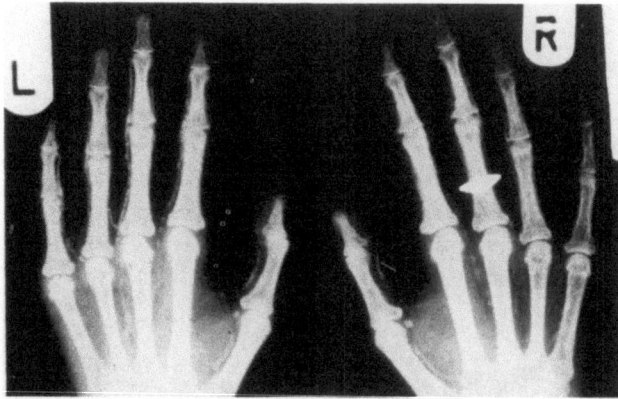

X-13. Subperiostial erosions (due to secondary hyperparathyroidsim) and calcification of the digital arteries in a 25-year-old woman with renal failure.

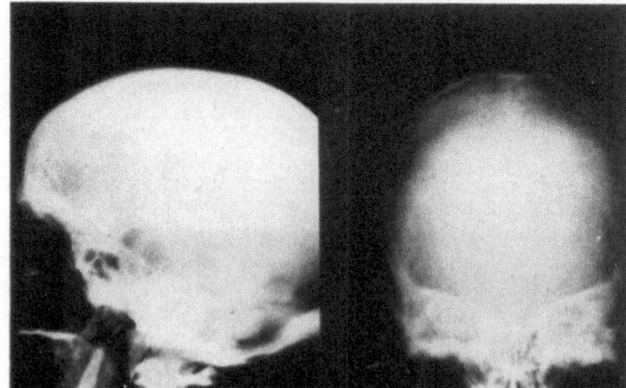

X-14. Calcification of the basal ganglia in an elderly woman with hyperparathyroidism.

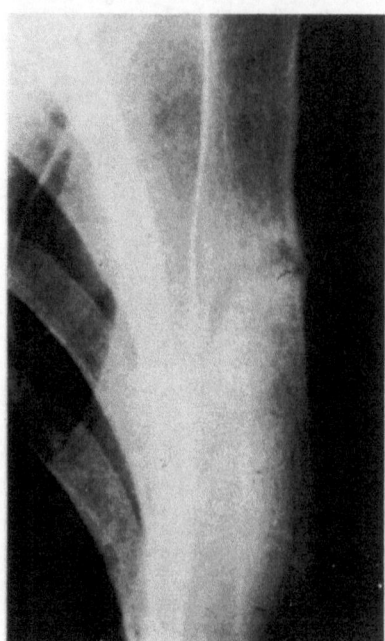

X-15. Looser's zone (pseudofracture) in the scapula of a 50-year-old woman with osteomalacia who presented with severe bone pain 10 years after a polyagastrectomy for a gastric ulcer. (Courtesy of Dr C.N. Pulvertaft.)

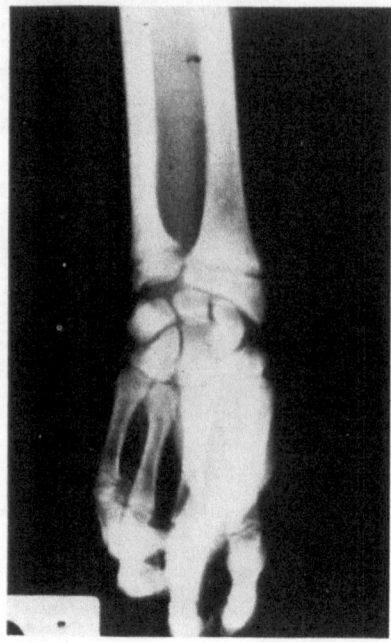

X-16. Rickets and tetany in a 14-year-old Asian boy with nutritional vitamin D deficiency.

TEST-YOURSELF QUESTIONS
Sections IV and V

42. Complete this table:

Hormone		Clinical syndrome
Growth hormone		
In adult	deficiency	
	excess	
In child	deficiency	
	excess	
Vasopressin	deficiency	
	excess	
Adrenocortico-steroid	deficiency	
	excess	
Thyroid	deficiency	
	excess	
Parathyroid	deficiency	
	excess	

43. Which of the following are typical features in acromegaly?

a) Low serum prolactin level
b) Low serum gonadotrophins
c) Raised growth hormone levels suppressed during an oral glucose tolerance test
d) Raised serum phosphate
e) Erosion of sella turcica on X-ray

44. Which of the following statements are true about the treatment of acromegaly?

a) Successful treatment leads to a reduction in mortality
b) Pituitary implantation of radioactive yttrium is the treatment of choice
c) Hypophysectomy is indicated when the optic chiasma is involved
d) Following hypophysectomy replacement therapy is usually unnecessary
e) Bromocriptine leads to a reduction in growth hormone levels

45. List four causes of growth hormone deficiency in a child.

46. Why is the diagnosis of growth hormone deficiency in the child important?

47. A 34-year-old woman is found to have hypopituitarism. What hormone replacement therapy might be required?

a) Thyroxine
b) Fludrocortisone
c) Growth hormone
d) Cortisol
e) Oestrogen/progesterone combination

48. Which of the following statements are true of undescended testes?

a) May be unilateral or bilateral
b) Surgical treatment should be postponed to five years
c) HCG is of no value
d) There is a risk of malignancy
e) Is associated with impaired testicular function

49. Match the response to LHRH in an LHRH test on the left with the conditions on the right.

a) No response
b) Low resting levels LH and FSH and sluggish response of LH and FSH
c) High resting levels LH and FSH and above normal response of LH and FSH

1) Hypothalamic cause
2) Primary anterior pituitary failure
3) Primary gonadal cause
4) Normal

50. A mother of an 8-year-old girl with precocious puberty asks you a number of questions about her daughter.

a) Does she need to have any tests done?
b) Can treatment be given?
c) Will she be excessively tall as an adult?

51. A patient with diabetes insipidus is receiving DDAVP. How can the dose be regulated?

52. i) Which of the following can lead to adrenocortical insufficiency?

a) Metastatic carcinoma
b) Amyloidosis

c) Autoimmune adrenalitis
d) Tuberculosis
e) Sarcoidosis

ii) Which is the commonest cause?

53. What is the commonest cause of Cushing's syndrome?

54. Match the results shown on the left with the disease on the right.

a) Raised ACTH
b) Suppression with large dose of dexamethazone
c) Loss of normal diurnal variation in serum cortisol

1) Cushing's syndrome
2) Ectopic ACTH syndrome
3) Adrenal tumour

55. Match the symptoms on the right with a disease on the left.

a) Tall stature with normal skeletal proportions
b) Immature looking child proportionately small
c) Tall child becoming short adult

1) Precocious puberty
2) Growth hormone deficiency
3) Gigantism
4) None of these

56. Why do patients with Cushing's syndrome get

a) acne?
b) impaired wound healing?
c) sodium and water retention?
d) osteoporosis?

57. What are four important symptoms diagnostically in hypothyroidism?

58. How can a diagnosis of hypothyroidism best be confirmed?

59. How can congenital hypothyroidism be avoided?

60. Match the clinical feature on the left with the disease on the right.

a) Pain and tenderness in the throat
b) General malaise and pyrexia
c) Dislike for fish
d) Short history
e) Strong family history

1) Dyshormonogenesis
2) Hypothyroidism
3) Subacute thyroiditis
4) Iodine deficiency
5) Autoimmune thyroiditis

61. Match the symptoms on the right with a disease or diseases on the left.

a) Carpal tunnel syndrome
b) Hypertension
c) Impaired glucose tolerance
d) Hyperprolactinaemia
e) Hyperpigmentation
f) Virilisation

1) Hypothyroidism
2) Cushing's syndrome
3) Acromegaly
4) Addison's disease
5) Adrenal hyperplasia due to 21-hydroxylase deficiency
6) Hyperaldosteronism
7) Phaeochromocytoma

62. Which of the following are typical features of hyperaldosteronism?

a) Moon face
b) Polydipsia
c) Muscle weakness
d) Paralysis
e) High circulating renin level

63. Match the findings on the left with the disease on the right.

a) Low Na, elevated K
b) Low Na, normal K
c) High sodium, low K

1) Cushing's syndrome
2) Adrenal hyperplasia due to 21-dihydroxylase
3) Adrenal insufficiency
4) Inappropriate ADH secretion
5) Hyperaldosterism

64. Match the clinical features on the left with the disease on the right.

a) Purplish red striae
b) Onycholysis
c) Atrial fibrillation
d) Delayed relaxation of ankle jerk
e) Muscle wasting

1) Thyrotoxicosis
2) Hypothyroidism
3) Cushing's syndrome
4) Addison's disease

65. Which of the following tests is typical of thyrotoxicosis?

a) Raised serum thyroxine level
b) Raised serum TSH level
c) Absence of response to administered TSH
d) Failure of suppression of thyroid after administration of triiodothyronine
e) Normal response in TRH test

66. Match the treatment on the left with a feature or features on the right.

a) Iodide
b) Carbimazole
c) Propranolol
d) Surgery
e) Radioiodine

1) Rapid action
2) Recurrence of thyrotoxicosis
3) Late hypothyroidism
4) Reduction in size of goitre

67. For each of the thyroid cancers on the left indicate the typical mode of spread on the right.

a) Papillary   1) Local
b) Follicular   2) Blood stream
c) Anaplastic   3) Lymph nodes
d) Medullary

68. Which features should alert you to the fact that a thyroid lump is malignant?

69. Which of the following may be associated with an ovarian tumour?

a) Menstrual irregularity
b) Hirsuitism
c) Enlargement of the clitoris
d) Deepening of the voice

70. Which of the following statements are true of diseases of the breast?

a) Breast cancer is usually associated with pain
b) Breast cancer may be hormone stimulated
c) Cancer is unlikely in a breast with fibroadenosis
d) Paget's disease of the nipple is relatively benign

71. In a woman complaining of infertility, how can you determine whether there is failure of ovulation?

72. Which of the following conditions are typically associated with pain in the testes?

a) Torsion
b) Hydrocoele
c) Epididymo-orchitis
d) Spermatocoele
e) Seminoma

73. Which of the following statements are true of male infertility?

a) The cause can usually be found
b) It may be associated with undescended testes
c) Evidence of androgen insufficiency is common
d) A high FSH suggests seminiferous tubular failure
e) Testicular biopsy may give an indication of the outcome of therapy
f) Bromocriptine may be indicated in some patients

74. Which of the following results are typical of (1) primary and (2) secondary testicular failure?

a) Plasma testosterone decreased
b) Plasma FSH decreased
c) Testosterone response to HCG decreased
d) LH, FSH response to LHRH increased

75. For each disease on the left indicate appropriate treatment on the list on the right.

a) Hyperparathyroidism   1) Surgery
b) Hypoparathyroidism   2) 1α-hydroxycholecal-
c) Osteomalacia   ciferol
d) Osteoporosis   3) Oral phosphate
e) Osteogenesis   4) Vit D
    imperfecta   5) Calcium
f) Paget's disease   6) Calcitonin
                       7) None of the above

76. Which condition would give the biochemical finding in A, B and C below?

Serum	A	B	C
Calcium	Reduced	Reduced	Normal
Phosphate	Raised	Low	Normal
Alkaline phosphatase	Normal	Raised	Raised

77. Match the findings on the left with the disease on the right.

a) Hypoplasia of the enamel of the teeth   1) Primary hyperpara-thyroidism
b) Tetany   2) Secondary hypopara-thyroidism
c) Pepper-pot skull on X-ray   3) Hypoparathyroidism
d) Renal stones   4) Pseudohypopara-thyroidism
e) History of osteomalacia
f) Short metacarpals   5) Osteogenesis imperfecta
g) Blue sclera   6) None of the above

78. Which is the one commonest cause of uric acid stones?

a) Gout
b) Myeloproliferative diseases
c) Uricosuric drugs
d) None of the above

79. In which of the following conditions is hypercalcaemia typically found?

a) Primary hyperparathyroidism
b) Secondary hyperparathyroidism
c) Tertiary hyperparathyroidism
d) Pseudohyperparathyroidism
e) Hypoparathyroidism

80. Which of the following may be associated with gynaecomastia?

a) Puberty
b) Digoxin therapy
c) Spironolactone therapy
d) Hyperthyroidism
e) Acromegaly
f) Cirrhosis of the liver

81. Which of the following features are typical of
i) insulin dependent diabetes?
ii) non-insulin dependent diabetes?
    a) Occurs in children
    b) Gradual onset
    c) Often symptomless
    d) Marked weight loss
    e) Relatively insensitive to insulin
    f) Low plasma insulin

82. What is the best simple test to confirm diabetes and what should be done if this gives equivocal results?

83. Match the patient on the left with the disease on the right.

a) Obese person with normal glucose tolerance test

b) Woman with normal glucose tolerance test but who had temporary diabetes while pregnant

c) Person with equivocal glucose tolerance test

d) Man with normal glucose tolerance test but excessive hyperglycaemia response to a glucocorticoid

e) Child with abnormal glucose tolerance test and weight loss and polyurea

1) Latent diabetic

2) Potential diabetic

3) Chemical diabetes

4) Clinical diabetes

5) Pre-diabetic

84. Which of the following are classical features of a diabetic retinopathy?
a) Microaneurysms
b) Haemorrhages
c) Exudates
d) New vessel formation

85. Which of the following statements are true of the chronic complications of diabetes?
a) Good diabetic control delays the progress of retinopathy
b) Diabetic peripheral neuropathy is mostly motor
c) Mononeuropathies may occur
d) Diabetic nephropathy usually presents with proteinuria
e) Diabetic gangrene is confined to maturity onset type of diabetes

86. Which of the following are characteristic of diabetic ketoacidosis?

a) Dehydration
b) Bradycardia
c) Blood glucose bicarbonate
d) Raised plasma bicarbonate
e) Normal initial serum K

87. Which of the following statements are true of hyperosmolar or aketotic diabetic coma?
a) The prognosis is more favourable than for diabetic ketoacidosis
b) It is much rarer than diabetic ketoacidosis
c) The serum bicarbonate is usually below 16 mmol/l
d) The blood glucose levels are usually less than 40 mmol/l

88. List three common causes of hypoglycaemia in a diabetic.

89. Which of the following statements are true of hypoglycaemia?
a) Symptoms are likely to occur when the plasma glucose concentrate falls below 2.2 mmol/l
b) Older individuals tend to tolerate low levels of blood sugar better than the young
c) β-blockers may mask the warning symptoms of hypoglycaemia
d) Physical aggression may be a feature
e) The patient typically has contracted pupils

90. Why is obesity of importance?

91. List four endocrine causes of small stature.

92. Which of the following could be attributed to ectopic hormone production by non-endocrine tumours?
a) Hyperpigmentation
b) Hypercalcaemia
c) Hypoglycaemia
d) Hypokalaemia
e) Polycythaemia

93. In which of the following situations is radiotherapy, either external or with radioisotopes, considered a useful treatment?
a) Hyperthyroidism due to Graves' disease in a 50-year-old woman
b) After surgery in a follicular carcinoma of the thyroid
c) Serous adenocarcinoma of the ovary
d) Seminoma of the testicle

94. Which of the following seriously affect patient compliance in taking medications?
a) The patient's inability to describe his or her drug regime

b) Age

c) Number of drugs

d) Frequency of dosage

e) The patient's perception regarding how necessary the drug is

95. Write down the major points in the instructions you would give a patient with newly diagnosed Addison's disease on starting replacement therapy?

96. Which of the following statements are true regarding contraceptive pills?

a) Progestogens alone are not very effective

b) The side effects are due to the progestogen content

c) They may be used to stimulate the start of normal periods in a patient with amenorrhoea

d) An important side-effect is increased blood clotting

97. What treatment is common to all diabetics?

98. Which of the following statements are true about diet in diabetes?

a) Alcohol may lead to hypoglycaemia on 'the morning after'.

b) An energetic young man might reasonably be given a dietary carbohydrate allowance of up to 1000 g/day

c) Fat consumption should account for about 30% of the energy intake

d) Diabetics should be discouraged from taking carbohydrate rich in fibre

e) Substituting polyunsaturated margarine for butter favours a low serum cholesterol level in diabetics

99. Which of the following drugs if prescribed along with a sulphonylurea may result in hypoglycaemia?

a) Large doses of salicylate

b) Thiazides

c) Glucocorticoids

d) Monoamine oxidase inhibitors

e) β-Adrenoreceptor blocking drugs

100. Which of the following statements are true about insulin therapy?

a) The use of home blood glucose monitoring allows much tighter diabetic control in insulin dependent diabetic patients

b) A sugar free urine implies a less strict standard of control in a pregnant diabetic

c) All new young insulin dependent diabetics should be treated from the start with highly purified insulins

d) If patients stop eating they should stop their insulin

e) Fat atrophy can occur at the site of insulin injections

ANSWERS TO TEST-YOURSELF QUESTIONS

Sections I, II and III

1   a – No; b – Yes; c – Yes; d – No; e – No (page 2)

2   Any of the following: primary causes due to aplasia, abnormalities of normal synthesis, surgery, radiotherapy, autoimmunity, infarction, haemorrhage, infiltration. Secondary causes due to lack of trophic hormone, decreased target organ sensitivity. (page 3)

3   a – 2; b – 1; c – 3; d – 4 (pages 6, 7 and 8)

4   a – Yes; b – No; c – Yes; d – No (page 11)

5   a – Yes; b – No; c – Yes; d – No (pages 7 and 8)

6   Any of the following: Increased gluconeogenesis; increased glycogenolysis; decreased protein synthesis; increased aminoacid uptake by liver; decreased aminoacid uptake peripherally; increased fat deposition on face, neck, trunk; increased uric acid secretion; increased sodium retention; increased potassium excretion. (Table 5 – page 17)

7   i) a; ii) b, c and d (Table 7 – page 21)

8   a – Yes; b – No; c – Yes; d – No (page 25)

9   Any of the following:
a – Decreases calcium excretion by kidney; Increases phosphate excretion by kidney; Increases production of 1,25 DHCC by kidney; Stimulates bone resorption
b – Inhibits bone resorption; Increases urinary phosphate excretion
c – Increases active transport of calcium in small intestine (pages 28 and 29)

10   a – Yes; b – Yes; c – Yes; d – No (pages 31 to 33)

11   Any of the following: oestrogen; placental lactogen; prolactin; oxytocin (page 36)

12   a – 2; b – 3; c – 1; d – 1 or 3; e – 4 (pages 39 and 40)

13   1 – Yes – Because is well below the 3rd centile in height. (Chart 3 – page 46)
2 – Yes – She is well above the 97th centile in weight. (Chart 6 – page 49)
3 – No – He is only about the 90th centile in the growth spurt. (Chart 7 – page 50)
4 – Yes – She has all parameters of puberty delayed. (Chart 4 – page 47)

14   a – Yes; b – No; c – No; d – No (pages 55 and 56)

15   a – Yes; b – No; c – No; d – No (page 59)

16   a – Yes; b – Yes; c – Yes; d – Yes (page 62)

17   a – Yes; b – Yes; c – Yes; d – Yes; e – Yes (Table 20 – pages 66 and 67)

18   a – 5; b – 4; c – 2; d – 3 (Table 28 – page 72)

19   a – No; b – No; c – Yes; d – Yes (See Section II)

20   a – Yes; b – Yes; c – No; d – No (See Section II)

21   a – No; b – Yes; c – Yes; d – No (See Section II)

22   a – No; b – Yes; c – Yes; d – Yes (Table 22 – page 68)

23   Any of the following: Graves' disease and exophthalmos; pituitary tumours and visual field defects or optic atrophy; diabetes mellitus and cataracts or retinopathy; hypoparathyroidism and cataracts. (page 72)

24   a – 2; b – 3 or 4; c – 3; d – 1 (Table 19 – page 74)

25   a – No; b – Yes; c – Yes; d – No

26   When the height falls below the − 3 standard deviation line or when the growth velocity over a year is abnormally slow. (page 78)

27   a – No; b – Yes; c – No; d – Yes; i) – b, c; ii) – c or d

28   a – 1, 3, 4, 5; b – 2, 3, 4, 5 (pages 84 and 85)

29   a – No; b – No; c – Yes; d – No (pages 86 and 87)

30   a – No; b – Yes; c – No; d – Yes (page 82)

31   a – No; b – No; c – No; d – No (page 88)

32   Any of the following: hyperthyroidism; hyperaldosteronism; hypothyroidism; diabetes mellitus; Addison's disease; Cushing's syndrome. (Table 36 – page 89)

33   Hyperthyroidism; diabetes mellitus (page 91)

34   Subacute thyroiditis; haemorrhage into a cyst or nodule

35   a – No; b – Yes; c – Yes; d – No (pages 92–94)

36   a – Yes; b – No; c – Yes; d – No (pages 95 and 96)

37   Compare your answers with Table 41 (page 98)

38   a – No; b – Yes; c – Yes; d – No; e – Yes (page 99)

39   a – No; b – No; c – Yes; d – Yes (page 101)

40   a – Yes; b – No; c – Yes; d – Yes; e – No (page 102)

41   Compare your answer with Table 44 (page 104)

Sections IV and V

42

Hormone		Clinical syndrome
Growth hormone		
In adult	deficiency	Hypopituitarism
	excess	Acromegaly
In child	deficiency	Hypopituitarism
	excess	Gigantism
Vasopressin	deficiency	Diabetes insipidus
	excess	No syndrome described
Adrenocortico-steroid	deficiency	Addison's disease
	excess	Cushing's syndrome
Thyroid	deficiency	Thyrotoxicosis
	excess	Hypothyroidism
Parathyroid	deficiency	Hypoparathyroid and pseudohypoparathyroidism
	excess	Hyperparathyroidism – primary, secondary, tertiary and pseudo

43   a – No; b – Yes; c – No; d – Yes; e – Yes (pages 110–111)

44   a – Yes; b – No; c – Yes; d – No; e – Yes (page 111)
45   The list should include: structural causes – hydrocephalus
     or tumour; irradiation; familial; injury at birth;
     idiopathic. (page 112)
46   The condition can now be treated with growth hormone.
     The earlier treatment is begun, the better the result. The
     diagnosis can easily be missed. (page 112)
47   a – Yes; b – No; c – No; d – Yes; e – Yes (page 111)
48   a – Yes; b – No; c – No; d – Yes; e – Yes (page 118)
49   a – 2; b – 1; c – 3 (page 119)
50   a – Yes. It is important to exclude a hypothalamic or a
     gonadal abnormality.
     b – Cyproterone acetate may be used to arrest the
     advance of puberty.
     c – No. The girl will not be excessively tall providing she
     is treated. (page 121)
51   She should receive sufficient to maintain a urine output of
     1–2 litres with a normal plasma osmolality. (page 123)
52   i) All of them; ii) c (page 124)
53   Administration of synthetic glucocorticoids (page 126)
54   a – 1 and 2; b – 1; c – 1, 2 and 3 (page 126)
55   a – 3 (page 111); b – 2 (page 112); c – 1 (page 119)
56   a – Virilising effect of adrenal androgens; b – Increased
     protein catabolism; c – Mineralocorticoid activity of
     excessive circulatory cortisol; d – Increased protein
     catabolism (page 127)
57   1 – Mental and physical slowness; 2 – Tiredness; 3 –
     Cold intolerance; 4 – Dryness of the skin and hair (page
     133)
58   By finding a low serum thyroxine and a high serum TSH
     (page 134)
59   By early diagnosis through screening programmes (page
     134)
60   a – 3; b – 3; c – 4; d – 3; e – 1; f – 1 (page 137)
61   a – 1 and 3 (pages 110 and 133); b – 2, 3, 6 and 7 (pages
     110, 125, 130, 131); c – 2 and 3 (pages 110 and 125); d –
     1, 2 and 3 (pages 113, 125); e – 4 (page 124); f – 2 and 5
     (pages 129 and 125)
62   a – No; b – Yes; c – Yes; d – Yes; e – No (page 131)
63   a – 3 and 2 (pages 123 and 129); b – 4 (page 122); c – 1
     and 5 (pages 125 and 130)
64   a – 3 (page 127); b – 1 (page 140); c – 1 (page 140); d – 2
     (page 134); e – 1 and 3 (pages 127 and 140)
65   a – Yes; b – No; c – No; d – Yes; e – No (page 140)
66   a – 1 (page 141); b – 2 (page 141); c – 1 (page 141); d –
     2, 3 and 4 (page 142); e – 3 (page 142)
67   a – 3; b – 2; c – 1 and 2; d – 1 (page 143)
68   Recent increase in size; Pain; Single nodule (pages 143
     and 144)
69   a – Yes; b – Yes; c – Yes; d – Yes (page 159)
70   a – No; b – Yes; c – No; d – No (page 163)

71   1 – No rise in basal body temperature; 2 – No secretory
     endometrium; 3 – Plasma progesterone < 20 mmol/l in the
     latter part of the menstrual cycle (page 158)
72   a – Yes; b – No; c – Yes; d – No; e – No (pages
     156–157)
73   a – No; b – Yes; c – No; d – Yes; e – Yes; f – Yes (page
     154)
74   1 – a, c and d; 2 – a and b (page 153)
75   a – 1 and 3 (page 145); b – 2 and 4 (page 146); c – 4
     (page 149); d – 7 (page 150); e – 7 (page 151); f – 6 (page
     151)
76   A – Hypoparathyroidism; B – Vitamin D deficiency; C –
     Paget's disease (page 147)
77   a – 3 (page 146); b – 3 (page 146); c – 1 and 2 (page 145);
     d – 1 (page 145); e – 2 (page 146); f – 4 (page 146); g – 5
     (page 151)
78   d (page 147)
79   a – Yes; b – No; c – Yes; d – Yes; e – No (pages 145 and
     146)
80   All of them (table 53 – page 164)
81   i) a, d, e and f; ii) b and c (page 165)
82   The 2-hour post-breakfast blood glucose. If equivocal
     proceed with a glucose tolerance test. (page 166)
83   a – 2; b – 1; c – 3; d – 1; e – 4 (page 167)
84   a – Yes; b – Yes; c – Yes; d – Yes (page 168)
85   a – Yes; b – No; c – Yes; d – Yes; e – No (pages 169 and
     170)
86   a – Yes; b – No; c – No; d – No; e – Yes (page 172)
87   a – No; b – No; c – No; d – No (page 172)
88   Too little to eat or food taken too late, excessive
     exertion, increased insulin dose. (page 173)
89   a – Yes; b – No; c – Yes; d – Yes; e – No (page 175)
90   Obesity is associated with an increased mortality and
     morbidity rate. (page 177)
91   Your list should include any of the following: growth
     hormone deficiency, delayed or precocious puberty,
     disorders of adrenal cortex, hypothyroidism or poorly
     controlled diabetes mellitus. (page 182)
92   a – Yes; b – Yes; c – Yes; d – Yes; e – Yes (page 187)
93   a – Yes; b – Yes; c – No; d – Yes (pages 190 to 193)
94   a – Yes; b – No; c – Yes; d – Yes; e – Yes (page 195)
95   Need for life-long treatment; Importance of *never* missing
     a dose; Instructions re higher morning than evening dose;
     Importance of increasing dose when ill; Obtaining medic
     alert bracelet; Attention to potential side effects. (page 196)
96   a – No; b – No; c – No; d – Yes (page 198)
97   Diet (page 199)
98   a – Yes; b – No; c – No; d – No; e – Yes (page 200)
99   a – Yes; b – No; c – No; d – No; e – Yes (page 203)
100 a – Yes; b – No; c – Yes; d – No; e – Yes (pages 205 and
     206)

INDEX